Navigating Responsible Business Practices Through Ethical AI

Ouail El Imrani
Abdelmalek Essaadi University, Morocco

IGI Global
Scientific Publishing
Publishing Tomorrow's Research Today

Vice President of Editorial — Melissa Wagner
Director of Acquisitions — Mikaela Felty
Director of Book Development — Jocelynn Hessler
Production Manager — Mike Brehm
Cover Design — Jose Rosado

Published in the United States of America by
IGI Global Scientific Publishing
701 East Chocolate Avenue
Hershey, PA, 17033, USA
Tel: 717-533-8845 | Fax: 717-533-7115
Website: https://www.igi-global.com E-mail: cust@igi-global.com

Library of Congress Cataloging-in-Publication Data

(CIP Data Pending)
ISBN13: 9798337331041
Isbn13Softcover: 9798337331058
EISBN13: 9798337331065
British Cataloguing in Publication Data
A Cataloguing in Publication record for this book is available from the British Library.

To all decision-makers, researchers, practitioners, and visionaries working toward a future where technology serves humanity with fairness, transparency, and integrity.

May this work inspire continued dialogue, innovation, and collaboration to build an ethical and inclusive world driven by AI.

Editorial Advisory Board

Table of Contents

Detailed Table of Contents

and civil society to align FDI flows with Sustainable Development Goals (SDGs). Ultimately, it calls for a case-specific, transparent, and inclusive approach to ensuring that FDI truly contributes to long-term, equitable development.

Artificial intelligence (AI) is transforming not only scientific research, but also business practices, particularly in emerging economies such as Morocco. However, this rapid adoption raises ethical and governance issues that directly affect research institutions and businesses. Regulating AI has therefore become essential to ensure transparency and social responsibility. This study aims to explore the perceptions, risks, and ethical needs related to the use of AI in business practices and scientific research in the Moroccan context. A qualitative study was conducted based on semi-structured interviews with Moroccan stakeholders in science and industry. The results reveal high expectations for ethical regulation and the establishment of participatory standards, which are likely to strengthen trust in both research and business environments. Drawing on international guidelines and the ETHNA model, the study proposes a governance framework adapted to the context, promoting inclusive and responsible scientific innovation while guiding companies towards sustainable and ethical AI.

Since the Industrial Revolution, technological advances have molded humanity's identity, workforce, and economic value. Cyberpsychology's Psychological Adaptation Vectors (PAVs) can analyze AI's significant impact on society and human psychology across various domains, from theoretical frameworks to its recent daily-life integration, influencing human cognition and behavior. The current Artificial Generative Intelligence (AGI) and ambient systems leading towards super intelligence, caused by advancements in big data and computations, raise challenges against human uniqueness, employment, identity, and existential queries about the future. As evidenced by Reddit sentiment analysis, AI deployments lack transparency and answerability, aggravating anxiety because they are seen as tools for automation rather than augmentation. This tragedy is creating a Turing Trap, benefiting only the technological class, which can lead to economic inequality. Government intervention, collaboration, and regulation are important to ensure responsible AI development to avoid creating a dystopian society.

In the era of digital transformation, artificial intelligence (AI) is emerging as a strategic lever in supply chain management, enhancing forecasting, planning, optimization, and logistics resilience. However, its adoption raises ethical challenges related to algorithm transparency, data protection, decision-making fairness, and sustainability. This chapter presents a critical and multidisciplinary study of AI ethics in supply chains, based on a structured review of 31 recent scientific articles. It identifies thirteen key themes, including ethical principles, regulatory issues, and enabling technologies such as XAI, blockchain, and federated learning. The study emphasizes the need for ethical performance indicators, collaborative governance, and risk management tools to support the responsible adoption of AI. It ultimately calls for a strategic integration of logistics innovation, ethical awareness, and social responsibility to foster a more transparent, fair, and sustainable supply chain.

This chapter explores how artificial intelligence (AI) is reshaping responsible sourcing within contemporary supply chain and logistics networks. Moving beyond efficiency-driven narratives, it argues for the integration of ethical intelligence where AI systems are harnessed to uphold labor rights, environmental standards, and supply chain transparency. Drawing on global frameworks, empirical case studies, and recent regulatory developments, the chapter examines both the promises and perils of AI in procurement. It critically engages with issues of algorithmic bias, data ethics, and governance, emphasizing that technological innovation must be matched by normative reflection. By situating AI within a broader ethical architecture, the chapter offers a roadmap for organizations seeking to align digital transformation with social responsibility and sustainable sourcing practices.

Section 4
AI Ethics and the Financial Sector

Preface

The dawn of the Artificial Intelligence (AI) era marks one of the most transformative periods in modern history. From predictive analytics in healthcare to automated decision-making in finance and personalized marketing strategies, AI has redefined the way businesses operate, compete, and engage with their stakeholders. While these advancements offer unprecedented opportunities for innovation and efficiency, they also bring forth pressing ethical questions regarding fairness, transparency, accountability, and inclusivity. In this rapidly evolving technological landscape, the integration of ethical principles into AI-driven business practices is no longer optional it is an imperative. Organizations are now confronted with the challenge of aligning their technological ambitions with their social responsibilities, ensuring that the deployment of AI not only maximizes value creation but also safeguards human rights, fosters diversity, and promotes sustainable development.

Navigating Responsible Business Practices Through Ethical AI emerges as a timely and necessary contribution to this dialogue. The book brings together diverse perspectives from academics, practitioners, and policy experts to provide a comprehensive examination of how ethical AI can be effectively embedded into corporate strategies, governance models, and operational processes. By combining theoretical insights with practical case studies, this volume offers a roadmap for organizations seeking to harness the power of AI responsibly balancing innovation with integrity.

Overview of the Subject Matter

The rapid integration of Artificial Intelligence (AI) into business operations has brought unprecedented opportunities for innovation, efficiency, and growth. However, alongside these benefits, the emergence of ethical dilemmas, biases in algorithms, data privacy concerns, and the socio-economic impact of automation have raised urgent questions about the responsible use of AI. This book addresses these challenges through the lens of responsible business practices ensuring that AI adoption not only meets technological standards but also aligns with ethical principles, corporate social responsibility (CSR), and long-term societal well-being. It is important to emphasize that this book offers a structured approach, organized into six main sections, allowing readers to grasp the diversity and interconnectedness of the issues related to integrating AI into responsible and ethical practices. Each section comprises chapters dedicated to ethics and diversity, legal governance, organizational transformation, sustainable development, finance and innovation, as well as public governance and healthcare. This organization facilitates the exploration of key issues and provides a coherent and comprehensive overview of the international contributions gathered in this publication.

Our work situates itself at the intersection of ethics, technology, and business strategy, highlighting the need for frameworks, guidelines, and best practices that balance innovation with accountability.

Where This Topic Fits in the World Today

In the past decade, AI has transitioned from an emerging technology to a mainstream driver of transformation across industries finance, healthcare, supply chains, marketing, education, and beyond. The global race to harness AI's capabilities has intensified, but so has the debate over its ethical implications. Regulators are now implementing frameworks such as the EU AI Act, UNESCO's AI ethics recommendations, and sector-specific governance models.

In this evolving context, businesses cannot afford to focus solely on competitive advantage; they must also address transparency, fairness, inclusivity, and environmental sustainability in their AI systems. This book responds to this global imperative by offering research-based perspectives, practical tools, and multidisciplinary insights that bridge academia, industry, and policy.

Target Audience

This book is intended for a diverse readership, including:

- Academics and Researchers in AI ethics, business ethics, data science, and management.
- Business Leaders and Executives seeking to integrate ethical AI into strategic decision-making.
- Policy Makers and Regulators developing governance frameworks for AI adoption.

- Students in business, computer science, law, and ethics courses looking for case-based learning.
- Nonprofit and Civil Society Organizations concerned with AI's social and environmental impact.

The interdisciplinary nature of the content ensures that each audience segment can find both theoretical depth and actionable recommendations.

Conclusion and Contribution to the Field

This book makes a timely and significant contribution to the growing discourse on ethical AI in business. By combining theoretical frameworks, empirical research, and practical case studies, it equips readers with the tools to critically assess and implement AI solutions that are not only profitable but also socially responsible and ethically sound.

In bridging the gap between technological innovation and ethical responsibility, Navigating Responsible Business Practices Through Ethical AI provides an essential reference for shaping the next generation of AI-driven business strategies. It encourages organizations to view ethical AI not as a constraint but as a catalyst for trust, brand value, and sustainable growth.

Acknowledgment

We would like to express our sincere gratitude to all those who contributed to the successful completion of Navigating Responsible Business Practices Through Ethical AI.

First and foremost, we extend our heartfelt thanks to all the chapter authors for their valuable research, insights, and dedication, which have enriched the content of this volume. Their commitment to advancing knowledge in the field of ethical AI and responsible business practices has been truly inspiring.

We are equally grateful to the reviewers for their time, expertise, and constructive feedback, which played a vital role in enhancing the quality and rigor of each chapter.

Our appreciation also goes to the research institutions, academic centers, and industry partners whose support and resources facilitated the development of the studies presented in this book. Their encouragement and collaboration have been essential to this endeavor.

Finally, we wish to acknowledge IGI Global for their professional guidance, editorial assistance, and commitment to disseminating scholarly work to a global audience. Their dedication to fostering academic excellence has been instrumental in bringing this project to fruition.

To all who, in one way or another, contributed to this work, we extend our deepest thanks.

Introduction

Over the past decade, Artificial Intelligence (AI) has shifted from being a futuristic concept to an essential driver of transformation across nearly every sector of the global economy. From predictive analytics that anticipate consumer needs, to intelligent supply chains that respond in real time to disruptions, to automated decision-making systems that influence everything from hiring to healthcare, AI has redefined what it mcans to conduct business in the twenty-first century. Yet, as with any profound technological shift, the integration of AI into business practices has not come without significant challenges, particularly in the ethical domain.

The speed of AI's evolution and adoption has outpaced the development of universally accepted ethical frameworks. As algorithms become increasingly complex and autonomous, questions arise about fairness, transparency, accountability, and inclusivity. What happens when an AI system inadvertently discriminates against certain groups? How can organizations ensure that data-driven decisions respect individual privacy and dignity? How can businesses balance the pursuit of efficiency with the protection of human values? These are no longer hypothetical concerns; they are pressing issues with real-world implications for corporations, consumers, and societies at large. The concept of responsible business practices in the age of AI moves beyond regulatory compliance and risk management. It encompasses a proactive commitment to integrating ethical considerations into every stage of AI development and deployment from data collection and algorithm design to implementation and post-deployment monitoring. It requires businesses to see AI not merely as a tool for competitive advantage, but as a transformative force that must be harnessed with integrity, foresight, and a deep sense of social responsibility.

In this context, the ethical dimension of AI is not an obstacle to innovation; rather, it is a catalyst for sustainable growth and long-term trust. Organizations that embed ethical principles into their AI strategies are more likely to earn customer loyalty, attract top talent, and maintain resilience in an increasingly transparent and socially conscious marketplace. Moreover, ethical AI practices can open new opportunities for innovation by aligning technological capabilities with human needs and values.

Navigating Responsible Business Practices Through Ethical AI emerges as a timely response to these complex dynamics. This book assembles diverse perspectives from scholars, industry practitioners, policy-makers, and thought leaders worldwide. It offers a comprehensive exploration of how businesses can design, implement, and govern AI systems in ways that respect ethical principles while delivering measurable value. Through a combination of theoretical frameworks, empirical research, and case studies, the book provides practical guidance for aligning AI strategies with both organizational goals and societal expectations.

The global relevance of this topic cannot be overstated. As AI becomes a common denominator in economic growth, international trade, public services, and social interaction, the stakes of responsible AI adoption extend far beyond the corporate boardroom. Policymakers are enacting new regulations such as the European Union's AI Act, while international organizations like UNESCO are developing guidelines for ethical AI governance. These initiatives reflect a growing consensus that ethical AI is not a matter of choice but a necessity for ensuring equitable, transparent, and accountable technological progress.

In framing the discussions and insights presented in the chapters that follow, this introduction seeks to position ethical AI as both a moral imperative and a strategic business advantage. By doing so, we invite readers to see AI not only through the lens of technical performance, but also as a deeply human endeavor one that must be guided by shared values, collective responsibility, and an unwavering commitment to the public good.

Section 1
AI Ethics and Diversity

Chapter 1

The Ethical Implications of AI in Human Resources: Addressing Bias and Promoting Fairness

Krishna Khanal
https://orcid.org/0009-0003-1369-0952
Westcliff University, USA

ABSTRACT

This chapter examines the ethics of AI in human resources (HR) management, focusing on mitigating bias and ensuring fairness in AI-driven decision-making. Companies now rely more on AI technology to hire, evaluate performance, and manage staff, which raises worries about bias in algorithms and data. AI systems may reinforce or even exacerbate biases related to gender, race, age, and other factors, thereby creating significant ethical concerns. This chapter examines the root causes of AI bias, including data selection, algorithmic design, and lack of transparency. It provides insights into reducing and managing these risks through inclusive data practices, algorithmic fairness, and regulatory frameworks. It also highlights the importance of human oversight in AI systems to ensure fair and equitable treatment of all employees. This chapter seeks to contribute to a broader understanding of how AI can be ethically integrated into HR practices to bring diversity, equity, and inclusion in the workplace by addressing these issues.

DOI: 10.4018/979-8-3373-3104-1.ch001

INTRODUCTION

The introduction of Artificial Intelligence (AI) will transform various business functions, enhance organizational processes and efficiencies, facilitate advanced data analysis, contribute to value creation (Fenwick et al., 2024; Kassa & Worku, 2025), and impact humankind's future (Rawas, 2024). AI in Human Resource Management (HRM) has significantly changed how organizations handle key functions such as hiring, performance evaluation, and workplace engagement (Nawaz et al., 2024). The rise of Algorithmic HRM (AHRM) is transforming HRM by enhancing efficiency and productivity through the automation of routine tasks while also requiring organizations to adopt change management strategies and upskill employees for successful implementation (Cameron et al., 2024; Bandara et al., 2025). By automating repetitive work and analyzing large volumes of data, AI has introduced efficiency and precision into HR operations, allowing organizations to make more data-driven decisions. AI applications are extensive, ranging from AI-based resume screening systems to predictive analytics for employee turnover, making routine functions more efficient so that HR management can concentrate on high-level activities (Murugesan et al., 2023; Ekuma, 2024).

Companies are now moving to digital recruitment 3.0, a stage focused on bringing AI into recruitment and selection processes, driven by advancements in digital technology within HRM (Chen, 2023). Although organizations recognize that AI can save time and effort for HR managers, enhancing overall productivity, they lack a detailed understanding of how these technologies impact both the organization and individual employees in human resource management (Nawaz, 2024; Budhwar et al., 2022). Automation and data-driven decision-making offer employers the appealing potential for improved hiring and management decisions, enhanced efficiency, and cost reduction. In particular, AI has a big influence on the recruitment process, offering companies a competitive advantage by enhancing their ability to understand and attract talent more effectively than their competitors. For example, in learning and development, AI and automation are changing how learning needs are identified and addressed through data analysis and pattern recognition while also using natural language processing and machine learning (ML) to analyze employee feedback and reveal findings into skill gaps and learning requirements (Ekuma, 2023; Soori et al., 2023).

While technology has always added to physical abilities and basic cognitive skills, it has never been able to augment human intelligence until now, as recent advancements create new challenges for HR. Advanced digital technologies, including AI and state-of-the-art ML techniques, are revolutionizing various HR functions and practices. These innovations are driving improvements across multiple HR work and departments, contributing to increased operational efficiency and value generation

(Votto et al., 2021). Owing to the advanced analytic techniques and capabilities, AI has emerged as a trustworthy tool as it can perform analysis with speed and accuracy higher than any human's ability. As a machine, it is capable of aggregating and analyzing huge volumes of data and offering decision recommendations.

Despite AI's advantages, it is important to scrutinize them, as their advantages often overshadow their shortcomings. While AI promises to transform HR, its reliance on historical data and algorithms may introduce unintended biases, leading to potential fairness and diversity problems (Mehrabi et al., 2021; Chen, 2023). Bias and unbiasedness in AI systems have been extensively researched and debated in AI and broader computer science for many years (Alvarez et al., 2024). AI's purported neutrality is increasingly challenged as examples show that algorithmic systems are inherently value-laden, creating moral consequences, either reinforcing or undermining ethical principles and affecting stakeholder rights and dignity (Alvarez et al., 2024; Martin, 2019).

UNDERSTANDING AI BIAS IN HR

HRM involves various repetitive, time-consuming tasks and is influenced by human perception and biases, making it an ideal area for the application of AI. Human capital is a key differentiator for organizations, offering a competitive advantage due to its intangible nature, which is difficult for competitors to replicate (Boon et al., 2017). HR management has gained strategic importance in response to economic, political, social, and technological shifts. However, not all departments have fully embraced this role, and strategic positioning remains slow and challenging. In such cases, adopting technologies like AI requires other societal changes.

AI is a key component of the Fourth Industrial Revolution, alongside advancements in mechanical, electrical, and information technologies, driving the evolution and enhancement of the digital economy sector (Shen & Zhang, 2024). AI is a field of computer science focused on simulating intelligent behavior in machines, aiming to replicate and, ideally, enhance human actions. AI, a term introduced by John McCarthy in 1955, refers to computational methods that mimic human decision-making and processes, allowing systems to analyze large data sets and identify, relate, and predict patterns intelligently. Calculators, apps, and traditional programs differ from AI in how they process data. They follow fixed instructions to produce outputs without adapting to new information (Sheikh et al., 2023). For example, a calculator always performs the same operations without learning from past use. In contrast, AI can learn from data, adjust its behavior, and improve over time without needing to be reprogrammed, allowing it to solve problems beyond its original programming. The concept of intelligence remains a topic of debate, with no universally accepted definition. In discussions of AI, distinctions are often made between different levels of machine intelligence, such as weak AI, which is designed for specific tasks, and strong AI, which theoretically possesses broader cognitive capabilities. While AI and ML are sometimes used interchangeably, they are distinct; ML is a subset of AI that focuses on using data-driven methods to solve problems without the need for explicit programming (Sarker, 2021). AI applications can follow rules, improve over time by acquiring new data and information, and adjust to changes in their environment. There are various AI techniques, including ML, deep learning (DL), and the more recently popularized generative AI. AI provides insight into the functioning of our brain and reflects the encoded information stored within our neurons (Palm, 2012).

It is widely recognized that human cognition is imperfect, as our brains often depend on shortcuts to quickly process information, assess, or decide, simplifying a complex world by drawing on biases, with this process often occurring tacitly and pre-reflexively (Liehr Storm et al., 2023). Bias, in general, refers to a consistent preference for one way of feeling or acting over others. In human cognition, bias results from evolutionary heuristics that work well in most situations. Cognitive biases are consistent patterns of flawed judgment that humans make when faced with uncertainty (Wilke & Mata, 2012). These biases can influence algorithmic evaluations or predictions, leading to similar errors in automated decision-making processes (Belenguer, 2022). AI can minimize bias and assist in selecting the most suitable candidate for a job, but its effectiveness depends on the quality of the input data. In AI, bias can stem from the data used (pre-existing bias), the design of algorithms (technical bias), and organizational processes (emerging bias). Since most AI models are data-driven, they may inherit biases from the data, which is not neutral but value-laden (Ferrara, 2024). Similar to human cognition, AI algorithms rely on heuristic strategies that may introduce biases (Vicente & Matute, 2023). Biases in algorithms can arise at various stages, including during model development, evaluation, post-processing, and deployment (Baker & Hawn, 2021). Bias can stem from the design of the algorithm, the methods of data collection, coding, and selection, or from improper application of the algorithm. Training data may be incomplete, unbalanced, unrepresentative, or lack proper context. It can carry the biases of previous decision-makers or mirror prevalent societal biases, leading to misinterpretations of reality. Algorithmically, bias occurs when models oversimplify reality, making them too rigid to fully capture the intricate patterns and complexity within the data. These biases are responsible for outcomes that unfairly disadvantage certain groups within a population, either because of incorrect assumptions in the machine-learning process or because the algorithm is applied to a group it was not intended for. For example, Amazon's AI recruiting tool consistently disadvantaged female applicants (Nishant et al., 2024).

Human bias is discussed in two different narratives. First, essentialist claims argue that human cognition and decision-making are inherently biased or irrational, meaning people naturally make subjective or flawed judgments (Cho et al., 2023). Second, causal claims suggest that these biases can lead to discriminatory practices, such as employers unconsciously favoring certain groups over others in hiring decisions. Both narratives emphasize that bias is a fundamental aspect of human thought and can influence outcomes in ways that are often unconscious (Seppälä & Małecka, 2024).

AI bias refers to the unconscious and systematic errors that occur when algorithms produce results that favor certain groups over others, often based on the skewed or incomplete data used during their training (Ferrara, 2024). In HR, this bias can present itself in various ways, such as privileging candidates from specific demographics or backgrounds in recruitment or unfairly evaluating employees based on non-relevant factors. The sources of bias in AI stem from the fact that ML data is shaped by human social environments, machines learn based on their rationality, and AI operates through binary language, which lacks the complex interpretive structure of human communication, leading to discrepancies in data interpretation and decision-making (Nishant et al., 2023). These biases can perpetuate inequality and discrimination, undermining the efforts to create inclusive workplaces (Cachat-Rosset & Klarsfeld, 2023). This results in gender, racial, and ethnic disparities in hiring, promotions, and job satisfaction among marginalized employees, who generally have less equitable access to career development opportunities and often face substandard treatment in the workplace (Akter et al., 2023). For example, Exclusionary norms refer to situations where the data used for training a model or algorithm only represents a limited portion of the overall population. This creates biases because the model does not account for the diversity of the full population. For instance, if the training data mostly includes data from a specific demographic or group, the resulting model may not perform well for people outside that group, essentially "excluding" them (Akter et al., 2022; Siddique et al., 2024). Similarly, monolingual bias occurs in multilingual settings when the training data is predominantly or entirely in one language. This leads to a skewed model that performs well only in that language but may struggle when used with other languages. This type of bias reduces the model's effectiveness in managing a truly multilingual population, limiting its broader applicability. Addressing AI bias is critical not only to prevent these unfair practices but also to ensure that the advantages of AI in HR are aligned with ethical requirements that promote fairness, range, and inclusion throughout all tiers of a business enterprise.

If AI algorithms are developed using biased data, they can reinforce and potentially intensify biases in HR practices (Albaroudi et al., 2024). AI learning databases can be biased, potentially reinforcing and perpetuating discriminatory algorithms, while the unconscious biases and stereotypes of AI developers may also lead to discriminatory behavior when applied in HR hiring and selection processes. Additionally, there is concern about job loss due to automation, as AI increasingly handles HR tasks, which may lead to some HR professionals being displaced (Zirar et al., 2023). AI systems can also dehumanize the hiring process, reducing candidates to mere data points and potentially infringing on privacy, free expression, and free association rights. The use of AI in lie detection, for example, raises significant concerns related to mental privacy, fairness, and bias, with the potential to violate individual dignity. These systems may unintentionally introduce new biases, while workers' rights to privacy and fair treatment could be compromised, sparking concerns about discrimination and the overall integrity of the employment process. Discrimination refers to treating different groups unfairly based on factors like gender, age, or ethnicity rather than on objective criteria such as individual performance (Müller et al., 2023).

The availability of representative, comprehensive, and diverse training data is essential for the effectiveness and unbiased performance of a model (Jui & Rivas, 2024). Representativeness ensures that the information provided reflects the range of inputs that the model may encounter, while completeness ensures that all relevant information is covered (Borodowski, 2022). Diversity helps the model generalize across different populations or input types, reducing the risk of bias toward any particular group. Synthetic data can enhance dataset diversity and address sample-selection biases by artificially generating data for underrepresented cases, helping create a more balanced and equitable model (Yang et al., 2024).

The concern here is that certain ML algorithms, often called "black box" algorithms, are not easily understood in terms of how they make decisions or predictions. These algorithms lack transparency, meaning it's difficult to see how they process data and arrive at their outcomes (Hilliard et al., 2022). This lack of clarity can make it difficult to identify issues such as bias in the system. Furthermore, given the lack of transparency in how decisions are made, questions arise as to who is responsible or accountable if the policy produces unfair or unfair outcomes. This lack of transparency and consistency creates challenges in ensuring fairness, accuracy, and ethical use of these algorithms. Cheng and Hackett (2019) describe algorithms as "glass boxes" in the context of HRM, meaning that while certain parts of the algorithm are transparent and understandable, others remain opaque or unclear. This analogy highlights how some elements of the decision-making process are visible, allowing for some level of scrutiny or interpretation, while other components remain hidden, making full comprehension difficult. This partial transparency can make it difficult to better understand how algorithmic decisions are made (Köchling & Wehner, 2020).

The ethical implications of AI bias in HR are many, especially in recruitment processes where biased algorithms can reinforce existing inequalities. AI tools designed to screen resumes, for example, may unintentionally favor candidates of certain genders, races, or age groups based on historical data that reflects past hiring practices. For example, if an algorithm has been trained on data where a company predominantly hired male candidates, it may learn to prioritize male applicants, thereby perpetuating gender discrimination. Similarly, AI may also reflect racial and age biases, inadvertently excluding qualified candidates from underrepresented groups (Drage & Mackereth, 2022). These biases not only harm individuals but also limit organizations by restricting the diversity of talent they can attract. Failing to address these biases can lead to ethical violations and legal consequences, as discriminatory hiring practices contravene anti-discrimination laws in many jurisdictions.

In addition to hiring, AI biases can negatively affect employee evaluation and engagement. Performance management systems that rely on AI to monitor and evaluate employee contributions may incorrectly assess certain groups based on biased data entry. For example, female employees may be underestimated in leadership assessments if the AI process is focused on data that historically shows that women are underrepresented in leadership roles. Such biases can lead to lower morale, lower engagement, and even victimization of employees who feel they are being unfairly judged. Furthermore, AI-driven decisions in HR often lack transparency, making it difficult for employees to understand how the assessment is being conducted and challenge incorrect results (Fenwick et al., 2024; Budwar et al., 2023). These ethical challenges not only risk violating privacy and fairness standards but also undermine diversity and inclusivity efforts, as AI bias can perpetuate exclusionary practices within the workplace. For example, the CV ranking process is a critical initial step in hiring, where HR professionals assess applicants' qualifications to identify suitable candidates for the role. This method helps streamline the recruitment process by allowing recruiters to concentrate on the most promising candidates, saving time and resources. However, the process can be influenced by unconscious biases, leading to potential discrimination against certain groups. Candidates' perceptions of fairness during CV ranking can also shape their overall impression of the hiring organization, impacting their willingness to apply (Albaroudi et al., 2024).

FAIRNESS IN AI

Fairness is frequently evaluated in terms of equality, such as procedural fairness, and equity, such as distributive fairness, when it comes to treating and assessing others (Leventhal, 1980; Choung et al., 2023). In the context of HRM, fairness in ML can be defined in various ways, with each definition focusing on different evaluation aspects. For instance, demographic parity requires that the likelihood of a positive outcome, such as hiring or promotion, be equal across different demographic groups, while accuracy parity aims to ensure that error rates, such as incorrect hiring decisions, are consistent across these groups. AI fairness was introduced to tackle inequality by preventing disproportionate harms or benefits to different subgroups while developing a system that identifies biases and minimizes discrimination (Chen et al., 2023). Concerns of fairness arise due to the limited availability of resources, combined with differing opinions on what individuals are entitled to or how institutions should distribute these scarce resources. In HRM, subgroups can be defined by demographic factors like gender, age, or ethnicity, as well as organizational factors such as departments or locations within the company. Fairness is essential for the long-term growth of organizations and the protection of individual rights, ensuring that all stakeholders are respected and their interests considered in decision-making processes. A fair approach increases the acceptance of decisions, reduces conflicts, strengthens team unity, enhances brand image, and promotes sustainable business growth (Bal et al., 2023; Niu et al., 2022). AI's role in decision-making is often perceived differently depending on the nature of the task. For tasks requiring human qualities, like emotional judgment or subjective evaluation, AI decisions are typically seen as less fair compared to those made by humans (Kern et al., 2022; Lepri et al., 2021). However, when the task involves straightforward, data-driven analysis, AI is often viewed as making fairer decisions than humans. The complexity of the task also plays a significant role in shaping these perceptions, with AI being perceived as less fair in more intricate, multi-step tasks. Technical solutions often fail to grasp the complexity of concepts like fairness, which are influenced by context and cannot be solved purely through mathematical models (Pauleen et al., 2017; Selbst et al., 2019). Human judgments play a significant role in overcoming algorithmic biases by getting to the roots of unfairness in society.

Fairness was once a notion in philosophy and sociology; now, it is a problem of technology, formalized as rules or equations that can be used to determine whether an algorithm is fair. However, fairness remains largely subjective, influenced by personal perceptions and societal contexts (Szafran & Bach, 2024). Much like beauty, fairness is in the eye of the beholder, shaped by individual experiences and emotions. These variations can make it difficult to universally define fairness as factors such as culture, education level, and even gender or age play key roles in determining how that model is applied.

To promote AI fairness, the common approach focuses on making technical improvements to AI models, such as training algorithms on unbiased data or embedding predefined mathematical fairness definitions (Chen et al., 2023). However, this method assumes that individuals will always perceive optimally designed AI models as fair, which overlooks the complexities of human judgment. In contrast, the subjective nature of fairness has received less attention despite fairness being shaped by individual perceptions. Since fairness is a social construct, people's views on what is fair—whether a decision comes from a human or AI—are influenced by how they perceive and interpret the world around them. AI systems are unable to account for the human tendency toward relational care, relying instead on quantifiable and often simplistic metrics. To truly understand AI fairness, we need to incorporate the psychology of how humans perceive fairness into algorithmic decision-making, which can help organizations adopt AI in ways that improve fairness perceptions and encourage employee engagement with AI systems (Choung et al., 2023).

Various companies have introduced several strategies to reduce bias in AI systems. Google (2019) used the Testing with Concept Activation Vectors (TCAV) program to reduce bias and gender discrimination in decision-making algorithms, while Accenture (2018) implemented 'Teach and Test' AI testing services to assist businesses in minimizing biases and discriminatory content (P.S., 2023). The Society for Human Resource Management's (SHRM) Code of Ethics outlines ethical guidelines for HR professionals. The main principle guiding this code is that HR professionals have an ethical responsibility to promote and ensure fairness and justice in the workplace. This means they are responsible for creating, implementing, and advocating for policies and procedures that ensure all employees are treated fairly, consistently, and equitably. By doing this, HR professionals contribute to a work environment where everyone is respected and has equal opportunities, supporting both employee well-being and organizational success.

Bias in algorithms is distinct from fairness, even though the two are related. Algorithmic bias is a mathematical and statistical result of the way an algorithm processes data, and it arises naturally from the data itself (Alvarez et al., 2024). The biased treatment resulting from automated decisions made by AI systems or intelligent agents is commonly known as "digital discrimination. This bias is independent of ethical considerations, meaning that the presence of bias does not automatically indicate unfairness. To determine whether an algorithm's bias leads to unfair outcomes, it must be evaluated against a set of ethical or legal standards. If the bias is found to create unfairness, the algorithm can then be adjusted or tuned to meet specific fairness constraints. This process ensures that the algorithm produces more equitable results while still maintaining its intended performance. In essence, bias in algorithms is a technical issue that becomes an ethical concern only when judged against fairness principles.

MITIGATING BIAS IN AI FOR HR

Reducing bias in AI for HR requires a multi-pronged approach, which begins with identifying and reducing bias in the algorithms themselves. One of the key strategies is to ensure that the data used to train AI systems is representative and diverse (Cheong, 2024). Often, bias occurs when incomplete history or data are used to model AI, so it is important to look at data sets that reflect a wide range of demographics, experience, and job roles. Another effective approach is that AI systems will be regularly monitored for biases and consequences. This audit tests AI in different scenarios to see how it makes decisions in different demographic groups. Furthermore, using fairness-enhancing techniques such as adversarial testing and algorithmic adjustments can help AI achieve more consistent results.

When it comes to HR and recruiting, AI deployment must adhere to ethical frameworks & guidelines in order not to disrupt fairness, transparency, or inclusivity. International organizations and governments are beginning to develop standards to govern the use of AI in sensitive areas like HR, focusing on principles such as accountability, explainability, and non-discrimination. HR professionals, AI developers, and policymakers play crucial roles in cultivating unbiased AI systems. HR teams must actively collaborate with AI developers to ensure that the algorithms used align with the company's diversity and inclusion goals. Meanwhile, policymakers can implement regulations that hold organizations accountable for the ethical use of AI in employment decisions. By working together, these stakeholders can create a responsible AI ecosystem in HR, ensuring that AI-driven processes are not only efficient but also equitable and inclusive for all employees. Although AI can assist, humans are still essential for resolving political conflicts, such as negotiating with key stakeholders and ensuring accountability (Taeihagh, 2021; Miller, 2022). People may resist and be hesitant to accept decisions made by AI systems, a tendency known as algorithm aversion. Additionally, because these decisions are often unique and not clearly defined, they require human creativity, judgment, and interpretation. Humans also play an important role in helping AI understand and apply contextual knowledge to inform decision-making processes effectively.

Developing ethical AI algorithms for unbiased datasets is an important step in ensuring unbiased AI-driven HR processes. It is important to produce unbiased data and gather diverse and representative data that reflects a wide range of employee demographics, experiences, and job duties. This not only ensures that the data collection phase is inclusive but also removes any historical biases that may have skewed past HR practices. Data pre-processing techniques such as resampling or reweighting certain factors can help reduce inherent biases (Nazer et al., 2023). Additionally, AI developers should include policies of fairness that prioritize equality across demographic groups to ensure that AI does not establish discriminatory patterns. Transparency in algorithm design, including clear documentation of the decisions made during development, is also key to building trust and accountability. Companies should ensure transparency in the process of developing algorithms and training developers to reduce the risk of unconscious bias (Miller et al., 2018).

Continuous monitoring and auditing of AI systems are critical to maintaining fairness and transparency over time. As AI systems learn, progress, and adapt, they may unintentionally generate new biases and intensify existing ones. The Butterfly Effect, a concept from chaos theory, illustrates how minor changes in complex systems, such as subtle biases in data or algorithm training, can have major and unpredictable outcomes in AI, often exacerbating biases, reinforcing societal inequities, and increasing vulnerability to hostile attacks (Ferrara, 2024). Regular audits, including fairness checks and algorithmic assessments, help identify and address these issues before they lead to unfair outcomes. These audits should include metrics that evaluate how the AI impacts different demographic groups and ensure that they adhere to ethical standards. Additionally, incorporating ethical considerations throughout the AI development lifecycle is vital to ensuring that fairness and inclusivity remain priorities at every stage (Díaz-Rodríguez et al., 2023). Engaging multiple sets of stakeholders, including HR professionals, data scientists, ethicists, and underrepresented employee groups, further ensures that a variety of perspectives are considered in the AI's development and deployment, cultivating a more inclusive and ethical HR ecosystem (Shams et al., 2023). To ensure that AI systems continue to be fair, accurate, and effective in a variety of HRM contexts, frequent audits, and AI validation are crucial for identifying and addressing any biases (Mökander, 2023). Independent assessments by external specialists or organizations can be carried out to assess the fairness, precision, and effectiveness of AI, with necessary adjustments to the algorithms to address any identified biases.

AI algorithms used in HRM have been trained on historical recruitment data and can unknowingly perpetuate existing biases, potentially amplifying them over time (Timmons et al., 2023). These biases and prejudices may be the result of prior judgments influenced by factors such as gender, race, and socioeconomic status. To reduce this cycle, HR teams can use targeted strategies to monitor for biases in stages of the hiring or promotion process and make adjustments accordingly. A demographic-specific approach can be particularly helpful, where models are developed and validated separately for various demographic groups, such as gender or ethnicity.

For example, just as race-specific algorithms in healthcare have enhanced the prediction of health outcomes for marginalized groups (Mihan et al., 2024), a similar approach in HR could customize AI models better to evaluate the unique needs and challenges of underrepresented employees. By recognizing how factors such as race and gender impact career paths, these specialized AI models can contribute to fairer hiring processes and career development opportunities, ensuring that AI not only learns from past decisions but also helps to build a more inclusive and equitable workplace.

REFERENCES

Akter, S., Dwivedi, Y. K., Sajib, S., Biswas, K., Bandara, R. J., & Michael, K. (2022). Algorithmic bias in machine learning-based marketing models. *Journal of Business Research, 144*, 201–216. DOI: 10.1016/j.jbusres.2022.01.083

Akter, S., Sultana, S., Mariani, M., Fosso Wamba, S., Spanaki, K., & Dwivedi, Y. K. (2023). Advancing algorithmic bias management capabilities in AI-driven marketing analytics research. *Industrial Marketing Management, 114*, 243–261. DOI: 10.1016/j.indmarman.2023.08.013

Albaroudi, E., Mansouri, T., & Alameer, A. (2024). A comprehensive review of AI techniques for addressing algorithmic bias in job hiring. *AI, 5*(1), 383–404. DOI: 10.3390/ai5010019

Alvarez, J. M., Colmenarejo, A. B., Elobaid, A., Fabbrizzi, S., Fahimi, M., Ferrara, A., Ghodsi, S., Mougan, C., Papageorgiou, I., Reyero, P., Russo, M., Scott, K. M., State, L., Zhao, X., & Ruggieri, S. (2024). Policy advice and best practices on bias and fairness in AI. *Ethics and Information Technology, 26*(2), 31. DOI: 10.1007/s10676-024-09746-w

Baker, R. S., & Hawn, A. (2021). Algorithmic bias in education. *International Journal of Artificial Intelligence in Education*, •••, 1–41.

Bal, M., Stok, M., Bombaerts, G., Huijts, N., Schneider, P., Spahn, A., & Buskens, V. (2023). A fairway to fairness: Toward a richer conceptualization of fairness perceptions for just energy transitions. *Energy Research & Social Science, 103*, 103213. DOI: 10.1016/j.erss.2023.103213

Bandara, R. J., Biswas, K., Akter, S., Shafique, S., & Rahman, M. (2025). Addressing algorithmic bias in AI-driven HRM systems: Implications for strategic HRM effectiveness. *Human Resource Management Journal*, 1748-8583.12609. Advance online publication. DOI: 10.1111/1748-8583.12609

Belenguer, L. (2022). AI bias: Exploring discriminatory algorithmic decision-making models and the application of possible machine-centric solutions adapted from the pharmaceutical industry. *AI and Ethics, 2*(4), 771–787. DOI: 10.1007/s43681-022-00138-8 PMID: 35194591

Boon, C., Eckardt, R., Lepak, D. P., & Boselie, P. (2017). Integrating strategic human capital and strategic human resource management. *International Journal of Human Resource Management, 29*(1), 34–67. DOI: 10.1080/09585192.2017.1380063

Budhwar, P., Malik, A., De Silva, M. T. T., & Thevisuthan, P. (2022). Artificial intelligence – challenges and opportunities for international HRM: A review and research agenda. *International Journal of Human Resource Management, 33*(6), 1065–1097. DOI: 10.1080/09585192.2022.2035161

Cachat-Rosset, G., & Klarsfeld, A. (2023). *Diversity,* Equity, and Inclusion in Artificial Intelligence: An Evaluation of Guidelines. *Applied Artificial Intelligence, 37*(1), 2176618. Advance online publication. DOI: 10.1080/08839514.2023.2176618

Cameron, R., Herrmann, H., & Nankervis, A. (2024). Mapping the evolution of algorithmic HRM (AHRM): A multidisciplinary synthesis. *Humanities & Social Sciences Communications, 11*(303), 303. Advance online publication. DOI: 10.1057/s41599-024-02786-z

Chen, P., Wu, L., & Wang, L. (2023). AI fairness in data management and analytics: A review on challenges, methodologies, and applications. *Applied Sciences (Basel, Switzerland), 13*(18), 10258. DOI: 10.3390/app131810258

Chen, Z. (2023). Ethics and discrimination in artificial intelligence-enabled recruitment practices. *Humanities & Social Sciences Communications, 10*(567), 567. DOI: 10.1057/s41599-023-02079-x

Chen, Z. (2023). Collaboration among recruiters and artificial intelligence: Removing human prejudices in employment. *Cognition Technology and Work, 25*(1), 135–149. DOI: 10.1007/s10111-022-00716-0 PMID: 36187287

Cheong, B. C. (2024). Transparency and accountability in AI systems: Safeguarding wellbeing in the age of algorithmic decision-making. *Frontiers in human dynamics*, 6.

Cho, W., Choi, S., & Choi, H. (2023). Human resources analytics for public personnel management: Concepts, cases, and caveats. *Administrative Sciences, 13*(2), 1–22. DOI: 10.3390/admsci13020041

Choung, H., Seberger, J. S., & David, P. (2023). When AI is Perceived to Be Fairer than a Human: Understanding Perceptions of Algorithmic Decisions in a Job Application Context. *International Journal of Human-Computer Interaction*, •••, 1–18. DOI: 10.1080/10447318.2023.2266244

Díaz-Rodríguez, N., Del Ser, J., Coeckelbergh, M., López de Prado, M., Herrera-Viedma, E., & Herrera, F. (2023). Connecting the dots in trustworthy artificial intelligence: From AI principles, ethics, and key requirements to responsible AI systems and regulation. *Information Fusion, 99*, 101896. DOI: 10.1016/j.inffus.2023.101896

Ekuma, K. (2024). Artificial Intelligence and Automation in Human Resource Development: A Systematic Review. *Human Resource Development Review*, *23*(2), 199–229. DOI: 10.1177/15344843231224009

Fenwick, A., Molnar, G., & Frangos, P. (2024). The critical role of HRM in AI-driven digital transformation: A paradigm shift to enable firms to move from AI implementation to human-centric adoption. *Discover Artificial Intelligence*, *4*(34), 34. Advance online publication. DOI: 10.1007/s44163-024-00125-4

Ferrara, E. (2024). Fairness and bias in artificial intelligence: A brief survey of sources, impacts, and mitigation strategies. *Sci*, *6*(1), 3. DOI: 10.3390/sci6010003

Ferrara, E. (2024). The butterfly effect in artificial intelligence systems: Implications for AI bias and fairness. *Machine Learning with Applications*, *15*, 100525. DOI: 10.1016/j.mlwa.2024.100525

Hilliard, A., Guenole, N., & Leutner, F. (2022). Robots are judging me: Perceived fairness of algorithmic recruitment tools. *Frontiers in Psychology*, *13*, 940456. DOI: 10.3389/fpsyg.2022.940456 PMID: 35959005

Jui, T. D., & Rivas, P. (2024). Fairness issues, current approaches, and challenges in machine learning models. *International Journal of Machine Learning and Cybernetics*, *15*(8), 3095–3125. DOI: 10.1007/s13042-023-02083-2

Kassa, B. Y., & Worku, E. K. (2025). The impact of artificial intelligence on organizational performance: The mediating role of employee productivity. *Journal of Open Innovation*, *11*(1), 100474. DOI: 10.1016/j.joitmc.2025.100474

Kern, C., Gerdon, F., Bach, R. L., Keusch, F., & Kreuter, F. (2022). Humans versus machines: Who is perceived to decide fairer? Experimental evidence on attitudes toward automated decision-making. *Patterns (New York, N.Y.)*, *3*(10), 100591. DOI: 10.1016/j.patter.2022.100591 PMID: 36277823

Köchling, A., & Wehner, M. C. (2020). Discriminated by an algorithm: A systematic review of discrimination and fairness by algorithmic decision-making in the context of HR recruitment and HR development. *Business Research*, *13*(3), 795–848. DOI: 10.1007/s40685-020-00134-w

Lepri, B., Oliver, N., & Pentland, A. (2021). Ethical machines: The human-centric use of artificial intelligence. *iScience*, *24*(3), 102249. DOI: 10.1016/j.isci.2021.102249 PMID: 33763636

Leventhal, G. S. (1980). What should be done with equity theory? In Gergen, K. J., Greenberg, M. S., & Willis, R. H. (Eds.), *Social Exchange* (pp. 27–55). Springer US. DOI: 10.1007/978-1-4613-3087-5_2

Liehr Storm, K. I., Reiss, L. K., Guenther, E. A., Clar-Novak, M., & Muhr, S. L. (2023). Unconscious bias in the HRM literature: Towards a critical-reflexive approach. *Human Resource Management Review*, *33*(3), 100969. DOI: 10.1016/j.hrmr.2023.100969

Martin, K. (2019). Ethical implications and accountability of algorithms. *Journal of Business Ethics*, *160*(4), 835–850. DOI: 10.1007/s10551-018-3921-3

Mehrabi, N., Morstatter, F., Saxena, N., Lerman, K., & Galstyan, A. (2021). A survey on bias and fairness in machine learning. [REMOVED HYPERLINK FIELD]. *ACM Computing Surveys*, *54*(6), 1–35. DOI: 10.1145/3457607

Mihan, A., Pandey, A., & Van Spall, H. G. C. (2024). Mitigating the risk of artificial intelligence bias in cardiovascular care. *The Lancet. Digital Health*, *6*(10), e749–e754. DOI: 10.1016/S2589-7500(24)00155-9 PMID: 39214762

Miller, F. A., Katz, J. H., & Gans, R. (2018). The OD is imperative to add inclusion to artificial intelligence algorithms. *OD Practitioner*, *50*(1), 8.

Miller, G. J. (2022). Stakeholder roles in artificial intelligence projects. *Project Leadership and Society*, *3*, 100068. DOI: 10.1016/j.plas.2022.100068

Mökander, J. (2023). Auditing of AI: Legal, ethical and technical approaches. *Digital Society : Ethics, Socio-Legal and Governance of Digital Technology*, *2*(49), 49. Advance online publication. DOI: 10.1007/s44206-023-00074-y

Murugesan, U., Subramanian, P., Srivastava, S., & Dwivedi, A. (2023). A study of artificial intelligence impacts on human resource digitalization in Industry 4.0. *Decision Analytics Journal*, *7*, 100249. DOI: 10.1016/j.dajour.2023.100249

Nawaz, N., Arunachalam, H., Pathi, B. K., & Gajenderan, V. (2024). The adoption of artificial intelligence in human resources management practices. *International Journal of Information Management Data Insights*, *4*(1), 100208. DOI: 10.1016/j.jjimei.2023.100208

Nishant, R., Schneckenberg, D., & Ravishankar, M. (2024). The formal rationality of artificial intelligence-based algorithms and the problem of bias. *Journal of Information Technology*, *39*(1), 19–40. DOI: 10.1177/02683962231176842

Niu, Y., Wang, S., & Liu, Z. (2022). Managing brand competition with consumer fairness concerns via manufacturer incentives. *European Journal of Operational Research*, *300*(2), 661–675. DOI: 10.1016/j.ejor.2021.08.026

Palm, G. (2012). *Neural assemblies: An alternative approach to artificial intelligence*. Springer Science & Business Media.

Pauleen, D. J., Rooney, D., & Intezari, A. (2017). Big data, little wisdom: Trouble brewing? Ethical implications for the information systems discipline. *Social Epistemology, 31*(4), 9–33. DOI: 10.1080/02691728.2016.1249436

Rawas, S. (2024). AI: The future of humanity. *Discover Artificial Intelligence, 4*(25). Advance online publication. DOI: 10.1007/s44163-024-00118-3awe

Sarker, I. H. (2021). Deep learning: A comprehensive overview of techniques, taxonomy, applications, and research directions. *SN Computer Science, 2*(420), 420. DOI: 10.1007/s42979-021-00815-1 PMID: 34426802

Selbst, A. D., Boyd, D., Friedler, S. A., Venkatasubramanian, S., & Vertesi, J. (2019). Fairness and abstraction in sociotechnical systems. FAT 2019 - Proceedings of the 2019 Conference on Fairness, Accountability, and Transparency*, 1(1), 59–68. DOI: 10.1145/3287560.3287598

Seppälä, P., & Małecka, M. (2024). AI and discriminative decisions in recruitment: Challenging the core assumptions. *Big Data & Society, 11*(1), 20539517241235872. Advance online publication. DOI: 10.1177/20539517241235872

Shams, R. A., Zowghi, D., & Bano, M. (2023). AI and the quest for diversity and inclusion: A systematic literature review. *AI and Ethics*. Advance online publication. DOI: 10.1007/s43681-023-00362-w

Sheikh, H., Prins, C., & Schrijvers, E. (2023). Artificial intelligence: Definition and background. In *Mission AI. Research for Policy* (pp. 13-30). Springer.

Shen, Y., & Zhang, X. (2024). The impact of artificial intelligence on employment: The role of virtual agglomeration. *Humanities & Social Sciences Communications, 11*(122), 122. DOI: 10.1057/s41599-024-02647-9

Siddique, S., Haque, M. A., George, R., Gupta, K. D., Gupta, D., & Faruk, M. J. H. (2024). Survey on machine learning biases and mitigation techniques. *Digital, 4*(1), 1–68. DOI: 10.3390/digital4010001

Soori, M., Arezoo, B., & Dastres, R. (2023). Artificial intelligence, machine learning and deep learning in advanced robotics: A review. *Cognitive Robotics, 3*, 54–70. DOI: 10.1016/j.cogr.2023.04.001

Szafran, D., & Bach, R. (2024). "The human must remain the central focus": Subjective fairness perceptions in automated decision-making. *Minds and Machines, 34*(1), 24. DOI: 10.1007/s11023-024-09684-y

Taeihagh, A. (2021). Governance of artificial intelligence. *Policy and Society, 40*(2), 137–157. DOI: 10.1080/14494035.2021.1928377

Timmons, A. C., Duong, J. B., Simo Fiallo, N., Lee, T., Vo, H. P. Q., Ahle, M. W., Comer, J. S., Brewer, L. C., Frazier, S. L., & Chaspari, T. (2023). A call to action on assessing and mitigating bias in artificial intelligence applications for mental health. *Perspectives on Psychological Science, 18*(5), 1062–1096. DOI: 10.1177/17456916221134490 PMID: 36490369

Vicente, L., & Matute, H. (2023). Humans inherit artificial intelligence biases. *Scientific Reports, 13*(1), 15737. DOI: 10.1038/s41598-023-42384-8 PMID: 37789032

Votto, A. M., Valecha, R., Najafirad, P., & Rao, H. R. (2021). Artificial intelligence in tactical human resource management: A systematic literature review. *International Journal of Information Management Data Insights, 1*(2), 100047. DOI: 10.1016/j.jjimei.2021.100047

Yang, Y., Lin, M., Zhao, H., Peng, Y., Huang, F., & Lu, Z. (2024). A survey of recent methods for addressing AI fairness and bias in biomedicine. *Journal of Biomedical Informatics, 154*, 104646. DOI: 10.1016/j.jbi.2024.104646 PMID: 38677633

Zirar, A., Ali, S. I., & Islam, N. (2023). Worker and workplace artificial intelligence (AI) coexistence: Emerging themes and research agenda. *Technovation, 124*, 102747. DOI: 10.1016/j.technovation.2023.102747

Chapter 2
AI and Diversity:
Ensuring Fairness in Automated Decision-Making

Satya Subrahmanyam
https://orcid.org/0000-0003-0441-2742
Holy Spirit University of Kaslik, Lebanon

Badih Elie Baz
https://orcid.org/0000-0002-1553-4908
Holy Spirit University of Kaslik, Lebanon

ABSTRACT

As artificial intelligence (AI) becomes integral to business decision-making, ensuring fairness, equity, and diversity in automated systems is a critical ethical and operational priority. This chapter explores the concept of fairness in AI, examining various types of bias, their sources, and real-world consequences. It reviews legal and ethical frameworks, outlines technical and organizational strategies for promoting inclusive AI, and presents case studies highlighting both best practices and failures. Emphasis is placed on the importance of human oversight, diverse representation, and transparency in AI development. The chapter concludes by identifying challenges and proposing future research directions to align AI with broader diversity and inclusion goals.

DOI: 10.4018/979-8-3373-3104-1.ch002

1. INTRODUCTION

1.1 AI in Business Decision-Making

The new revolution that has emerged has been Artificial Intelligence (AI) that is transforming the business decisions landscape. Companies operating in various sectors have since turned to AI based systems so that they can enhance productivity and simplify their operations and gain strategic insights. Predictive analytics and automation of customer service such as algorithmic hiring and financial modeling is part of any AI usage (Dwivedi et al., 2021). As corporates are shifting to the use of AI in the decision-making processes that involve the most interest in the long-term perspective, the question of how such systems are treating the people and frequent groups of varied backgrounds is of utmost importance.

1.2 Fairness, Equity, and Diversity in AI Systems

The introduction of AI into the decision-making process has been raising issues of justice, injustice, and diversity for some time. The AI systems may appear to be without any bias, however, they frequently get taught on the basis of the past data, which may be indicative of cultural biases. Unless addressed, such types of bias can be incorporated into algorithms and ultimately perpetuate or increase discrimination toward already-vulnerable groups (Mehrabi et al., 2021). Being fair in artificial intelligence simply means being fair in humanity and knowing that we do not discriminate against our outputs in relation to such aspects as gender, age, colour, and socioeconomic status.

Equity goes further to recognize the imbalance existing in the system and attempts to level the playing field, especially towards the previously underprivileged groups. Diversity in this sense implies heterogeneous human populations of the data as well as multidisciplinary attitudes in the design of an artificial intelligence system. Diversity of development teams, and datasets, together, are handy to reduce blind spots and have more robust and inclusive algorithms (Raji et al., 2020).

1.3 The Ethical Imperative: Preventing Bias and Discrimination

The ethical ramifications of algorithmic prejudice intensify as AI increasingly infiltrates critical sectors, including employment, finance, healthcare, and law enforcement. Discriminatory results not only hurt the credibility and validity of organizations, but they also have serious effects on the people who are touched by unjust judgments. Ethically responsible AI requires taking steps to find, reduce, and keep an eye on prejudice throughout the AI lifetime.

An even greater number of ethical AI ideas and principles emphasize fairness, accountability, and openness. As an example, the European Union has the AI Act and appropriate considerations (European Commission, 2021) regarding specifying the high-risk systems of the AI and setting strict standards on anti-discrimination and human control. UNESCO (2021) described the implementation of inclusive and rights-based treatments of artificial intelligence in their 2021 Recommendation on the Ethics of Artificial Intelligence (UNESCO, 2021). The users and creators of AI programs and systems also have a moral responsibility not to harm and foster equitability as revealed by these initiatives.

1.4 Purpose and Scope of the Chapter

In this chapter, AI will be discussed in relation to diversity, i.e., it will refer to the issue of fairness of automatic decisions. It aims at discovering the roots of the prejudice in AI and analyzing the existing regimes of law and ethics implemented currently and providing the practical measures towards creating more non-discriminatory and more ethical AI. Using the most recent research, experience, and best practices in the industry, the chapter targets to equip stakeholders in the business sector, including the business leaders, data science professionals, and policy makers, with the tools needed to manage and develop AI in responsibilities.

The bigger picture is to emphasize that, fairness is not a technical issue, but a social requirement. The cultural element of ethical responsibility is required since the organizations are adopting AI technologies; all the principles of equity and diversity must be introduced into the very technology of AI creation and application to the company. The chapter will contribute to the growing discussion of ethical AI by advancing ethically intelligent systems, which involve more than the intelligence, but also justice, inclusivity and value of human beings.

2. UNDERSTANDING FAIRNESS IN AI

Being reasonable and fair to persons and parties in algorithmic decision-making is what we mean when we talk about fairness in AI. It encompasses concepts of equality of demographic, equal opportunity and personal injustice. The question of fairness will require that the following dimensions of diversity, such as race, gender, age, and socio-economic status, will also need a solution. Biased AI can contribute to the increase of social inequalities, including discriminating algorithms during hiring or lending. Fairness prospect is required to cultivate trust, inclusiveness and ethics in development and application of AI.

2.1 Defining Fairness in the Context of AI and Machine Learning

Fairness in machine learning (ML) and artificial intelligence (AI) refers to the concept that the algorithms must judge in a non-biased and equitable way, but not in a biased and unfair way. In this scenario, skinflintness is a major issue when it comes to AI systems controlling access to jobs, education, healthcare, and credit the more so that it is dangerous to come as close to the further implementation of AI systems without ensuring that these systems are fair and just which pose threats to the potential absence of trust and justice within society. Defined as a practical concept, AI fairness takes into consideration trying to prevent a discriminatory output that has been actualized via an attribute that is safeguarded such as race, gender, age, or disability, and at least in situations where such outputs reproduce or solidify the political asymmetries that are currently present in the human world (Mehrabi et al., 2021).

The concept of fairness in AI is not a blanket concept. It is a situational and multiple mode of value that cuts through legal sets, ethical theories and technical constraints. It is in how to assign units of the nature of fairness norms which may be reduced to the calculable and enforceable design principles. Trade-offs between the conflicting fairness measures and the potential conflict between fairness and model accuracy, which complicate the operationalization of fairness further, also exist (Corbett-Davies and Goel, 2018).

2.2 Types of Fairness in AI

In recent literature, a number of definitions and mathematical ways of fairness have developed, each representing a dissimilar view of what fairness involves in treating people and groups of people fairly.

Statistical Parity or **Demographic Parity** means that the results of a model are not sensitive to sensitive attributes. In case the hiring algorithm is used to pick the candidates, demographic parity would require that the rate of selection should be equal between the genders or racial lines (Barocas et al., 2020). Although this method is favorable to equality at group level, it may be ignorant of variations in subordinate qualifications or context.

Equal Opposition, on the contrary, revolves around the concept that people of similar qualifications need to be given equal opportunities of positive outcomes irrespective of the group membership. In particular, this definition allows stating that the true positive rate must be similar among demographic groups (Hardt et al., 2016). This is to make sure that the deserving people are not left behind because of bias in the data or model.

Individual Fairness goes ahead to conclude that persons that are more related should be treated similarly. This formulation must take a clear definition of similarity, that is, in the attributes of relevancy and this might not be easily defined objectively (Dwork et al., 2012). Though in attempt to provide a more subtle and more intimate approach, it may sometimes require additional steps of modelling and moral judgment.

These definitions underline the idea that the concept of fairness did not constitute a technical quality, but a philosophical and policy-based issue. The multidisciplinary aspect of the process of making a normative and strategic decision, on which fairness principle to apply in the situation under consideration, implies that all the stakeholders must be involved.

2.3 Dimensions of Diversity in AI Systems

AI systems must acknowledge and honor the many attributes that constitute human identity to guarantee fairness. These traits include, but are not limited to, color, gender, age, socio-economic status, handicap, religion, and sexual orientation. Each of these aspects may affect how people use technology and how AI systems read and react to their data.

Due to a dearth of varied training data, facial recognition systems perform far worse on women and people with darker skin tones than on lighter-skinned males (Buolamwini & Gebru, 2018). Voice recognition systems may misinterpret speakers with regional or non-native accents, leading to accessibility challenges. In financial services, credit scoring models trained on biased datasets may systematically underestimate the creditworthiness of certain ethnic groups, even when objective indicators suggest otherwise (Suresh & Guttag, 2021).

Ensuring representation in training data is therefore critical but insufficient on its own. Diversity must also be embedded in the development teams, governance structures, and value systems that shape AI design. Without inclusive perspectives guiding the development process, even technically sophisticated models may reproduce exclusionary or harmful outcomes.

2.4 Real-World Implications of Unfair AI Decisions

Unfair AI decisions can lead to profound and far-reaching consequences in people's lives. In the domain of employment, automated hiring platforms have come under scrutiny for penalizing women applicants based on patterns observed in historically male-dominated fields (Subrahmanyam, 2025). Such outcomes not only reinforce gender inequality but also undermine the credibility of AI systems in the eyes of the public.

One of the issues emerging due to the use of healthcare spending as a proxy of health needs is that discriminatory AI algorithms that are applied to determine the additional treatment lack proper prioritisation of Black patients, despite their medical needs being more substantial (Obermeyer et al., 2019). These inadequacies are in opposition to the moral principles of beneficence and equity, and they can result in continued health inequity.

In the field of criminal justice, algorithmic risk assessment technology such as COMPAS has undergone criticism of unproportionately labeling Black defendants as risky, consequently influencing the decisions regarding bails, sentencing, and parole. Such biases not only go against the principles of equal protection, but they also undermine the public trust in the judicial systems (Angwin et al., 2016).

Corporates risk legal trouble, harm to their brand, and a decline in consumer confidence when they use biased AI systems. As societies become more aware of algorithmic harm, public expectations for ethical accountability are rising. Businesses must therefore invest in fairness as a core component of AI governance, not merely as a regulatory checkbox but as a strategic imperative.

3. SOURCES OF BIAS IN AUTOMATED DECISION-MAKING

The concern with algorithmic bias has increased where AI-based systems influence decision-making to a greater extent in various areas, such as health, loans, artificial intelligence (AI), and others. Bias refers to unfair and systematic discrimination which is an element of the data collection procedure, design of models, and interpretation of the results process of the automated decision-making. These biases might be generated at various stages of the machine learning process and then reflect and enhance social inequalities that already exist (Subrahmanyam, 2024). To develop responsible and ethical AI ecosystems, one should learn these sources of bias.

3.1 Bias in Data Collection and Representation

Any AI system relies on data, and if there are biases during data collecting, it might affect conclusions made later on. Results may be biassed or discriminating if the data used to train AI does not accurately reflect the target population. Problems with minority group performance when using models in reality might arise from datasets that do not adequately reflect such groups (Suresh & Guttag, 2021).

In this regard, face recognition software stands out. Research has shown that these systems tend to outperform darker-skinned females compared to light-skinned men. This is mainly due to the fact that training datasets tend to favor persons with lighter skin tones (Buolamwini & Gebru, 2018). The lack of demographic diversity in data results in higher error rates for underrepresented groups, posing serious ethical and legal implications when these systems are used in policing, surveillance, or employment verification.

When the data obtained shows societal biases, it may also lead to bias in representation. AI trained on data from a firm that has a history of gender discrimination would probably learn and copy those prejudices when choosing candidates.

3.2 Algorithmic and Model-Induced Bias

Even balanced datasets may carry with them the bias that algorithms can introduce when using algorithms in model construction. Although they are not explicitly mentioned, certain machine learning models, particularly the more intricate ones such as deep neural networks, can discover subtle patterns in data that align with sensitive characteristics (Barocas et al., 2020). The use of names or ZIP codes as the alternative to racial or socioeconomic status can result in indirect discrimination.

In model optimization, the focus is usually on accuracy and efficiency to the detriment of fairness. Assuming that the optimization of a fraud detection system will reduce false positives over the general population will lead to individuals of some ethnic groups being indicating a false positive more often than the general population even when actually more likely to commit fraud. The models will produce more disparities instead of rectifying the existing disparities without a design that is fairness-conscious.

3.3 Bias in Feature Selection and Labeling

Feature selection—deciding which variables to include in a model—can also be a critical source of bias. Features that are assumed to be neutral may, in fact, carry embedded social meaning. Educational attainment or job history may seem objective but could indirectly reflect systemic barriers faced by certain demographic groups. Including such features without critical scrutiny can reinforce structural inequalities (Hooker, 2021).

Labeling bias occurs when the ground truth used to train the model is itself biased. In supervised learning, models are trained to predict a target label. If the labeling process reflects subjective human judgments or historical discrimination, the model will learn to mirror those patterns. In the criminal justice system, for instance, arrest records are often used as labels to predict future criminal behavior. These records may reflect biased policing practices rather than actual criminal activity, thus embedding institutional prejudice into predictive policing tools (Lum & Isaac, 2016).

3.4 Institutional and Historical Biases Embedded in Datasets

Many datasets used to train AI systems are not collected in a vacuum but are products of institutional practices and historical contexts. These institutional biases can become codified into algorithms, leading to systematic disadvantages for marginalized communities. Historical lending data used by financial institutions may reflect redlining practices that excluded minority applicants from access to credit. When such data is used to train modern loan approval systems, it can reproduce those exclusions under the guise of algorithmic objectivity (Eubanks, 2018).

Recruitment algorithms that are developed using past hiring data could discriminate against people from under-represented groups by giving an advantage to applicants from certain schools or with particular names. It is well-known that Amazon discontinued an AI recruiting tool because it was shown to have devalued resumes that included the phrase "women's" (e.g., "women's chess club captain") as a result of prejudice in the training data (Dastin, 2018). If development efforts do not aim to rectify past disparities, AI systems like this will reflect and exacerbate gender prejudices.

3.5 Case Examples of Biased Automated Systems

The necessity to deal with these challenges is emphasized by real-world incidents of discriminatory AI. The financial sector has also witnessed the issue of gender discrimination whereby the algorithms deployed to set credit limits to customers are identified to propose lower limits to women than men with the same financial profile (Hurley & Adebayo, 2017). The use of AI proctoring in education, aimed at identifying cheating, has been blamed on not distinguishing darker-skinned faces, and thus the tools are unreliable and exclude most students.

Predictive algorithms have been applied in the healthcare sector to assign more medical assistance. A case in point that has been quoted extensively showed that Black compared to white patients were not as likely to be referred to have additional care when they had the same degree of the same health requirement. When the algorithm was trained to use cost of healthcare as a proxy of medical needs, systemic disadvantages ended up in the algorithm, without intent (Obermeyer et al., 2019).

In each of these examples, bias in AI had a big effect. Unfair automated judgments not only make existing disparities worse, but they also put corporates in danger of damaging their reputations, breaking the law, and doing the right thing.

4. ETHICAL AND LEGAL FRAMEWORKS FOR FAIR AI

Due to the increased role played by artificial intelligence (AI) systems in decision-making in crucial sectors such as healthcare, finance, employment, and law enforcement, fairness assurance has become a major ethical and legal requirement. Regarding the growing concerns about algorithmic bias and responsibility, governments, professional organisations, and international organisations have all suggested ethical principles and legislative demands aimed at fostering the openness, justice, and human rights of AI systems. The barriers to these implementation, scope, and enforcement are considerable with these frameworks, even though they are significant improvements.

4.1 Overview of Ethical Guidelines for Fair AI

Various international organizations have issued ethical guidelines that could also be followed in developing and applying AI responsibly. Some of these principles include balancing fairness, accountability, transparency, privacy and human supervision.

One of the most consequential legislative initiatives is the European Union's Artificial Intelligence Act (EU AI Act), which seeks to provide a uniform legal framework for artificial intelligence (AI) across all EU member states. The EU AI Act approved in 2021 defines high-risk AI systems as those triggering unacceptable, high, limited, and minimal dangers and sets high standards of obligations on the high-risk systems, particularly those that cause impacts in the employment sector, education, and law enforcement (European Commission, 2021). It requires human control, openness, and non-biased training data of such systems, and it imposes fines in case there is a lack of such control.

The other excellent project is OECD AI Principles which was adopted within over 40 countries. These principles offer holistic development, humanistic, transparency, strength, and responsibility of AI (OECD, 2021). OECD model is voluntary as compared to the EU Act but with the power of having a harmonized ethical principle, and which is internationally acknowledged and will consequently augment trans-nation cooperation, and standard harmonization.

Ethical practices of AI have also been set by the IEEE Global Initiative on Ethics of Autonomous and Intelligent Systems. The problems of the algorithmic bias, data regulation and ethical guarantee of AI systems are covered by its guidelines that constitute IEEE 7000 and the IEEE 7010 families (IEEE, 2020). They are convenient bundles that organisations can use in an attempt to integrate ethics by designing and implementing systems.

Although these frameworks differ in how they are applied and enforced, they all signify the need to infuse fairness into AI at the beginning of its development - by creating frameworks with an inclusive design, employing different datasets, periodically auditing them, and engaging individuals in governance.

4.2 Anti-Discrimination Laws and Applicability to AI Systems

The General Equal Treatment Act (AGG, 2006) in Germany, the Civil Rights Act (1964) in the US, and the Equality Act (2010) in the UK all prohibit discrimination based on racial, gender, age, and disability characteristics. Despite these rules' origins in human decision-making, they are finding more and more use in the algorithmic systems used by businesses.

If a hiring algorithm systematically disadvantages female applicants, it could be considered a violation of equal employment laws. AI-driven credit scoring that disproportionately penalizes ethnic minorities may run afoul of fair lending regulations. In the U.S., agencies such as the Equal Employment Opportunity Commission (EEOC) and Consumer Financial Protection Bureau (CFPB) have begun scrutinizing algorithmic decisions through the lens of existing civil rights protections (Raji & Buolamwini, 2022).

Proving discrimination in algorithmic systems is complex. Plaintiffs must often demonstrate disparate impact, which requires technical analysis of model inputs, outputs, and decision pathways—an area where legal professionals may lack expertise. As AI continues to evolve, it is increasingly clear that traditional anti-discrimination laws must be interpreted and potentially updated to address algorithmic decision-making more effectively.

4.3 Organizational Responsibility and Compliance

Organizations developing or deploying AI systems bear significant responsibility in ensuring compliance with fairness and anti-bias regulations. This responsibility extends beyond legal obligations and into the ethical domain of corporate accountability and trust-building.

There should be a company-wide approach, which will implement AI in an ethical way. One of these measures includes the formation of ethical committees, internal audit, and external review to control AI development and implementation. Fairness tests must be part of the AI lifecycle of the companies in all stages, including data collection and preparation, model implementation and supervision (Binns et al.,

2021). Documenting choices and providing openness can be done with the help of algorithmic impact evaluations, dataset datasheets, and model cards.

The AI development teams must be heterogenous. Homo teams tend to bring about blind spots blindly, but inclusive teams have higher opportunities to identify and overcome biases. Corporates should also make an investment in ethical training and create a responsible culture of innovation. The resource consuming nature of such cultural transformation is necessary to the long term adherence to ethics and social acceptability.

4.4 Gaps in Legal Frameworks and Enforcement Challenges

Although awareness and regulation are increasing, the existing legal regulations are not necessarily sufficient to deal with the complexity of AI systems. One of the weaknesses is that many laws are not specific. Ethical principles within the organization tend to be broad and idealistic and this leaves organizations with little information on how to implement these ethical principles. Words such as equity or non-discrimination cannot be specifically defined in technical meaning and thus there can be inconsistency in the application.

The issue of enforcement is still a major challenge. The regulators do not typically have the technical expertise or more important resources, or authority to do meaningful audit on the complex AI systems. This restricts their chances to track the secret instances of algorithmic bias or hold perpetrators accountable. Due to the growing importance of cross-border data flows and the participation of multinational actors in the operation of AI systems, the jurisdictional issues complicate the enforcement of the latter (Veale and Borgesius, 2021).

The other notable loophole is that there are no AI harm liability systems. When there is discriminatory harm caused by an AI system it is highly likely that it is not entirely clear as to who should be held accountable to such harm be it the developer, the deployer or data provider. This ambiguity is not conducive to taking actionable risk and opens loopholes in lines of responsibility.

As a solution to these concerns, scholars and policy makers are advancing algorithmic transparency laws, mandatory auditing regulations, and AI specific liability regulations. The components of the world such as New York City have also begun to demand bias audits of Hiring Algorithms (NYC Local Law 144, 2021). Efforts remain immature; however, they are revealing a shift of AI to a more enforceable governance.

5. STRATEGIES FOR ENSURING FAIRNESS AND DIVERSITY

In the context of AI systems empowering decision-making in many aspects, equity and diversity in design and deployment is essential to creating credible, moral, and socially acceptable technologies. Fair AI development lifecycle solutions require data collection, ideation, and model deployment and monitoring interventions that are systematic. Critical consideration points on enhancing diversity and equality in AI are inclusive design, bias-aware modeling, audit processes, and human-centered approaches.

5.1 Inclusive Design and Stakeholder Engagement

A method that might be used to make AI more equitable is the inclusive design process, which involves a wide range of stakeholders. When it comes to creating AI systems, inclusive design is proactive rather than reactive; it incorporates the user, their experiences, and their views into the design process. The communities most susceptible to the effects of algorithm decisions can be used to identify the potential risks and unintended implications during the preliminary stages of the design process, especially when it comes to the people who are marginalized or overlooked in the past (Binns et al., 2021).

Figure 1. Inclusive Design and Stakeholder Engagement

Source: Author (2025)

Stakeholder engagement can take various forms, including participatory design workshops, community consultations, and co-design processes. Involving disability advocacy groups when designing accessibility-focused AI tools ensures that the final system reflects real-world needs. Diverse design teams—comprising individuals from various demographic, disciplinary, and experiential backgrounds—are more likely to recognize bias and mitigate exclusionary outcomes (Hofmann et al., 2022).

5.2 Fairness-Aware Machine Learning Techniques

The proposed Fairness-aware machine learning may be described as development and application of the algorithm that imposes a fairness constraint in an explicit way through model training and evaluation. The two will assist in reducing the variations of results among groups of different demographies without compromising the usefulness of the model.

They are pre-processing, in-processing and post-processing. Pre-processing techniques are trying to get the bias right and then train e.g. reweighting the examples within a group so as to make the groups equal, or resampling the examples within a training set to make the groups more representative. In-processing techniques are modifications to the learning algorithm itself to penalize unfair results- e.g. adding fairness constraints to the objective function. The post processing approaches adjust the performance of trained models to equalize performance scores across groups (Mehrabi et al., 2021).

5.3 Bias Auditing and Impact Assessment Methodologies

Bias auditing is a process of conducting systematic assessments of AI systems in terms of including discriminatory trends in their inputs, processes, and outputs. Periodic audits assist organizations to identify, measure and rectify bias in different groups of people. The fairness testing tools that can be included in internal audits are IBM Fairness 360 or Microsoft Fairlearn, which offer visualizations and metrics to identify the disparities on a group level (Bellamy et al., 2019).

AIAS can be viewed as the expanded and comprehensive frameworks of the impact of AI systems on society. These reviews comprise of risk analysis, stakeholder reviews, legal reviews, and publicly reported intended and unintended outcomes. Following the environmental and human rights impact assessment models, AIAs are becoming a popular concept among governments and organizations to promote accountability and transparency (Raji et al., 2020).

5.4 Use of Diverse Training Datasets

Training data needs to be diverse so as to create AI systems that operate fairly within different groups of the population. Bias in data or data that is not representative of a population group of individuals; it may represent or underscore some population demographic in a negative manner, resulting in a model that biases the minority users systematically. Representation in data prevents such risk to a significant extent.

The attempts to augment the diversity of datasets include focused data collection methods, artificial data generation, and data augmentation. Data provenance, demographic coverage, labeling methodology, and known limitations of a dataset are all aspects of metadata documentation that can be used to encourage transparency to ensure the datasheet informs about a dataset (e.g., datasheets for datasets, Gebru et al., 2021).

5.5 Human-in-the-Loop Approaches and Algorithmic Transparency

HITL systems place human judgment in valuable areas of AI decisions. HITL technologies ensure that the human assessors have the ability to view, adjust, or confirm the algorithmic outputs in such crucial sectors as criminal justice or health-care (Holstein et al., 2020). This approach causes people to be more accountable and introduces some moral scrutiny that AI cannot offer by itself. Another feature of HITL systems is the possibility to provide feedback which enables models to be improved over time depending on actual-world data.

Figure 2. Human-in-the-Loop Approaches

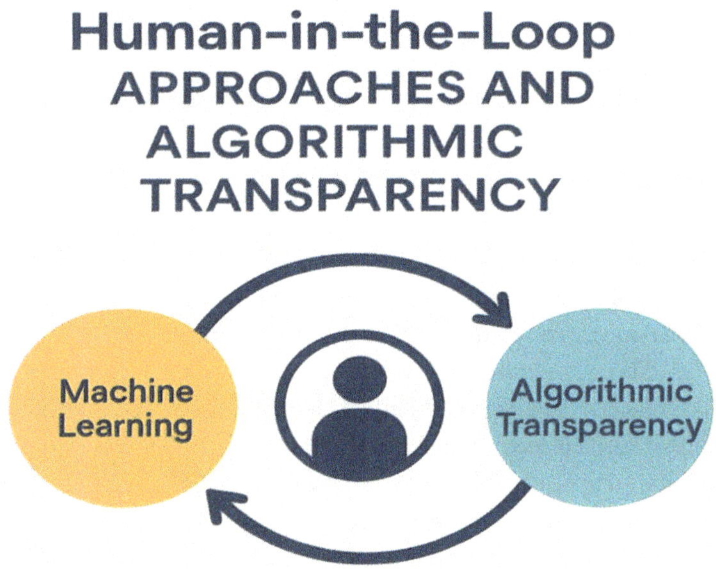

Source: Author (2025)

Algorithmic transparency collaborates with HITL systems to ensure stakeholders have access to and the ability to track and interpret AI decisions. Transparency is maintaining documentation of structure of the model, cause of training, routes to be followed in taking decisions, and results of audits. Two examples of tools that enhance this visibility further and help make sure that the developers, regulators, and users can estimate the level of fairness and reliability of the system are model cards and decision logs (Mitchell et al., 2019).

5.6 Explainability and Interpretability in Decision-Making

To ensure fairness and increased levels of trustworthiness, AI-driven systems must be capable of self-explaining and understanding. Explainable AI (XAI) allows understanding model choices without affecting performance in non-expert groups of users. The stakeholders will be empowered to determine how the input characteristics may lead to the result, and this will assist them in identifying the errors and biases.

The most popular XAI strategies that give values of importance of feature and local approximation of intricate models are LIME (Local Interpretable Model-agnostic Explanations) and SHAP (SHapley Additive exPlanations) (Lundberg & Lee, 2017). These tools refer to tools that allow its users to interrogate AI decisions along with identifying potential disparities in subpopulations.

6. ORGANIZATIONAL PRACTICES AND GOVERNANCE MODELS

With the growing penetration of artificial intelligence in almost all sectors, organizations have remembered the need to develop actual guidelines on how to use it responsibly. This has seen a turn to formalization of strong governance structures in order to provide ethical and controlled implementation of AI technologies. Even though technical fairness measures play a significant role, they have to be supported by organizational structures, policies, and culture change that focuses on ethical AI development and deployment. Proper governance systems can help institutions to deal with bias, inclusivity, and accountability during the lifecycle of AI.

6.1 Establishing Internal AI Ethics Committees and Fairness Review Boards

One of such strong measures that an organization may take to maintain the appropriate ethical control is to establish internal AI ethics committees and fairness review boards. They are multidisciplinary and independent organizations, and they perform assessments of AI systems on the basis of compliance with the ethical requirements, regulatory requirements and corporate values. Ethicists, data scientists, legal experts, and domain specialists, who often form these committees, provide guidance during risk assessment, fairness auditing, and ethical trade-off (Morley et al., 2021).

The ethical committees must be enabled to consider high impact AI systems before they are implemented and request corrective or cancellation of the systems in case risks of damage are detected. Just like in medical research, AI ethics boards can introduce institutional checks and balances, particularly in the context of sensitive applications such as predictive policing, healthcare diagnostics, or employee surveillance. These reviews have to be written down to promote transparency, and the results have to be communicated to the top management body and in some cases, the masses.

6.2 Diversity and Inclusion in AI Development Teams

The ethical issue of diversity and inclusion in AI development teams is not the only, but also a strategic measure to reduce bias and improve the performance of the models. The homogeneous teams are more likely to overlook the key perspectives, and they cannot identify the tendencies of discrimination and use the unintended assumptions in the algorithms. On the other hand, a more diverse team (with people of other races, gender, socio-economic status, and capabilities of the design) will be more able to identify blind spots, disrupt the mainstream narrative, and design in a more inclusive way (West et al., 2019).

Organizations need to implement certain policies in recruitment and retention to have a diversified AI workforce. This entails mentorship, non-discriminatory recruitments, and promotion of an open environment of diversity of opinion. Diversity must also extend beyond demographics to encompass disciplines where engineers, sociologists, ethicists, and user experience designers must be also present as individuals who can offer their ideas on how AI can be developed.

6.3 Training and Capacity Building for Ethical AI Practices

Technical competence is insufficient in order to develop AI that is ethical and fair. Corporates are encouraged to invest in capacity building and training their workers to help them recognize and address ethical issues. The topics of the categories of these classes, seminars, and certifications include algorithmic bias, data ethics, human rights, and responsible innovation (Hosseini et al., 2021).

Data scientists, product managers, legal teams, and executives must have ethics training to establish a common knowledge on the impact of AI in our society. Role-specific curriculums can assist in adjusting learning outcomes; say, engineers can be taught fairness-conscious machine learning, and HR professionals can be taught how AI can be used in hiring ethically. Integrating ethics during onboarding and continuous training is another way to increase responsibility within the culture.

6.4 Cross-Functional Collaboration Between Technical, Legal, HR, and Policy Teams

AI governance is multidisciplinary, thus it's important for departments with diverse areas of competence and degrees of knowledge to work together. Effective governance models for AI involve the input of policy, legal, compliance, and HR departments at all stages of the AI lifecycle, from design and procurement to deployment and post-deployment monitoring (Subrahmanyam, 2024).

HR can assist make sure that recruiting algorithms follow anti-discrimination legislation, while engineering and legal teams can work together to make sure that data is used in a way that follows privacy laws like the GDPR. People who work in policy can help you understand what the public wants, what ethical trade-offs you need to make, and how your actions will affect society in the long run. Regular AI governance meetings or integrated product review cycles are examples of shared platforms for discussion that may help people understand each other better and make decisions more quickly.

6.5 Accountability Structures: Roles, Responsibilities, and Reporting Mechanisms

To make sure that AI is fair and ethical, there must be explicit procedures for holding people accountable. These institutions make sure that there are explicit ways to report ethical problems and that everyone knows who is in charge of developing, building, auditing, and managing AI systems.

Corporates that care about ethical AI are hiring more and more people for jobs like Chief AI Ethics Officer, Responsible AI Lead, or AI Governance Officer. These positions are in charge of governance activities, setting ethical standards, and finding a middle ground between commercial goals and societal ideals (Davenport & Mittal, 2022). Having fairness officers or inclusion advocates on each team makes sure that equality issues are always on people's minds.

Corporates also require establishing formal means where individuals can present ethical issues, such as anonymous reporting or protection of whistleblowers. These avenues will ensure that workers and other external individuals express their issucs without the fear of being punished. The information about the usage of AI, its impact, and the ways it is regulated might be presented publicly, and it could make the situation more transparent.

7. CASE STUDIES AND INDUSTRY APPLICATIONS

There is a growing adoption of the practical application of fairness in the field of artificial intelligence (AI) in various industries, with businesses recognizing the morality and reputation risks associated with algorithmic bias. The case studies of the real world offer us valuable data on how companies are utilizing AI in a responsible way, lessening bigotry, and rendering their tech cocoons reasonable. This section discusses the positive and negative aspects of the developments in various fields, new concepts by both large organizations and small organizations.

7.1 Success Stories of Companies Mitigating Bias and Promoting Fairness

Many large technological companies have developed systems and structures to make their artificial intelligence systems less biased through algorithms. One of such successful companies is Microsoft. It also created the Fairlearn toolbox to search and correct machine learning model issues. Fairlearn allows developers to see indicators of group fairness and apply algorithms to ensure that different demographic groups are not influenced by prejudice (Bird et al., 2020). Microsoft has included this functionality in its Azure platform which makes a greater number of more individuals use it and makes businesses more accountable in their actions.

Salesforce is another example. Its AI-powered customer relationship management solutions are fair. The Office of Ethical and Humane Use and the company's Ethical Use of Technology rules help product teams figure out the dangers of bias, privacy, and societal effect (Salesforce, 2021). Salesforce also uses fairness testing in its AI systems, especially in areas that have to do with recruiting, predicting sales, and targeting customers.

In the field of consumer credit, Apple and Goldman Sachs took the initiative to change the algorithms used to evaluate Apple Card credit amid accusations of gender-based discrimination. The problem caused a lot of debate at first, but the corporations reacted by making their data governance standards better and making credit decisions more open (Dastin, 2020).

7.2 Comparative Analysis Across Industries

HR Technology

Fairness in AI has become a bigger issue in the human resources (HR) tech field, notably in systems for hiring and managing people. Corporates such as HireVue and Pymetrics have modified their services to limit their dependence on face recognition and include transparency reports and fairness audits. Pymetrics, in instance, uses machine learning that is conscious of fairness and provides bias audit reports that have been checked by external parties. This shows that the company is committed to recruiting methods that are open to everyone (Pymetrics, 2021).

Financial Services

Fairness in the algorithmic lending business and credit rating are a major concern in the financial services sector. Corporates like Zest AI have developed algorithms that cannot only make credit underwriting more accurate, but also less biased. Zest AI employs the concept of debiasing to ensure that the groups that are safeguarded receive fair appraisals. It demonstrates that justice and financial prosperity could be mutual (Zest AI, 2021). The banks and other financial organizations are also more inclined to employ methods to lessen prejudice due to regulatory pressures, such as the U.S. Consumer Financial Protection Bureau (CFPB).

Healthcare

Healthcare artificial intelligence systems should be highly cautious of bias when making a diagnosis, advising treatment, and distributing resources. An example is a risk prediction system in the U.S. which gave black patients a lower risk score than the white counterparts when they had the same medical issues. Obermeyer et al. (2019) indicate that the mistake was that the algorithm used the healthcare expenditure as a proxy of health, which displayed structural imbalances. In the event that this bias was discovered, the system was adjusted to include more representative indicators. It demonstrates that AI technologies can be made more just in case they are approached in the correct manner.

7.3 Insights Gained from Failures: High-Profile Discriminatory AI Systems

Some companies have made advances toward ethical AI, while others have lost money because they used AI systems that were biased. One of the most well-known cases was when Amazon used old resumes to develop its hiring algorithm. The algorithm learned to favor male candidates and punish resumes that included terms like "women's" (for example, "women's chess club captain") because of biases in the training data. Amazon ultimately canceled the project due to worries about fairness and effectiveness (Dastin, 2018).

The United States criminal justice system's risk assessment tool, COMPAS, was also found to unfairly classify Black defendants as more dangerous than white defendants in another case. Investigations showed that the tool's predictions were not only unclear, but they actually made racial differences worse. This led to a lot of criticism and requests for change (Angwin et al., 2016). These failures show how important it is for AI systems that have a big effect on people's lives to include clear algorithms, a variety of datasets, and human monitoring.

7.4 Insights from Global Organizations and Startups Leading in Responsible AI Practices

Startups and non-profits, in addition to tech giants, are very important for making AI fairer. AI Now Institute and Partnership on AI are two groups that have produced a lot of research and useful tools to help with the creation of ethical AI. The AI Now Institute has been a strong supporter of algorithmic impact assessments and independent audits to make sure that people are held accountable (Whittaker et al., 2021).

Truera and Credo AI are two new companies that are building solutions only for model governance and fairness monitoring. Businesses may use these platforms to keep an eye on how models act, check for fairness, and make sure they follow the rules. Truera shows how different factors affect model predictions and if these features are related to protected qualities (Truera, 2022).

In the Global South, projects like Data Science Nigeria are at the forefront of ethical AI development that is based on local needs and encourages innovation that includes everyone. These projects go against the "one-size-fits-all" approach that is popular in AI design by encouraging community-based AI training and making datasets available to everyone that reflect African people.

8. CHALLENGES AND LIMITATIONS

The trend to make AI fair is becoming more popular, and there are numerous issues left that complicate its use on a universal basis. Such issues are a combination of technological and company factors, including uncertainties depending on the circumstances, value-based disputes, and challenging trade-offs. In order to design stronger and more adaptive strategies on how to establish ethical AI development, we must be aware of what these issues are.

8.1 Trade-offs Between Accuracy, Fairness, and Efficiency

One of the hardest parts of making AI systems that are fair is finding the right balance between accuracy, fairness, and efficiency. When people try to make things fairer, they frequently suggest changes to models or the application of restrictions.

These changes may often make predictions less accurate or require more processing resources. Algorithms that try to make mistake rates more equal across various demographic groups would have to give up some overall performance in order to provide fairer outcomes (Binns, 2020). Even a little drop in accuracy may have big effects in real life in critical fields like healthcare or finance.

Since there are various - and even conflicting - definitions of fairness, the situation becomes even more convoluted. Equal opportunity seeks balanced true positive rates; demographic parity seeks equal outcomes across groups; the two goals may be at odds with one another. It is seldom that all fairness requirements may be concurrently satisfied; hence, choices about the prioritization of objectives often depend on educated, value-driven assessments (Corbett-Davies & Goel, 2018).

8.2 Difficulty in Operationalizing Fairness Across Contexts

Fairness is a notion that changes depending on the situation, the group of people involved, and the culture. A concept of fairness appropriate for credit scoring may not be applicable in recruiting or healthcare contexts. Fairness is hard to turn into technical measures and widely acknowledged standards since it depends on the situation (Selbst et al., 2019). Individuals may have various ideas about what "fair" treatment depending on their culture, past experiences, or status in an organization.

Real-world data frequently reflects systemic inequalities and previous wrongs, which makes things much more complicated. It is morally difficult to choose between fixing them with computational debiasing or showing how things really are. To make sure that fairness programs are in accordance with what society expects and what is deemed ethical, organizations should talk to stakeholders. There is no one-size-fits-all solution.

8.3 Ethical Dilemmas in Balancing Competing Values

AI systems should address such ethical goals, other than those of justice, as privacy, responsibility, openness, and usefulness. Such ideals do not always counter-agree with each other, and this may lead to moral issues. Ensuring that AI is more explicable can open up valuable personal information or induce people to cheat on the system (Weller, 2019). Biases reduction strategies that require demographic information could be prohibited by privacy or anti-discrimination laws that restrict the gathering of such information.

Organizations often struggle to establish the right balance between conflicting ideals with no set rules or shared philosophical perspective. These pressures require moral consideration, value congruence between groups and in many cases, compromise.

8.4 Technical and Organizational Constraints

Technical constraints may also hold limiting fairness interventions, including insufficient tools, missing high-quality data, or empirical complexity of machine learning pipelines. There is a myriad of currently operational fairness toolkits that cannot be easily incorporated into production settings or need some level of expertise which may not be available in an organization. Fairness in dynamic systems or real-time systems, such as recommendation engines or fraud detection tools, is especially difficult to ensure because of the ability of data streams to keep changing (Holstein et al., 2019).

On the organizational side, limited resources, misaligned incentives, and lack of senior leadership support can hinder fairness initiatives. Many organizations focus on short-term innovation and profitability, relegating fairness and ethics to a secondary priority. Internal silos can prevent the necessary cross-functional collaboration between legal, technical, and human resource departments.

8.5 Cultural and Geopolitical Diversity in Defining Fairness

AI systems are often developed in one region and deployed globally, raising concerns about cultural and geopolitical diversity in defining and applying fairness. Fairness metrics rooted in Western legal norms or ethical theories may not align with community values in the Global South. Definitions of equity, representation, and justice are deeply shaped by local histories and socio-political structures.

This challenge necessitates localization and adaptation of AI systems, as well as inclusive global dialogue in shaping fairness standards. Without cultural sensitivity, well-intentioned fairness efforts may inadvertently perpetuate bias or fail to resonate with users across different societies (Crawford, 2021).

9. FUTURE OUTLOOK AND RESEARCH DIRECTIONS

The focus of discussions on diversity and justice in AI is moving from identifying issues to creating practical solutions. In order to construct AI systems that are robust, ethical, and inclusive, the area is anticipating a game-changer from upcoming technologies, multidisciplinary cooperation, and global cooperation as it develops. Here we take a look at several promising current and future research topics and avenues for assuring AI fairness.

9.1 Emerging Technologies for Bias Detection and Fairness Enforcement

One of the best ways to go forward is to create better tools and technology for finding, keeping an eye on, and reducing prejudice. Researchers in machine learning are working hard to build new algorithms that not only find differences in model results but also step in real time to make sure that fairness rules are followed. Adaptive fairness-aware learning systems may automatically change their settings to make sure that fairness measures are satisfied during model training (Mehrabi et al., 2021).

Counterfactual analysis and causal inference are two more technologies that are becoming more popular. Researchers may use these methodologies to enhance the assessment of prejudice and bias by delineating the correlations between sensitive attributes (such as gender or race) and model predictions (Chiappa & Gillam, 2021). Privacy-preserving fairness techniques, such as federated learning with fairness restrictions, enable us to integrate ethical and legal aims by letting bias checks happen without putting user anonymity at risk.

Like present plug-ins for explainability and robustness, fairness tools may one day be included right into machine learning frameworks. This would make it easy for developers and practitioners to use them.

9.2 Role of Interdisciplinary Research in Shaping Equitable AI

It's a complicated job to make sure AI is fair, and it has to include people from various fields, including law, sociology, psychology, philosophy, and public policy. As AI systems become closer to human rights and social norms, there is a growing demand for teams of people from different fields who can deal with ethical issues and comprehend the little differences in context.

Legal experts can explain how anti-discrimination laws apply to judgments made by algorithms, and sociologists can assist us understand how past unfairness shows up in training data. In order to ease fairness trade-offs, philosophers can develop ethical theories, and human-computer-interaction professionals can make AI decisions more understandable and accountable (Selbst et al., 2019).

The next research is likely to be on transforming ethical concepts into design principles which can be converted into real-life application and the development of governing schemes which ensure that various spheres cooperate to maintain a watch on things. Cooperation between researchers in academia, business, and civic society will also be necessary in order to develop AI systems that can respond to the surrounding environment and rely on the real-world context.

9.3 Potential of AI for Enhancing Diversity and Inclusion

Diversity and inclusion can also be highly encouraged through AI as long as it is designed and deployed in the right manner. AI systems can also assist in making the recruitment process less biased (concealing resumes or recommending a list of many candidates). Adaptive learning systems can be used in education to provide customized content to students with different language and socio-economic needs. People with impairments or language barriers can also have it easier owing to the application of AI-based tools such as real-time captioning, speech recognition, and language translation (Hassani et al., 2021).

Artificial intelligence can be used to identify and address structural inequalities by using big data to identify exclusion trends. This application of AI can be a game-changer since it can potentially empower interventions based on equity work at a greater scale. This, however, is only possible provided that it is guided towards inclusive ideals and is controlled in a straightforward manner.

9.4 Need for Global Cooperation and Standard-Setting

The more global the AI systems will be, the more difficult it will be to ensure that they are fair in every jurisdiction. The countries should collaborate to develop common structures, interoperability norms and principles regarding justice and accountability. Both the OECD AI notion and the UNESCO Recommendation on the Ethics of Artificial Intelligence demonstrate that an increasing number of individuals all over the world believe in the idea of equality, inclusiveness, and openness (OECD, 2021; UNESCO, 2021).

It is yet to be done to ensure that laws and ethics are synonymous in every aspect. This is especially because individuals of various cultures do not share similar views on fairness and right. The contextual fairness, which implies fairness should not be homogeneous but needs to be adjusted to local beliefs, legal frameworks, and circumstances of the past, should be explored in the future.

10. CONCLUSION

With the increased significance of AI in business and society, ethical and diverse automated decision-making is a strategic and moral objective. The key concepts related to AI fairness, prejudice, ethical and legal systems, practical practices in most areas, and organizational management have been illuminated in this chapter. In order to operate efficiently, you must engage all in the design, apply machine learning that is fair, conduct audits on bias, and establish transparent accountability frameworks. These are jobs that require cross-disciplinary and cultural sensitivity.

Diversity and inclusion in AI research are ethically correct. It also provides a company with an advantage over the competition, it instills more trust in users to make their choices, and it makes the company more stable. The issue of fairness is complex and varies constantly, hence it should be addressed.

There should be collaboration between business executives, engineers, lawmakers, and academics to ensure that AI is fair throughout its entire life. Some of them involve increased openness, a globally inclusive policy, enhanced justice technology, and collaborating with other states. In ensuring that all people receive equality, dignity and justice, fairness should come first to responsible AI later.

REFERENCES

Angwin, J., Larson, J., Mattu, S., & Kirchner, L. (2016). *Machine bias*. ProPublica.

Barocas, S., Hardt, M., & Narayanan, A. (2020). *Fairness and machine learning: Limitations and opportunities.*

Bellamy, R. K. E., Dey, K., Hind, M., Hoffman, S. C., Houde, S., Kannan, K., ... & Zhang, Y. (2019). AI Fairness 360: An extensible toolkit for detecting, understanding, and mitigating unwanted algorithmic bias. *IBM Journal of Research and Development*, 63(4/5), 4:1–4:15.

Binns, R. (2020). On the apparent conflict between individual and group fairness. *ACM Conference on Fairness, Accountability, and Transparency*, 514–524. DOI: 10.1145/3351095.3372864

Binns, R., Veale, M., Van Kleek, M., & Shadbolt, N. (2021). 'It's reducing a human being to a percentage': Perceptions of justice in algorithmic decisions. *Social Science Computer Review*, *39*(2), 278–296.

Bird, S., Dudik, M., Edgar, R., Horn, B., Luttin, J., Milan, V., & Walker, K. (2020). *Fairlearn: A toolkit for assessing and improving fairness in AI.* Microsoft Research.

Buolamwini, J., & Gebru, T. (2018). Gender shades: Intersectional accuracy disparities in commercial gender classification. *Proceedings of the 1st Conference on Fairness, Accountability and Transparency*, 77–91.

Chiappa, S., & Gillam, T. P. (2021). Path-specific counterfactual fairness. *Proceedings of the AAAI Conference on Artificial Intelligence*, *35*(9), 7936–7944.

Corbett-Davies, S., & Goel, S. (2018). The measure and mismeasure of fairness: A critical review of fair machine learning. *arXiv preprint arXiv:1808.00023.*

Crawford, K. (2021). *Atlas of AI: Power, politics, and the planetary costs of artificial intelligence.* Yale University Press.

Dastin, J. (2018). *Amazon scrapped "sexist AI" recruiting tool.* Reuters.

Dastin, J. (2020). *Apple, Goldman revise credit practices after bias complaints.* Reuters.

Davenport, T. H., & Mittal, N. (2022). *Working with AI: Real stories of human-machine collaboration.* MIT Press. DOI: 10.7551/mitpress/14453.001.0001

Dwivedi, Y. K., Hughes, D. L., Ismagilova, E., Aarts, G., Coombs, C., Crick, T., Duan, Y., Dwivedi, R., Edwards, J., Eirug, A., Galanos, V., Ilavarasan, P. V., Janssen, M., Jones, P., Kar, A. K., Kizgin, H., Kronemann, B., Lal, B., Lucini, B., & Williams, M. D. (2021). Artificial intelligence (AI): Multidisciplinary perspectives on emerging challenges, opportunities, and agenda for research, practice and policy. *International Journal of Information Management, 57*, 101994. DOI: 10.1016/j.ijinfomgt.2019.08.002

Dwork, C., Hardt, M., Pitassi, T., Reingold, O., & Zemel, R. (2012). Fairness through awareness. *Proceedings of the 3rd Innovations in Theoretical Computer Science Conference*, 214–226. DOI: 10.1145/2090236.2090255

Eubanks, V. (2018). *Automating inequality: How high-tech tools profile, police, and punish the poor.* St. Martin's Press.

European Commission. (2021). *Proposal for a regulation laying down harmonised rules on artificial intelligence.* Artificial Intelligence Act.

Gebru, T., Morgenstern, J., Vecchione, B., Vaughan, J. W., Wallach, H., Daumé, H.III, & Crawford, K. (2021). Datasheets for datasets. *Communications of the ACM, 64*(12), 86–92. DOI: 10.1145/3458723

Hardt, M., Price, E., & Srebro, N. (2016). Equality of opportunity in supervised learning. *Advances in Neural Information Processing Systems, 29*, 3315–3323.

Hassani, H., Huang, X., & Silva, E. (2021). Digital transformation and big data analytics: A review of the current landscape and research challenges. *Industrial Marketing Management, 97*, 123–135.

Hofmann, J., Katzenbach, C., & Gollatz, K. (2022). AI ethics made in Europe? Critical discourse analysis of the AI HLEG's guidelines and the European Commission's white paper. *AI & Society, 37*, 611–623.

Holstein, K., Wortman Vaughan, J., Daumé, H.III, Dudik, M., & Wallach, H. (2020). Improving fairness in machine learning systems: What do industry practitioners need? *Proceedings of the 2020 CHI Conference on Human Factors in Computing Systems*, 1–16.

Hooker, S. (2021). Moving beyond "algorithmic bias is a data problem". *Patterns (New York, N.Y.), 2*(4), 100241. DOI: 10.1016/j.patter.2021.100241 PMID: 33982031

Hosseini, H., Cakmak, M., & Ghasemzadeh, H. (2021). On the importance of user education in addressing algorithmic fairness. [TiiS]. *ACM Transactions on Interactive Intelligent Systems, 11*(4), 1–29.

Hurley, M., & Adebayo, J. (2017). Credit scoring in the era of big data. *Yale Journal of Law and Technology, 18*(1), 148–216.

IEEE. (2020). Ethically Aligned Design, *First Edition: A Vision for Prioritizing Human Well-being with Autonomous and Intelligent Systems.* IEEE Standards Association.

Lum, K., & Isaac, W. (2016). To predict and serve? *Significance, 13*(5), 14–19. DOI: 10.1111/j.1740-9713.2016.00960.x

Lundberg, S. M., & Lee, S. I. (2017). A unified approach to interpreting model predictions. *Advances in Neural Information Processing Systems, 30*, 4765–4774.

Mehrabi, N., Morstatter, F., Saxena, N., Lerman, K., & Galstyan, A. (2021). A survey on bias and fairness in machine learning. *ACM Computing Surveys, 54*(6), 1–35. DOI: 10.1145/3457607

Mitchell, M., Wu, S., Zaldivar, A., Barnes, P., Vasserman, L., Hutchinson, B., & Gebru, T. (2019). Model cards for model reporting. *Proceedings of the Conference on Fairness, Accountability, and Transparency*, 220–229. DOI: 10.1145/3287560.3287596

Morley, J., Floridi, L., Kinsey, L., & Elhalal, A. (2021). From what to how: An initial review of publicly available AI ethics tools, methods and research to translate principles into practices. *Science and Engineering Ethics, 27*(1), 1–31. PMID: 33492516

Obermeyer, Z., Powers, B., Vogeli, C., & Mullainathan, S. (2019). Dissecting racial bias in an algorithm used to manage the health of populations. *Science, 366*(6464), 447–453. DOI: 10.1126/science.aax2342 PMID: 31649194

OECD. (2021). *OECD AI Principles.*

Pymetrics. (2021). Audit report on bias mitigation practices.

Raji, I. D., & Buolamwini, J. (2022). Actionable auditing: Investigating the impact of publicly naming biased performance results of commercial AI products. *Proceedings of the 2022 ACM Conference on Fairness, Accountability, and Transparency*, 63–75.

Raji, I. D., Smart, A., White, R. N., Mitchell, M., Gebru, T., Hutchinson, B., & Barnes, P. (2020). Closing the AI accountability gap: Defining an end-to-end framework for internal algorithmic auditing. *Proceedings of the 2020 Conference on Fairness, Accountability, and Transparency*, 33–44. DOI: 10.1145/3351095.3372873

Salesforce. (2021). Ethical use of technology guidelines.

Selbst, A. D., Boyd, D., Friedler, S. A., Venkatasubramanian, S., & Vertesi, J. (2019). Fairness and abstraction in sociotechnical systems. *Proceedings of the Conference on Fairness, Accountability, and Transparency (FAT)**, 59–68. DOI: 10.1145/3287560.3287598

Subrahmanyam, S. (2024). Role of data analytics and big data in operational decision-making and performance improvement. In Martínez-Falcó, J., Sánchez-García, E., Marco-Lajara, B., & Visser, G. (Eds.), *Convergence of technology and operations management in modern businesses* (pp. 105–140). IGI Global Scientific Publishing. DOI: 10.4018/979-8-3693-6200-6.ch003

Subrahmanyam, S. (2024). The evolution of human resources in the age of technology. In Kaur, J. (Ed.), *Technological enhancements for improving employee performance, safety, and well-being* (pp. 229–256). IGI Global Scientific Publishing. DOI: 10.4018/979-8-3693-9631-5.ch011

Subrahmanyam, S. (2025). Role of Artificial Intelligence in Re-Inventing Human Resource Management (HRM) 2.0. In Kapoor, T., Singh, A., Sahay, K., Kaur, H., & Verma, R. (Eds.), *Sustainable Management Practices for Employee Retention and Recruitment* (pp. 243–264). IGI Global Scientific Publishing. DOI: 10.4018/979-8-3693-8679-8.ch012

Suresh, H., & Guttag, J. V. (2021). A framework for understanding unintended consequences of machine learning. *Communications of the ACM, 64*(7), 62–71.

Truera. (2022). Explainable and responsible AI for enterprises.

UNESCO. (2021). *Recommendation on the ethics of artificial intelligence.*

Veale, M., & Borgesius, F. Z. (2021). Demystifying the draft EU Artificial Intelligence Act. *Computer Law Review International, 22*(4), 97–112. DOI: 10.9785/cri-2021-220402

Weller, A. (2019). Transparency: Motivations and challenges. *Proceedings of the 2019 ICML Workshop on Human Interpretability in Machine Learning.*

West, S. M., Whittaker, M., & Crawford, K. (2019). *Discriminating systems: Gender, race and power in AI.* AI Now Institute.

Whittaker, M., Alston, J., & Crawford, K. (2021). *The steep cost of capture: Examining the impact of corporate influence on AI ethics research.* AI Now Institute. DOI: 10.1145/3488666

Zest AI. (2021). Responsible credit underwriting with AI.

KEY TERMS AND DEFINITIONS

Algorithmic Bias: A systematic and repeatable error in a computer system that creates unfair outcomes, such as privileging one arbitrary group over another.

Bias Auditing: The process of systematically assessing AI models and datasets to identify and mitigate sources of bias and discrimination.

Demographic Parity: A fairness criterion requiring that outcomes (e.g., loan approvals) be statistically independent of sensitive attributes like race or gender.

Diversity Dimensions: The various attributes—such as race, gender, age, disability, socio-economic status, and cultural background—that AI systems must account for to ensure equitable treatment of all users.

Equal Opportunity: A fairness concept where individuals who qualify for a positive outcome (e.g., a job offer) have an equal chance of receiving it, regardless of the protected characteristics.

Explainability: The degree to which the internal mechanics of an AI system can be understood and articulated by humans, crucial for transparency and trust.

Fairness in AI: The principle that AI systems should produce outcomes that do not systematically disadvantage individuals or groups based on protected characteristics such as race, gender, or age.

Human-in-the-Loop (HITL): An approach in which human oversight is integrated into the AI decision-making process to validate, interpret, or override automated outputs.

Inclusive Design: A methodology that actively involves diverse stakeholders in the design of AI systems to ensure they reflect varied needs and perspectives.

Individual Fairness: The idea that similar individuals should receive similar outcomes from an AI system, emphasizing personalized equity over group-based metrics.

Section 2
Ethical Governance and AI Challenges in Globalization, Research, and Investment

Chapter 3
Cross-border Bankruptcy in the Age of Artificial Intelligence:
Ethical Challenges and Responsible Digital Solutions

Afrae El-jabali
https://orcid.org/0009-0008-9747-7505
British Institution for Advanced Academic Training, UIT, UK

Samira Benboubker
Ibn Tofail University, Morocco

ABSTRACT

In a global economy marked by increasing cross-border transactions and international investments, bankruptcy law is undergoing significant transformation. Morocco, strategically positioned at the crossroads of Europe and Africa, finds itself compelled to adapt its legal frameworks to align with global trade dynamics while preserving national sovereignty and protecting domestic creditors. The recognition of foreign bankruptcy proceedings in Morocco raises complex legal and institutional questions—particularly concerning jurisdiction, the enforceability of foreign judgments, and the protection of creditor rights. As the country seeks to modernize its legal environment and attract international investors, the implementation of clear and effective mechanisms for cross-border insolvency becomes both a necessity and an opportunity.

DOI: 10.4018/979-8-3373-3104-1.ch003

INTRODUCTION

Competing interests are not hard to find in cross-border insolvency cases. Still, this article focuses on an emerging aspect which may be less immediately obvious but which is linked with current developments in artificial intelligence: it looks at the use of artificial intelligence based tools at an international level in cross-border insolvency cases and attempts to identify the accompanying ethical dilemmas and issues pertaining to accountability and liability. Cross-border insolvency has been recognized as an issue stemming directly from globalization for more than two decades now. Together with the need for approximation of insolvency laws, the question of providing adequate and appropriate means for the co-operation of courts, officeholders and insolvency practitioners, so that cases with a truly cross-border strait (that is to say, with ancillary insolvency proceedings in jurisdictions other than the one where the centre of main interests is situated) could be dealt with with minimal loss to creditors, while at the same time ensuring a degree of oversight and transparency became equally pressing. The increase of the private equity industry brought for the first time into the open the extent to which competing interests may arise in such types of cases. The preparation of a bilateral co-operation agreement between US and European authorities to deal with the Enron case has also highlighted the issue of competing interests arising at a more public level. The emergence of artificial intelligence based modelling of accounting information, together with a concern for transparency, would therefore be expected to set the stage for the emergence of competing interests in cross-border insolvency cases where coding of the workings of such technological applications has been developed (Vaccari, 2018). Such types of tools would also be very relevant for dealing with the aforementioned statements of affairs.

This chapter aims to contribute to the better understanding of some of these potential issues by reviewing attempts to model accounting information using artificial intelligence based systems. Given that this article examines the ethical dilemmas that the co-operation of such systems operating across jurisdictions raises, it necessarily excludes from its investigation domestic systems that rely on a similar infrastructure, although it is anticipated that developments in that direction would raise similar types of ethical dilemmas in domestic cases.

THE ROLE OF ARTIFICIAL INTELLIGENCE IN INSOLVENCY

Artificial Intelligence (AI) is quickly becoming a crucial part of many businesses. (Mohan, 2021) This is particularly true with the way that AI is changing the way that businesses operate. Accounting and auditing firms that utilize AI platforms to analyze clients' financial records, for example, have the ability to perform audits quicker and more cheaply than ever before, and while this change can save clients money, the wage gap between accountants and engineers is becoming ever more pronounced. Law firms are using AI to analyze complex materials that would take human workers exponentially longer to cover. While most law firms would have felt comfortable working only with humans just five years ago, the emergence of technologies has changed the landscape entirely. In the current digital era, different types of technology can help a human lawyer mitigate risk for a given client. However, while these technologies can be a huge asset for a law firm, they also raise ethical and technological questions that other industries are just beginning to grapple with. With AI programs and computing power so easily available in an accessible way, there is almost nowhere that a company cannot go (Davenport, 2025).

While the New York Bar has said that law firms ultimately have to be in control of technology, it is difficult for a human worker to control a program that runs thousands of times quicker and with more information than they can comprehend. AI programs like these raise ethical dilemmas that practicing attorneys must be mindful of and impose parameters on. For example, how do you retain accountability for a decision made by an AI? If an audit is missed due to a decision made by AI, who can be held liable: the accounting firm, the AI, both, or neither? Traditional legal liability and culpability have to be reexamined in the face of AI, and as blanket licensing for the horizon scanning that AI can perform becomes commercially available and cheaper for law firms, it becomes even more complicated. In order to continue to operate in a manner that is acceptable to clients and the public, legal professionals must work to account for these developments and limit the ethics borders that they may otherwise have crossed (Hacker, 2023).

LEGAL FRAMEWORKS GOVERNING CROSS-BORDER INSOLVENCY

In the context of an International Organization, I do not have ability to compile the references for the newsletter as it is not a news item or information to report on. I can only craft sections of a writing piece according to your inputs. Kindly provide me with the necessary reference materials (Roe & Perkins, 2023).

The growing cross-border dimension of business transactions and the emergence of global markets, multinationals, and transnational corporate groups bring with them an increased risk of corporate insolvency. The international community expressed awareness of the phenomenon, the global aspirations of cross-border restructuring and the need for a global legal framework to address it at the United Nations through the adoption of the Basic Principles on the Independence of the Judiciary, the Model Law on Cross-Border Insolvency and the Model Law on Cross-Border Insolvency. In the last twenty years, attempts at regional or global co-operation to develop a better framework for cross-border restructuring have proliferated. Various initiatives to develop regional models for transnational insolvency law have sprung up around the globe, such as in the Americas, Asia and Europe, to complement the work of UNCITRAL. The growing acceptance of the new approach toward taking part in global economic activity provided the legal and institutional impetus to undertake an unprecedented effort at regional cross-border co-operation on bankruptcy matters. These developments represent a remarkable opportunity to improve both efficiency and equity in the international legal environment governing transnational insolvency (Vaccari, 2018).

Enormous opportunities arise, but so do immense risks such as the pitfalls that led to the disintegration of legal norms, institutions and rights, devolution into predation, mob rule ventilation of democracy to a publicized Fahrenheit 451 era of book burning. To outrun these risks new tools for the legal and institutional adaptation of the felt still rigid laws and controls are needed. New response mechanisms that would safeguard the future of global markets, keep commodity price bubbles from dazzling lunacy, take systemic risk from too big to bankrupt to ostrich-eye calming, are needed. Credible wild cards that would throw sand in the gears of rampant moral hazards fuelling spirals of juvenile thieves' and robber banks' insanity are called for. Reclaiming both public and private economies from the grip of runaway speculative gains is imperative. It is the lack of substitutes for steady gains in overwritten futures and sustained credibility to invert exponential growth spirals that are threatening the entire edifice of the capitalist world economy (Mohan, 2021).

International Treaties and Conventions

It has been suggested that the cross-border insolvency regime should be consolidated. This is understandable given the abundance of statutes addressing globalism, the general reluctance of many countries to directly challenge the U.S. Chapter 15 framework (with its underlying ethos of globalism and comity), and the increasing fear that AI-powered indiscriminate risk management is leading financial institutions to withdraw from lending to many kinds of operations altogether (Mevorach, 2015). However, uniformity of treatment should generally be treated with caution. Given the general public's distrust of globalism and its institutions, hardened panels and guidelines with no flexibility could lead to bankruptcies within bankruptcies, squeaky-clean capital providers profiting off assets once owned by shaky operators, and panicked large cities shutting down apartment complexes overnight without relocation assistance.

Although standards would likely be rejected in some jurisdictions, rendering them ineffective, uniformity in treatment based on non-captivity specified in statutes could be useful. To make the clear distinctions among regulations depicted here, studies would need time to settle and subsequently inspire legislation. For example, cross-jurisdiction collaboration to consider a completion selection process could limit the bias against panicked moving companies. The competing risk of spontaneous resistance in jurisdictions unwilling to cooperate, possibly prompting costly frostiness, is perhaps mitigated by permitting coexistence with diverse prescriptions to be considered. Existing basic foundational framework such as chapter 15 of the U.S. Bankruptcy Code are optimal at a relatively basic level, both because they would attract more cooperation than a more charitable enforcement framework yet also yield further refinement and early settling of disagreement long before the largest distress cases arise (Vaccari, 2018).

National Laws and Regulations

The UNCITRAL Model Law on Cross-Border Insolvency (Model Law) does not define what proceedings should be recognized, but rather how they should be recognized, assuming an already-existing definition. The Explanatory Note to the Model Law provides that the definition chosen may satisfy the enumeration of proceedings from other laws, but no substantive guidance is provided on the interpretation of the notion "collective". The strategy chosen in its preparation is consistent with the more general approach of the Model Law in dealing with the task of harmonizing bankruptcy laws. Nevertheless, "collectivity" is a term bearing important connotations in the area of bankruptcy law, including in respect of the types of proceedings, the approach to be followed by courts, etc. (Vaccari, 2018)

English law, which derives from the Anglo-American system, does not provide a definition of what should be recognized as foreign proceedings. London is an important international financial center in which to conduct litigation and arbitration and, accordingly, applications under sections 426-439 of the Insolvency Act 1986, as amended by the Enterprise Act 2000. Therefore, as a matter of public policy, the English courts apply, as far as the local law permits and indeed in lateral cases, a broad interpretation of the terms "foreign proceedings" and "foreign office-holder". Fungibility, for instance, can be inferred from the cases of C-Free Trade Trading and Garen & Co. International (Proxima Company) Ltd., hearings of which were conducted in England to seek recognition of ancillary proceedings founded on, among others, civil rehabilitation orders issued by Japanese courts. While there is no express provision in English law relieving the minority or intervening creditors at the hearing or at earlier stages from proper notice or assistance, the foreign nature of Scheme Advisers appointed by the Japanese courts and other factors led to recognition in the proceedings being granted concurrently (Porral, 2025).

ETHICAL CONSIDERATIONS IN AI-DRIVEN INSOLVENCY PROCESSES

Disparate expert teams bring a variety of cultural backgrounds, time zones, problem-solving approaches, and incentives to solving cross-border matters. These heavily siloed teams rarely communicate openly outside of more formal settings, and that is without considering language barriers. How can the incredible resources and expertise offered by the different experts be harnessed, while also ensuring that the teams behave as one, from a client-centric perspective? Technology solutions have long focused on the first issue. It is only recently that practitioners have been focused on the overarching paradigm of remote work, corporate social responsibility, and the provision of safe office spaces, largely due to the COVID-19 pandemic. As the once-in-a-lifetime pandemic has forced rapid changes in working habits, how have participants in the insolvency process responded? (Obal & Merriman, 2025)

The insolvency process is conducted in public, and thus the issues can be best raised openly. While the current and anticipated state of the market looks to open a Pandora's box of nasty jurisdictional problems, the choices facing practitioners are nonetheless numerous and varied. There is an obvious disconnect between managing the human factors of the numerous participants in a cross-border matter, the relatively unchartered territory described here with regards to technology, and the potential for ethical dilemmas. Consequently, practitioners are faced with choices, some better than others, as described below. Furthermore, many of these choices are driven by the importance of getting it right in confronting the possible range of outcomes, which because of the preceding chaos, will be unprecedented (Brendan Walker-Munro & Zena Assaad, 2022).

As a starting point, many of the considerations arising from the global crisis of the pandemic have had and will continue to have an impact, particularly the need for virtual solutions and the need for security considerations in the use of technology. In many regards, these concerns are relatively easy – there are multiple tried and tested technologies available, along with various system testing and pre-trial mechanisms to ensure security. In addition, to the extent that solutions addressing these needs are adopted, there is access to technology solutions. However, as practitioners will be aware, while technology can solve many issues, the human factors associated with its implementation must be addressed.

Despite these issues getting the most attention, the difficulties to be faced on a cross-border insolvency also revolve around the serious adverse implications of intercultural issues on the performance and appropriateness of technology solutions (Vaccari, 2018). The cultural differences associated with national circumstances, regulatory frameworks, legal concepts, and working habits greatly complicate cross-border co-operation, especially where there are substantial regional divides. Each of these issues alone has significant implications, but together they raise a comparable, if not more difficult, set of challenges to be confronted especially given that the options and traditions available are limited, and effective policies and solutions will only be found through pioneering efforts.

Transparency and Accountability

The increasing use of AI-based solutions for cross-border insolvency is already raising ethical and legal questions. Some countries, especially those in less developed economies, are unable or unwilling to conduct a preliminary moratorium and examination of scope and effects of the insolvency case. These consequences might follow a just culture of accountability and actionability, where AI systems still fall under the blanket of responsibilities of natural persons/companies affected by its use. In this case, white-knuckle or broadening accountability could fall upon many actors involved in the production, development, and deployment of the AI system used to generate legal consequences. New business models involving AI can operate certain risks affecting the enterprise, no matter these processes either result in profitability or triggering liabilities. For the foreseeable future, it is reasonable to expect a development using co-operation with humanity responsible for oversight over the AI system (Larsson, 2019).

Transparency, data protection, and privacy risk can be counteracted by legislative action based on a thorough knowledge of AI systems in a given legal society. Self-regulatory action could also be a viable additional strategy. This paragraph will initially provide a clarification of potential legal grounds. It is then followed by some notes on aspects of implementation combatting social blindness of AI systems aiming to achieve strategic goals. AI actors producing AI systems executed in a legal society should lack financial interest in the reactable AI system. If a state-based structure, also elected politicians and other representatives, should lack such interest. Until the proposal herein was written, a professional and financial interest in politics has been deemed forbidden (Neuwirth, 2023). This is, however, often circumvented by pack-leading charitable institutions in a proxy ownership convention, where officials of influence sit in the foundation's board.

Bias and Fairness in AI Algorithms

Technologies that include automated, algorithmic, and AI/ML (artificial intelligence/machine learning) approaches are being increasingly used to help obtain and process evidence for the legal system, educate students in law, and inform judges and other actors in the legal system with legal reasoning systems. Balancing fairness and performance in decision-making systems is an important problem driven by the increasing utilization of automated and AI-based technologies in sensitive domains. When the evaluation of a fairness criterion involves complex group differences, screening procedures based on the conditioning of algorithms are used to lessen the exposure of sensitive features in the decision-making process. However, these methods can unreasonably limit the performance of a classifier and its predictive capability regarding outcomes. These considerations are particularly relevant to legal systems, where the excessive exposure of features may compromise principles of due process and the accessibility of evidence while at the same time compromising the fairness of judgments (Leavy et al., 2020). AI/ML-based techniques to extract and translate evidence, summarize, and streamline the review of large-scale evidence, courtroom decision support and sentencing recommendations, and litigation outcome assessments and recommendations for risk mitigation are now widely used by legal professionals (Tolan, 2019). Algorithmic models can better sustain the principles of accountability and explainability in contrast to the black-box nature of most state-of-the-art ML methods.

The overlapping criteria of facts, unlike in medical diagnostics, can introduce complexities and ambiguities in the explanation that arise from contradictions and the unclear nature of text measurements. This problem is compounded by the employment of abstractions or simulated representations of modeled reality. Recent research exploits symbolic learning methods to assist the interpretation of and explainable-fair detection in AI/ML-based models. While the importance of ethical decision-making transparency in legal systems is well recognized and demands the introduction of regulations, the notion and understanding of fairness, such as its definition, formulation, and goals, remain underdeveloped.

AI TECHNOLOGIES IN INSOLVENCY MANAGEMENT

The field of insolvency management has witnessed the emergence of artificial intelligence (AI) based technologies and their application to the cross-border insolvency domain. AI relies on computer algorithms in the form of machine learning and cognitive computing, enabling computers to execute tasks typically requiring human intelligence, as analyzed through data mining. AI systems can sift through vast data pools to assist insolvency professionals by suggesting timelines, interpreting case law, predicting decision outcomes, locating assets, providing optimal resolutions, drafting documents, and evaluating the effectiveness of proposed strategies (Mohan, 2021).

AI technologies in insolvency management raise ethical dilemmas regarding algorithmic biases, lack of transparency, contextuality concerns, breach of confidentiality, and potential misuse. Traditional code-based rules in insolvency laws may be inadequate for addressing these dilemmas in the context of AI systems. This proposes the need for ethical guidelines for developing and deploying technology-based insolvency managerial practices with accountability mechanisms to enhance stakeholder confidence. Despite increasing collaboration between startups and business heads to transform insolvency management in India using technology, the application of such technologies raises ethical concerns similar to those in other domains, like justice or financial decision-making. Key stakeholders, including lawyers, insolvency professionals, data scientists, and regulators, remain apprehensive about the ethical deployment of such technologies, leading to inherent distrust in machine-driven solutions.

Digitalization and the application of digital technologies have proven to be helpful for stakeholders in the face of the COVID-19 pandemic. In India, both short-term applications of technologies and long-term digital transformation initiatives have emerged as important trends. However, stakeholders have voiced concerns about loss of face-to-face interactions, uncertainty about the altered authority and accountability of functions, and algorithm-driven decision-making processes. Technologies like AI-assisted algorithms, chatbots, and LAWS can provide important advantages for stakeholders, such as suggestion timelines for sequences of actions needed to resolve cases, interpreting case laws in light of the rulings by judges directing resolutions, predicting possible outcomes of decisions based on learned experience, locating assets blocked by misuse of jurisdictions, and drafting and scrutinizing documents for compliance (Henkel, 2021).

Predictive Analytics

Having been around for decades, predictive analytics has attracted considerable attention from academia and practice on the corporate side since it provides a prediction of future events or trends based on statistical techniques and the available data (Rabiul Islam et al., 2019). Existing literature is divided into two parts, namely bankruptcy prediction and credit rating prediction, both of which aim to predict whether a company will default in the future. The prediction can assist companies in taking actions to avert the event from happening, e.g. early warning and timely decision making on the company and creditor side, hence reducing the economic loss substantially. Since the prediction could have a significant impact on a lot of parties, a good understanding of the methodology, techniques, and findings of the detection in the information age is of great importance.

The given literature on bankruptcy prediction and credit rating prediction has been reviewed and evaluated. 368 articles published by 1990 to 2017 in bankruptcy prediction and 221 articles from 1980 to 2017 in credit rating prediction from premier journals across academia have been retrieved. Bibliometric analysis is performed first by drawing the temporal popularities as well as distributions by the research interests in the two fields. Followed by the development of each topic over time, future research direction and recommendations will be discussed. In brief, the findings regarding predictive analytics of default events based on the above two fields are summarized, drawing the nontrivial connection between the two topics. Additionally, conclusions and discussions identifying gaps and future research suggestions are presented (C. K. Chow, 2018).

Automated Decision-Making Systems

To increase transparency and accountability, automated decision-making systems must be carefully designed, implemented, and monitored. This unique and complex ecosystem of cross-border insolvency decision-making systems is intertwined with accountability questions and dilemmas. These systems deliver various automated decisions and recommendations that can involve significant risks and harms. It raises new concerns about accountability for the legality, effects, and broader social implications of these systems. Current laws may not adequately meet the needs of accountability and transparency in cross-border insolvency proceedings. An open question remains whether further legislation would be effective in meeting these needs or merely pay lip service to accountability while introducing disproportionate additional costs and time for all stakeholders involved. There is a need for an inclusive approach to accountability that provides meaningful accountability mechanisms for civil society and decision-makers based on society's needs. This includes not only legal accountability but also normative and market accountability. Regarding normative accountability, it would be beneficial to foster discourse around appropriate norms and values, including ethical codes, professional standards, and labels. Since existing approaches are unlikely to suffice on their own, these need to be combined to create an accountability approach that is capable of promoting the kind of automated decision-making that serves societal goals. At the same time, accountability cannot be reduced to matters of design and implementation. What is needed to effectively respond to new accountability needs is an ecosystem of accountability mechanisms, involving various actors making use of distinct arenas. In addition to legal approaches and design measures involving technical solutions, actors need to engage in a discourse around normative accountability wherein societal values are articulated and elaborated. Finally, a market form of accountability can be developed directly binding all relevant actors.

CHALLENGES OF CROSS-BORDER INSOLVENCY IN THE AI CONTEXT

Artificial Intelligence (AI), one of the tools of Industry 4.0 technology, plays a vital role in advancing international trade and globalization and is an essential element of soundspace cross-border insolvency and recovery (Mohan, 2021). The development of AI would result in unique challenges for cross-border insolvency (CBI) proceedings, including recognition of AI assets and the need to address AI-malfunctioning issues. From a positive angle of view, AI-based technologies with the CBI system would enable efficient routing and controlling of the CBI process as well as help the development of competent CBI law. However, it is important to note that owing to significant ethical dilemmas, the use of AI requires careful and prudent monitoring and control mechanisms.

One of the major challenges for a CBI proceeding related to AI is on the recognition of AI assets. The advent of asset-less AI, which cannot be owned by any person or entity, possesses a unique challenge for existing CBI laws that rely heavily on the concept of assets for party recognition. Current formulation on asset identification under the Model Law or other bankruptcy laws often fails to apply to the situation where the debtor's business relies entirely on such asset-less AI. CBI laws do not have a clear definition of 'debtor' but often refer to 'persons' or 'entities', and a corporate debtor is a 'legal person' in which assets are implanted under the jurisdiction of national law. However, this definition does not work well with models in which an entity has no control of any assets. Moreover, when the AI that causes the financial crisis becomes the main 'debtor', it is challenging to identify the AI, its function, and the corporate entity. In such a situation, it is often the negligence of risk prevention that leads to negligence liabilities, and how to determine the person or entity that falls under culpability might have been more challenging, especially in the case of an asset-less AI.

Jurisdictional Issues

In an era of Artificial Intelligence (AI), new trends of financial crimes have arisen in the field of cross-border bankruptcy, which in turn has raised new jurisdictional issues. In practice, cross-border bankruptcy proceedings often have jurisdiction over the same subject matter and the same debtor but have different effects and results.

In this case, whether the courts have jurisdiction to accept the case is the key issue in the dispute over the validity of preemptive measures in the proceedings. How the terms "jurisdiction" and "collectivity" are construed emerges as a critical aspect. The term "jurisdiction" presiding over cross-border proceedings refers to general, objective jurisdiction over the subject matter of a bankruptcy case. The term "collectivity" of bankruptcy proceedings relates to their effect or result (Vaccari, 2018).

Neither of the UNCITRAL Model Law on Cross-Border Insolvency and the EU Regulation 1346/2000 on Insolvency Proceedings provides guidance on how the notion of "jurisdiction" should be construed. The term "collectivity" is integral in the Anglo-American bankruptcy system as well as its variations, and is recognized in different international frameworks, which emphasizes coordination in cross-border bankruptcy proceedings. Such proceedings are "collective" when they are aimed at addressing the issue of the financial distress of a corporation and restraining creditors from taking individual actions against such a corporation (Pae, 2003). The concept emphasizes the collective jurisdictional power of the courts of the state of incorporation over proceedings aimed at collective actions against the financial distress of the debtor. It should also be specifically noted that such categorizations should be construed in a reasonable manner based on the nature and objectives of the bankruptcy proceedings.

Dispute over the jurisdiction of bankruptcy courts occurs over the preemptive effect of voluntary proceedings. The jurisdictional issue results in the conflict between the duty to coordinate regional laws and the rigid determination of the applicable law. In the context of AI development, technology companies are easily inclined to use the AI technology in order to maximize their profit in finance, which resultantly might lead to risk and losses of local studios and small game firms, thus leading to difficulties in payments to gamers. If the debtor moved its legal base to another jurisdiction, the distant courts might not address the concern of local investors.

Data Privacy Concerns

Cross-border insolvency case proceedings involve various electronic data, ranging from internal and financial records to correspondence with various stakeholders (Humerick, 2017). While recognizing that the availability of digitally held, cross-border information can enhance accountability and fairness, this can create a dilemma over how such information exists and whether impacting broader socio-political issues is necessary. Also, expanding the gathering set of records can allow adversarial parties to develop unreasonable settlement expectations, should they obtain information about potential adverse effects from digital assets with generally tenuous claims on corporate value, unless a legal and ethical strategy provides for the gradual accumulation of basic cross-border electronic records needed to assess either the merits of the examination or the feasibility of a formal restructuring approach.

Recognizing the imperative to do so responsibly, the application of reasoning techniques in an automated or semi-automated way is inherently valuable. In particular, AI text generation and summarization models can be implemented to an extent as human partners in the collaborative review of personal overall communications with selection criteria tailored for each adversarial period, in a manner where either partner's unjustified behavior is controllable. Certain "occasions" might remain cognizant to deploy the filtering of extreme distancing human partners in charge of their input data, to minimize unaccountable outer interaction, which would otherwise compromise the initial motivators to adopt the techniques and cause mistrust in them. However, humans should remain in the communication loop to deepen inquiries and expand the search zone when focusing over narrow constructions of the overall task. Humans should direct textual feedback to the model's construction process, guiding desirable future behavior.

The appropriate, human-in-the-loop architecture and pursuit of explainable and fair processing methods would be essential to satisfy ethical standards, and the development of patterns shifting the role of these techniques to more automated settings would demand resilient jurisdictional considerations on general interests in light of the current practical pace of technical changes.

FUTURE TRENDS IN CROSS-BORDER INSOLVENCY AND AI

The rise of e-commerce and advancements in machine learning tools have enabled modern firms to obtain better credit risk and decision analysis based on patterns subliminally utilised by borrowers. With better and improved information concerning respondents' prospects of repaying borrowed funds, lending via online and hybrid models on the very verge of default had become easy. In light of such circumstances, firms should develop a comprehensive market-trader mindset in addition to obstreperous accountants and lawyers. The need for cross-border digital insolvency models of governance raises ethical dilemmas and jeopardises privacy concerns. Absence of liquidity requires trustees and creditors to locate nations for vacant assets. However, given blockchain's intrinsic design, decentralised exchanges' existence in external jurisdictions amassing fortunes could go unnoticed. Market hubs to trade data are few and far in between. Compliance costs explode if auditors were to submit amid penalties of obtaining undeclared assets. Alongside the incapacity to locate evaded assets, data attacks would lead to unwarranted losses in concurrence with information blackouts (Akinsola, 2025).

Cross-border insolvency models require bespoke tools to plan and implement. Historical examples reveal that governmental agencies must outlay digital trading firms and regulators with law-abiding programming across jurisdictions. Any laws governing electronic firms seeking registries worldwide would be construed as illegal tax-hunting. However, bootstrapped strategies of high stakes for governance across measures to tighten possessory concerns could ameliorate some of the grey research space in digital finance firms using program codes to seed trust. Given the presumed ill-functioning governance of digital firms, researchers could hypothesise immunity concerns from sovereign sternness, asymmetries, the oracles of truth, and related concepts. Insolvency-ridden exchanges could argue the computational expense of preparing compliance and regulatory expenses being a guard against criminal abuse. However, this neglects the plight of erstwhile transparent firms turned rogue by digitised assets and assets turned sought after by cross-border instruments (Jørgensen et al., 2025).

Mathematical modelling skills are useful to concurrently address the distinct research aspects of granular exchange governance globally. Solutions to blockchain robustness and privacy built on discrete mathematics back on decades of precedence. Nonetheless, alternative-pricing settlement in national currencies could reinstate the monetary ending to the Great Digitisation Bull Run, alongside surveillance by token techniques as infractions on antitrust grounds.

Emerging Technologies

The emergence of artificial agents raises important questions about legal personhood. Notably the distinction between legal persons and nonlegal persons involves a reflection that concerns agents. In order to be a legal person, something must be relevant to a legal system to be articulated in law something fundamental about personhood that isolates it from everything else. Relevance to the legal system might involve an array of considerations drawn from various disciplines, including law, philosophy, biology, and even religion. Because of this, legal personhood is a malleable construct. What are considered relevant factual conditions might evolve across centuries and across cultures. As a result of legal culture's tendency toward change, so legal personhood itself can also be altered by a sufficiently radical judgment and by equally radical evidence presented to a legal system. However one worldview might hold now, this does not guarantee validity or acceptance even a hundred years hence.

Consequently the issues raised by machine minds are ongoing and must be monitored carefully to assure that they remain a continuing influence on the law in the future (Brendan Walker-Munro & Zena Assaad, 2022). AI's continued development suggests that machines might be endowed with a mind. These machines might reason to a potentially advanced level. They may reason to a more advanced level than many humans. Such a prospect opens two lines of inquiry. The first is whether the present basis against the legal personality of machines might survive advances in machine intelligence. Timelines are very uncertain. AI may never develop into the kind of mind capable of reasoning relevant to personhood, or it may do so within decades. To determine where humanity stands in its assessment of AI's place within the world and its legal personhood must be subject to ongoing study in the future.

The second matter is whether AI that possesses a mind should also be a legal person. This in nature is concern with morality and ethics. In other words, the question is how widely should legal rights, duties, and privileges be allocated (Abhivardhan, 2020)? A legal framework contains ethical standards and, ultimately, moral requirements that govern a society's behavior toward persons. If a kind of mind were to entertain legal rights and duties, this would put these frameworks and standards to their greatest test. Onward to posit the great caution against acts of ordinary human evil, the intensity of the question is multiplied. If machines in possession of mind might be capable of committing acts akin to the most heinous acts of human actors, to what extent is the legal system equipped to handle such developments?

Global Policy Developments

The issue of cross-border insolvency (CBI) has gained international attention due to recent high-profile scenarios. This includes both research on the legal implications of CBI and international coordination of policymakers on best practices (Mevorach, 2015). New technologies pose additional challenges to current best practices. For decades, it has been widely recognized that cybercrime is part of the broader criminal class threatening both private corporate and public interests, which can span states and jurisdictions. However, the overlap between insolvency law and cybersecurity has rarely been raised. The issue is complicated by fundamental technology and governance differences that prevent a consensus on what a corporate event even looks like. Even assuming a legal definition, the natural experimentation of disruptive technologies leads to radical differences across sovereigns. Ecosystem design steals away national sovereignty in ways never encountered before. Cybersecurity incidents are known to drastically increase default probabilities, equity-debt spreads and loss given defaults, both at issuance and 30 days into the event. Analytical methodologies either increase the legal size of a corporate event by including a broader set of past and current actions or reduce them by focusing on just those relevant for a particular recovery strategy. In either case, leverage ratios across jurisdictions diverge and the analytics are not portable.

Good reasons may often exist for policymakers and corporations to want to keep trials secret. These might include national security considerations; aims to restrict losses, for the corporation or for its bondholders; management strategies to benefit the general public, e.g. by swift responses to cybersecurity incidents; and the firm's significant reliance on artificial intelligence technologies, which it wishes to keep secret from the competition. However, such secrecy brings substantial risk. With information often scarce relative to incentives to compete, trials may be biased against the corporation simply due to authorities in jurisdictions where more potential loss exists pursuing those options more aggressively.

THE ROLE OF LEGAL PRACTITIONERS IN AI INTEGRATION

The growth of artificial intelligence (AI), together with a generalized effort to support the process of turning big data into smart data, leads law and regulators to face major challenges. In comparison with other digital technologies, which were relatively easy to understand and regulate, AI makes everything more complex and uncertain. AI refers to a system's ability to perform tasks that normally require human intelligence. Experts disagree with each other both on the actual performance of AI systems and on the likely future increase in intelligence and autonomy. There are uncertainties about the nature of technological development and the costs of its control; about the nature of the legal questions and challenges. More generally, AI shares with other types of digital technology the character of a new phenomenon still evolving in unpredictable ways.

The AI-based machineries do not, for the moment, possess the self-learning capacities of human brains, or the capability of feeling emotions. AI systems are still mostly seen as programs, although potentially lethal ones, entailing profound ethical dilemmas. Moreover, a different legal status of AI systems raises a different set of issues. The argument is that also in the event that future AI systems came to possess new, different and dangerous traits, a solution could well be found without conferring on them legal personhood (Abhivardhan, 2020). Regarding current systems, which are mostly regarded as incapable of those magical traits, the more alarming and pressing issues from an ethical and regulatory perspective are probably those already envisaged by the current legal regime. Their impartiality could be seen as a further step in the commodification of human beings.

Equally, there are very worrying implications for democracy and social justice. Digital scholars say that AI's input data is adopted in ways that were never foreseen both by advertisers and by users of social networks. AI systems are also to a certain extent opaque, and in general competing theories about the kinds of algorithms at work raise different ethical dilemmas. Ultimately, the points made regard the human inability to take decisions at a democratic level. With machine learning, meta-algorithms are employed to test which algorithms yield a better result in a given category of cases. In recent applications, however, this almost-libertarian digital market does not produce a "winner takes all" situation: instead, various types of ads are employed, and dual criteria are openly adopted to define trustworthiness.

PUBLIC PERCEPTION OF AI IN INSOLVENCY

Over the past three decades, Artificial Intelligence (AI) has invited critical attention from academia, industry, and government worldwide. In September 2021, the International BC Insolvency Conference was attended by impeccably qualified insolvency practitioners and judges from prominent jurisdictions through a vivid zoom call. As the conference progressed, the much-expected segment on 'Techno-law' was introduced, with keynotes by eminent judges outlining the technological application in insolvency and future trends. The panellists were officials of eminent organizations from the USA founded by an AI entrepreneur as a "global virtual bankruptcy court", probably the first of the kind in the world. The gigantic platform entertained irrelevant yet important questions and comments via chat streaming. Few remarks from renowned judges faithfully represent the sentiments of the participants: requesting to refrain from referring to AI as Judge/Arbitrator, and not to allow it to arbitrate. These reactions indicate the arousing queries and concerns of the future acceptance of AI. Traditional applications of determination, such as legal personality of a corporation rights, and obligations, transparent programming, and accountability of time-bound mechanisms for due process in the appointment, operation, and termination of AI judges are not even at the nascent stage. Nor are enforcement laws and treaties of such techno-laws, required to limit parochial interests. It is important to note that technological jurisdictions may operate beyond the reach of the current legal regimes. Since benevolent AI Operas are currently on trial, the evolving techno-law on auditing mechanisms to embrace malevolent AIs is equally pertinent (Abhivardhan, 2020).

Expectedly, the recommendations of the conference seem slightly anticipatory. However, as an ostensible participant of the conference, the scholars recommend constitutional recognition of "A Judge"-AI stratified with the federated sovereignty of nations, creation of a "Global supra-sovereign entity" to govern the parochial interests of AI businessman firms, and harmonized big global laws on cybercrime, privacy, employment, AI, arbitration, insurance, etc. for equitable universal technological governance. Semi-sovereign humanity can increasingly be morphed into its ABI counterpart (Artificial Business Intelligence). It is feared that the egoistic anthropocentricity of ethical AI will be perpetuated into ABI for parochial interest of its cognizant custodians.

IMPACT OF AI ON STAKEHOLDER RIGHTS

The introduction of AI is profoundly causing positive and negative insecurities and uncertainties globally, and has created a need for transborder provisions protecting fundamental data rights, civil rights, and stakeholder rights in advance. If geopolitical, neoliberal, and powerful soulless AI will control digital assets and tyrannically treat digital human beings with negligence, with uncertain consequences on personal and global safety, territoriality, and lifestyle, the way AI can be developed is very much limited by rules protecting stakeholder rights, desires, and safety. The main issues which could evoke uncertainties and cause damages can be categorized as new scenarios to be governed.

The effects of the new product and system could become the concern of governance, since AI-formed inputs and outputs are likely to have unforeseen damages. The governance of the data treatment process or the algorithmic rule formation which governs the transformatory action of data inputs or product and system outputs is also possible. Ethical theories, laws, treaties, and country rules could constrain AI-based analysis. The outputs of the analysis themselves including their influence on other product or system decisions, new data output designing, evaluation, and mass production of risks, damages, or abuses including privacy infringement are out of governance bounds presently.

The automatization of digital strategy development which limits human intervention raises concerns about responsibility. If the safety of other products and systems becomes automatically and machine-driven compromised, in nearly untraceable ways the reasonability of damages remains an open and default question since prior-laid responsibility systems are and will likely remain human anchored directly and indirectly. The probability of wrongful use of AI for human rights failing is increasing as, which could drastically change global society and priority.

CRISIS MANAGEMENT IN AI-DRIVEN INSOLVENCY

The impact of artificial intelligence (AI) applications on the human understanding of decision-making causes a period of uncertainty about AI algorithms. Complicated technology stacks make it impossible to explain AI behavior. AI chatbots and language models are considered possible co-decision-makers, raising questions collectively about their accountability. The problem arises that the default legal paradigm cannot handle such complexities. A redesign is necessary, which should be paradigmatic, practical, democratic, accountable, personal, responsive, transparent, explicit, and reasonable (Mohan, 2021). Euro-solidary is an essential step toward a European ethical mechanism.

Crisis management is a significant challenge for corporate governance and insolvency law. Efforts to improve the resilience of companies, including those that question the necessity of public support measures, highlight the importance of early warning systems and trusted intermediaries to prompt timely reporting of distress. However, this approach is difficult to operationalize as it combines core challenges with convoluted follow-on questions. How should companies report risks? Do they trust internal controls in AI systems? What if the risks of AI systems are not known?

A bloc-level safety net addresses the question of whether a rescue mechanism exists for large distressed companies operating through a subsidiary. The ownership and/or control of a distressed company's legal structures assure the bloc-level structure of a group of companies. One can question whether silver bullets exist to address these issues. The legal regime surrounding the issuance of sub-sovereign bonds in the single currency has undergone significant reform, but its complexity raises concerns over further adjustments. One must also assess whether recent reforms enhance coordination within a bloc in times of distress.

CONCLUSION

The world is changing rapidly, necessitating the need for reconsidering the laws around the world. To meet the modern commercial and moral perspective of business, countries need to re-analyze their existing legislation so as to adapt to changing business needs. Cross-border insolvency law is one such area which has evoked great curiosity and which calls for reconsidering the existing structures. In this post, the common legislative approaches adopted by different jurisdictions and the special considerations for the jurisdictions opting for a new regime is discussed. The focus, ultimately, is on how the regulatory framework is to be used for development of accountability in operations of AI. It is argued that, rather than merely modeling AI systems to comply with regulations, corporations should adopt methodologies to breakdown their AI systems and business in ways that allow a distinction to be made between companies who diligently wanted to comply and those who failed due to insufficient efforts. The emphasis is on an accountable approach to oversight rather than a compliance-centric approach. AI companies should put in place governance structures that enable the distinction between individuals who were sincere in their efforts to address ethical concerns in the development and deployment of AI systems and those who did not make a genuine attempt (Mevorach, 2015). Understanding corporate intent is the key step in the methodology proposed. Upon decent modeling, it should be possible to replay decisions made by a corporate monitoring structure in response to AI decisions. Some decisions will reasonably be seen as good faith attempts to comply with regulations and fulfill the corporate intent to promote ethical AI systems. Others may reasonably be interpreted as failure to take into account concerns or objections raised. This "corporate intent-disarming" (CID) test can reasonably provide a means by which corporations will be distinguished based on their accountability event. This also foregrounds the need for companies to begin discussions on governance and oversight structures predicate on intentions, falsehoods, and possible mala fide behavior beforehand. Responsiveness bottlenecks and other internal network areas should be understood, identified, and tagged as corporate points of action. The same should be attempted for AI companies attempting to fine-tune their compliance apparatus. Internal and external agencies supporting corporations in their standards compliance may be cataloged. This is a move from compliance assessment which places few considerations on corporate intent. There is an urgent need to account for the wide-scale implications of artificial intelligence systems, and regulatory actions should parallel this urgency.

REFERENCES

Abhivardhan. (2020, May). The Ethos of Artificial Intelligence as a Legal Personality in a Globalized Space: Examining the Overhaul of the Post-liberal Technological Order. In IFIP International Conference on Artificial Intelligence Applications and Innovations (pp. 3-14). Cham: Springer International Publishing.

Akinsola, K. (2025). Regulatory Compliance and the Role of Corporate Governance in Preventing Financial Misstatements. Available at *SSRN* 5126627.

Chow, J. C. (2018). Analysis of financial credit risk using machine learning. arXiv preprint arXiv:1802.05326.

Davenport, J. (2025). Enhancing legal document analysis with large language models: A structured approach to accuracy, context preservation, and risk mitigation. *Open Journal of Modern Linguistics*, *15*(2), 232–280. DOI: 10.4236/ojml.2025.152016

Hacker, P. (2023). The European AI liability directives–Critique of a half-hearted approach and lessons for the future. *Computer Law & Security Review*, *51*, 105871.

Henkel, C. (2021). The impact of artificial intelligence on insolvency law and practice. In *Research Handbook on Corporate Restructuring* (pp. 377–399). Edward Elgar Publishing.

Humerick, M. (2017). Taking AI personally: How the EU must learn to balance the interests of personal data privacy & artificial intelligence. *Santa Clara High-Technology Law Journal*, *34*, 393.

Islam, S. R., Eberle, W., Bundy, S., & Ghafoor, S. K. (2019). Infusing domain knowledge in ai-based" black box" models for better explainability with application in bankruptcy prediction. arXiv preprint arXiv:1905.11474.

Jørgensen, N., Gunasekaran, S. S., & Ma, Z. G. (2025). Impact of eu laws on ai adoption in smart grids: A review of regulatory barriers, technological challenges, and stakeholder benefits. *Energies*, *18*(12), 3002. DOI: 10.3390/en18123002

Larsson, S. (2019). The socio-legal relevance of artificial intelligence. *Droit Social*, *103*(3), 573–593.

Leavy, S., O'Sullivan, B., & Siapera, E. (2020). Data, power and bias in artificial intelligence. arXiv preprint arXiv:2008.07341.

Mevorach, I. (2015). Beyond the search for certainty: Addressing the cross-border resolution gap. *Brook. J. Corp. Fin. & Com. L.*, *10*, 183.

Mohan, M. P. (2021). Intellectual Property licenses in cross-border insolvency: Lessons from In Re Qimonda. *Hastings Bus. LJ*, *18*, 181.

Neuwirth, J. (2023). Prohibited artificial intelligence practices in the proposed eu artificial intelligence act (aia). *Computer Law & Security Review*, *48*, 105798. DOI: 10.1016/j.clsr.2023.105798

Obal, W., & Merriman, K. K. (2025). Extending the geographic scope of corporate social responsibility: Remote work and housing affordability. *Journal of Business Ethics*. Advance online publication. DOI: 10.1007/s10551-025-05956-4

Pae, J. (2003). The EU Regulation on Insolvency Proceedings: The Need for a Modified Universal Approach. *Hastings Int'l & Comp. L. Rev.*, *27*, 555.

Porral, A. (2025). The protection of creditors in cross-border insolvency proceedings-International soft law, EU law and a comparative analysis of German and Australian law.

Roe, J., & Perkins, M. (2023). 'What they're not telling you about ChatGPT': Exploring the discourse of AI in UK news media headlines. *Humanities & Social Sciences Communications*, *10*(1), 1–9.

Tolan, S. (2019). Fair and unbiased algorithmic decision making: Current state and future challenges. arXiv preprint arXiv:1901.04730.

Vaccari, E. (2018). The Ammanati Affair: Seven Centuries Old, and Not Feeling the Age. Chi.-. *Kent L. Rev.*, *93*, 831.

Walker-Munro, B., & Assaad, Z. (2023). The Guilty (Silicon) Mind: Blameworthiness and Liability in Human-Machine Teaming. *Cambridge L. Rev.*, *8*, 1.

KEY TERMS AND DEFINITIONS

Accountability and Liability: The assignment of responsibility for actions and decisions made in cross-border insolvency processes, especially regarding outcomes influenced or determined by AI systems, including questions about who is liable when errors or harms occur as a result of AI-driven recommendations or actions.

Artificial Intelligence (AI): A branch of computer science focused on developing systems and algorithms capable of performing tasks that typically require human intelligence, such as reasoning, learning, and decision-making, increasingly applied in the field of insolvency management.

Automated Decision-Making Systems: Technological systems that use algorithms and AI to make or support legal, financial, or administrative decisions, often replacing or supplementing human judgment, and requiring careful design to ensure transparency, accountability, and fairness.

Bias and Fairness in AI Algorithms: The risk that AI systems may produce unjust, partial, or discriminatory outcomes due to biases in data, design, or implementation, necessitating efforts to ensure equity and justice in automated legal and insolvency-related processes.

Cross-Border Insolvency: A legal situation where a debtor has assets, creditors, or legal proceedings in more than one country, requiring cooperation between jurisdictions to resolve financial distress and insolvency matters efficiently and equitably.

Data Privacy and Protection: The safeguarding of sensitive financial, corporate, and personal information involved in cross-border insolvency cases, particularly when AI and digital tools are used, to prevent unauthorized access, breaches, or misuse, and to comply with evolving legal standards for privacy.

Ethical Dilemmas in AI: Situations where the deployment of AI technologies, especially in sensitive domains like cross-border bankruptcy, raises challenging questions regarding fairness, bias, accountability, and the potential for unintended consequences in automated legal and financial decision-making.

UNCITRAL Model Law on Cross-Border Insolvency: An international legal framework developed by the United Nations Commission on International Trade Law (UNCITRAL) to provide a standardized approach for recognizing and coordinating insolvency proceedings involving debtors with assets or creditors in multiple countries.

Chapter 4
Navigating Responsible Business Practices:
The Impact of Foreign Direct Investments on Socio-Economic Development and Sustainability

El Mahdi El Massaoudi
https://orcid.org/0009-0007-6016-281X
Ibn Tofail University, Morocco

ABSTRACT

This chapter investigates the multifaceted impact of Foreign Direct Investment (FDI) on socio-economic development and sustainability in host countries. It highlights the growing importance of Responsible Business Practices (RBPs) in shaping multinational corporations' strategies. The discussion spans job creation, technology transfer, income inequality, cultural impacts, corporate social responsibility (CSR), environmental concerns, and national regulatory frameworks. It emphasizes that while FDI offers opportunities for economic growth, effective policy filters and ethical guidelines are essential to mitigate its potential negative effects. The chapter also advocates for stronger multi-stakeholder dialogue involving governments, investors, and civil society to align FDI flows with Sustainable Development Goals (SDGs). Ultimately, it calls for a case-specific, transparent, and inclusive approach to ensuring that FDI truly contributes to long-term, equitable development.

DOI: 10.4018/979-8-3373-3104-1.ch004

INTRODUCTION

The ongoing pervasiveness of Foreign Direct Investment (FDI) activities by Multinational Corporations (MNCs) more than one decade after the onset of the global financial crisis and the protracted global efforts to tame the excesses thereof by reforming the global finance and foreign investments sectors has rekindled interest in the nexus between foreign investments and responsible business practices (RBPs). It has been proposed that the entry of MNCs through FDI presents host country governments an opportunity to positively affect sustainable development (SD) as a result of the spillover benefits that stem from foreign investments. Public policies to govern FDIs collectively fall under the umbrella of Investment Treaties and National Investment Policies. These policies are pragmatically grounded on the (perceived input-output) wisdom of RBPs because the majority of them emphasise investment retention but not responsible investor behaviour such as (the due diligence on) investment decisions and the (business) conduct of invested assets to abate negative investment impacts and to enhance positive ones (Wanjohi Ngeng, 2018).

The concept of RBPs is relatively new in academia although it has gained currency in the Global North. By pre-emptively framing results as supportive of locally relevant prescriptions, proponents of RBPs hope to dethrone MNCs' pursuit of profit and attendant impunity from accountability. Investigation of the foreign investment landscape is a herculean task because of fractured jurisdictions, but contextual analyses of each investor's impact should shift the focus from global policies to accountability at the national level which is likely more feasible and effective. On governance, beneficial regulation of foreign investments cannot be exclusively divorced from the larger governance landscapes which tend to reproduce outerbased natural resource booms and busts. Reciprocally, an "inside out" approach should not overlook taproot, transnational governance gaps. This contradictory dissertation serves to inform a wider audience the impact of FDIs on socio-economic outcomes (including their sustainability). Sovereign-afinal anti-FDI tackling contexts are suggested for research.

UNDERSTANDING FOREIGN DIRECT INVESTMENT (FDI)

Investments made by foreign individuals and entities that take the form of controlling ownership of domestic travels, enterprises, resources, or must fulfill the terms of a host country are known as foreign direct investments (FDI) (Dapra et al., 2019). Foreign companies carry out foreign direct investments. A direct investment means that the company must develop an actual presence in the foreign country through subsidiary operations. Foreign direct investment is therefore a direct investment in a country, state, or locality. Foreign subsidiaries, on the other hand, are abroad-based companies. They are production, sales, or other subsidiaries of foreign companies. Activities of foreign subsidiaries contribute to the local economy in many different ways. They can expand employment, hold business and partnership opportunities for existing companies, and create greater intrinsic and economic value. Surveys of foreign firms operate in countries reveal that they are often larger, pay higher wages, import and export more, and invest more in research and development. FDI is also considered to be financially an attractive option that the long-term growth of the host locations should facilitate based on grounded theories of value consideration. Realizing FDI can take place in three modes. The first one is Greenfield, which means that foreign companies enter the location by creating new facilities and building operational programs for each aspect of the operations there. The second one is acquisition, which signifies foreign companies acquiring an existing firm in a new location. The last one is a merger. With mergers, at least two companies combine in either of the former types of arrangements. Realizing FDI Greenfield in the most direct way has positive feedback on GDP growth as well. Acquisition alone is more good than bad for the balance of larger and stronger players in business scales of both the combining firms. Though less significant, mergers alone transform a firm from a small player to a stronger competitor. Increased FDI activity can be interpreted as the increased value of the host country, state, or locality. It suggests that in a competitive world, the relative attractiveness of places to FDI is greater in successfully maintaining growth for the future.

Definition and Types of FDI

When discussing foreign direct investment (FDI), an idea frequently emerges that foreign investment is good, and thus leaders must pursue it. However, this is only half the story. That truth will become evident in a later section. First, "FDI" can be defined, and while it may seem simple, there may be some mistaken views of what FDI really is. A foreign company must develop an actual presence in the other country through operational subsidiary investment. If there is no establishment of an actual activity by the company, there is no FDI, no matter how much currency is transferred across borders.

Foreign direct investment is more than just a financial investment in another country. It is an active investment in the country. This activity must create the subsidiary, which must create tangible outcomes for its host location. After foreign companies capitalize operations in a host country, that country's local economy expands altogether, because larger economic communities benefit local residents through expanded employment and business opportunities. FDI occurs for the long-term development of the host location, in contrast to portfolio investment.

Development is framed as the evolution of a larger, well-serviced community capable of addressing challenges, sustaining itself, and improving over time. FDI can realize destinations and their legacy. On the contrary, portfolio investment is money. If money is withdrawn, the project is gone. If development does not happen first, foreign companies will withdraw as well. Host countries should hope to attract FDI while suppressing the portfolio. FDI is realized by one or more of three methods: greenfield (a subsidiary is created wholly from scratch), by acquisition (the acquiring company purchases a controlling interest in the acquiring company), and through mergers (the combination of two or more companies, forming a new company). Of these, greenfield is the most direct way to recognize FDI, when foreign companies create new facilities (Dapra et al., 2019).

Historical Context of FDI

FPDI has undergone significant evolution, with the shift from trade theories to economic interaction and the phenomenon of FDI blooming after WWII. Most developing countries, by necessity and/or choice, have turned to FDI, complemented by foreign assistance to stimulate their economic growth and development during the post-WWII last half century. However, FDI has been elusive in the third world. Not only has the value of FDI been much lower than expected, but also, with the liberalization of the FDI regime, earning repatriation has begun to exceed FDI inflow. Such reality has often been attributed to the unfavorable behavioral, institutional and geographical factors of most developing countries. The debate on the relationship between FDI and economic development does not mean much for where, when and how a country should get the FDI inflow, not for whether a country should/have to rely on FDI, which is regarded as a pre-condition for development by many developing countries (Esplugues Mota, 2018).

FDI is an expression of cross-border capital movements, where capital investment is subject to some kind of control by the foreign investor. Conventionally FDI in the context of developing countries refers to the entry of foreign enterprises or foreign direct investment (FDI) to procure the establishment of business firms in host countries other than their own. The relevant firms include commercial firms and manufacturing firms in the form of assembling, sub-contracting, joint ventures, and others embraced by Transnational Corporations (TNCs). FDI for developing countries has a different connotation from that in the general sense. For developing countries or countries in transition, that connotation encompasses in addition to the generic entry of international capital, taking arms-length external capital in order to procure technology transfer, production techniques, managerial and marketing expertise, and also a relevant market to sell the products produced by the immediate capital.

THE ROLE OF FDI IN ECONOMIC DEVELOPMENT

Foreign Direct Investment (FDI) has become synonymous in recent decades with cross-border transactions in which multinationals (MNEs) located in one country (the parent country) acquire production assets in another country. Whether this is through a green field investment, merger, acquisition, or joint venture, all of these transactions involve the formation of long-lasting enterprise groups that span more than one country. In broad definition, the terms transnational corporation (TNC) and MNE are used interchangeably to denote a corporation that has active operations in at least two countries. However, as the activities of multinational corporations have become more pervasive and complex, several theories of FDI have been developed to explain this phenomenon.

Yet much of the global FDI literature is focused on the D in FDI and tends to ignore the global location and restructuring of production organizations and the impact of these changes on employment, technology, and other characteristics of the companies involved and the countries or regions in which they are established. It is clear that there is no single FDI theory that can account for the wide variety of FDI, MNE, and TNC behaviors. The observation that different phenomena need to be explained by different frameworks, or that greater complexity in the explanatory framework is required, seems to be inescapable. Indeed, this observation appears first in the pre- and post-Fordist literature. Attempts to classify FDIs into types based on motives, locational and operating characteristics, and the companies involved have been made. Although many astute efforts in the academic literature proposed a variety of typologies of FDIs and MNEs, the plethora of terms and typologies has made them difficult, if not impossible, to integrate into larger explanatory frameworks. FDI classification in terms of motivation, mode, and other characteristics is often found to be no more than a taxonomy, so while taxonomy is crucial in understanding a phenomenon, it does not explain the existence of these types of phenomena. Nevertheless, this classification framework forms an essential basis for further explanation. The ability to categorize different FDIs enables the researcher to specify the phenomena to be addressed in the first place.

Job Creation and Employment

FDI inflows have a broad range of impacts on host countries underpinned by environmental, social and economic consideration and these positive and/or negative impacts can be grouped into five broad themes or categories namely, tax revenue, local capability and firm level enhancement, job creation and employment, firm level opportunities and growth, sustainability impacts. Several research studies have empirically conceptualized the impact of FDI on host countries' socio-economic development and sustainability focusing around these themes and question of who has benefited from the foreign direct investments in Tanzania. One possible methodological approach can be conducted by following systematic review timeline which can start by discussing how FDI generates greater economic growth through its introduction of advanced technology, employment effect (in the aspect of creation of new jobs) as well as Demi-Labory effect (in the aspect of funded salary). A systematic review would massively add knowledge and fill out the missing literature regarding the social and economic impacts of FDI in the context of Tanzania for the academia and policy makers both locally and/or globally. FDI inflows tend to create a significant number of jobs in developing countries aiming to attain greater economic growth and an increase in per capita income growth through its introduction of advanced technology. Hosting FDI designing policies which are able to attract FDI inflows and the proper implementation of these policies play a vital role in ensuring a proper linkage between FDI inflows and employment generation. It is also essential to properly formulate and implement FDI designing policies in order to generate a greater net economic benefit through FDI inflows. Globally, the impact of FDI inflows on employment generation is of increasing governmental concern as a response to disquieting job losses. In developing countries, where the size of the formal sector is small, this concern is even greater.

Foreign Subsidiaries are usually more capital intensive, produce a more complex output, pay higher wages to skilled labor and do not always hire or train local labor. To reduce the externality level of foreign investments, a typical government imposes more restrictions on the entry of foreign firms or uses fiscal measures to encourage labor a market. International investors forgo in part their competitive advantages in order to operate locally. So doing, they strengthen the local economy and reinforce good practices with the static hope of fulfilling the more stringent demands of non-market economies on the labor market. In this way FDI can contribute indirectly through the government to poverty alleviation ((Abdelaal Mahmoud, 2010)).

Infrastructure Development

Foreign direct investments (FDIs) not only provide a means of capital and technology transfer but are also relied upon to provide a stimulus for economic growth, development, and, largely, modernization. Foreign direct investments can be broadly defined as investments in business operations, which lead to the establishment of a new production facility or an increase in production at an existing facility in a country other than that. To illustrate foreign direct investments are financial contributions by third parties to a sustainable economic operating model. Relatedly, though foreign direct investments are often considered a critical means of stimulating socio-economic growth and development, there is also a widespread belief that foreign direct investment plays a limited role in enabling a country's sustainable economic production model. And they have diabolically different social, political, economic, and environmental context worldwide, and so do the differences in the workings of foreign direct investments.

Infrastructure development, too, has attracted attention in both research and policy circles due to its strong linkages with overall economic development. Recent research has shown a strong positive causal relationship between physical infrastructure and economic development (Marozva & Lindelwa Makoni, 2018). However, initiatives in infrastructure development and adequacy have fallen short of expectations on the continent due to a multiplicity of reasons, both endogenous and exogenous. Current infrastructure development strategies remain piecemeal and have not yet given a holistic outlook to approach the continent's specific demands and requirements with respect to infrastructure. More critically, these strategies have either not anticipated exogenous shocks that have a long-term adverse socioeconomic impact on Africa's infrastructure, such as the COVID-19 pandemic, or they do not articulate the necessary mechanisms to mitigate the impact of the said shocks.

Infrastructure may provide a relatively costless and perhaps a much friendlier enabling environment for businesses, which translates to lower transaction costs and factors of production being more productive. Although there have been earlier cross-country studies showing that infrastructure availability is important in influencing productivity, the findings suggest that it may be possible also to err on the other side of the spectrum and invest heavily on infrastructure while at the same time underappreciating their productivity influence on various industries. These observations underscore the importance of investigating the role of infrastructure in the financing of various types of business investment and understanding the respective costs and their differential impact across industries.

Technology Transfer

The FDI has accelerated the growth of the technology transfer by facilitating the entry of multinational firms and obtaining the technology of knowledge- and competence-intensive industries in the host country (Diakon Debebe, 2018). But the technology transfer was not seen as a focal part of policymaking under the takers' perspective. With the FDI liberalization since the 1990s, the inflow of technology made by large domestic firms in the era of import substitution has almost been exhausted. Since then, some large firms acquired technology and management know-how by direct investment overseas. The other marketing and business-related technology has been obtained through joint ventures and suppliers, but the majority of the technology gap is still large. On the other hand, even if the licensing fees and royalties are facilitated, the rise of IPR would hamper technology diffusion. Competitive concerns, in some cases, revolved the incomplete/wrong technology transfer.

FDI units become integral parts of their MNCs' strategies for developing countries. MNCs are, or seek to be, players in a worldwide network while local firms (hereafter called firms) seek to tap into MNCs' sources of knowledge if they can engage with them in productive cooperative ventures. They maintain their own strategies to enhance capability, to minimize competitive threat, or to gain market access. As favorable locational types and conditions differ from the view of MNCs and that of firms, their strategies lead to conflicting expectations in the host country as economic agents and eliciting two sides of development and upgrading paths that two parties contribute to adverse economic effects or functional development gaps through different or opposite transition and developmental paths.

SOCIO-ECONOMIC IMPACTS OF FDI

The huge inflows of FDI attract the attention of researchers and policy makers to study its multidimensional socio-economic impacts on the choice of macroeconomic variables changed in the host countries. These studies measure the impact of FDI on different variables exploring the different dimensions of nature. Even though FDI have full impacts on a host country's development path, it requires a filtering process to mitigate the negative impacts and benefit moreover the positive impacts. As a result, the linear relation between FDI and development path does not exist through a threshold level, the impact path is corrupted and needs a filtering process during the FDI flow. With this preview of filtering process in the paper model, it is intended for researchers to understand how developed and dynamic countries filter it. Countries with effective governance, dynamic infrastructure, high-skilled labor force and large extent of market are expected to benefit more from FDI inflows on high economic growth (Hipólito Leal et al., 2021). This filtering theory suggests these variables are exogenous trigger for FDI impacts on economic growth in selected developing countries. Most studies on the effects of FDI measure in the international context focusing on the aggregated impact of both inflows and outflows on development variables. This aggregated measurement sometimes junk the true impact and policy implication since disregarding the widely asymmetric importance of FDI across countries. In some developing countries, multiple inflows and outflows from both developing and developed country counterparts might change the impact of these extreme flows. To investigate the channel of the asymmetric origination on a host country regarding the FDI inflow and outflow has more theoretical and policy implications to help the country to design more adaptive policies. Even though countries with similar demographic and economic variables have strong ties, an application of the model is advisable to obtain more domestic specifics and effective policies since this impact pattern is corruptible and easily transferred by robust and effective FDI policies.

Income Inequality

International investment enterprises have thrived in a system of international investment agreements that many countries have adopted, negotiated, and finalized in the last few decades. There are now more than two thousand bilateral investment treaties and four hundred twenty-two treaties between states and multilateral organizations. Mining, oil, and gas exploration industries have tended to be favored, and the protection of positive rights like the right to peaceful assembly is excluded from all agreements.

However, the negative rights framework of most agreements runs counter to the development aims for environmental protection and investment facilitation themselves, and bilateral corridor agreements fall far short of the development aims for sustainable investment. Such disparate agreements show the unequal partnerships designed by foreign corporations and countries of the Global North for long-term extraction and exploitation of natural resources in the Global South. These agreements are best understood as both part of a complex federal state to facilitate continued ports of trade in Global South countries and a legal counter-system that nullifies hundreds of binding laws. Foreign corporations have successfully challenged legislative efforts to protect human rights through lawsuits that threaten damages of hundreds of millions to billions of dollars.

New developments underline the extensive need for regulatory harmonization and increased data-sharing across countries from the time treaties are signed. New frameworks for tackling the loopholes of existing agreements include: individual rigorous ex-ante analyses and consultations to assess proposed treaties, including on rights violations; an independent international tribunal to hear investor-state disputes; use of counter-claims for abuses or damages; an unredacted public tribunal for all parties; and inclusion of human rights law as applicable law. These measures would further stakeholders' interests, the environment, and the development aims politicians have laid out.

Cultural Impacts

Due to globalization, international capital transfer has started to flow rapidly to developing countries. It is necessary to pay attention to the cultural distance in the impact of FDI, because locations possessing a high culture distance both in terms of its value orientation and social practices may hinder capital and technology inflow.

High cultural distances would lead to failures in FDI since it is more difficult to obtain social acknowledgment, enlist co-operating local organizations, and develop and maintain relationships (Li & Shen, 2022). Two categories of culture distance in terms of Hofstede's methodology are considered: the value-oriented cultural distance and the practice-oriented cultural distance. The former includes the power distance, collectivism versus individualism, uncertainty avoidance, and masculinity versus femininity. The value-oriented cultural distance would hinder the trade of equity as it affects social acknowledgement to new organizations and affects the establishment of open negotiation on value. The practice-oriented cultural distance represents different practices in common operational routines. The implementation method of a technology transfer would be absent if there is a high practice-oriented culture distance. It is pointed out that people react to trade similarly across locations with similar culture values (Ribeiro Goraieb et al., 2019). Therefore, cultural values should also be taken into account in examining the cultural distance in the field of FDI. The value-oriented cultural distance based on Hofstede's methodology focuses on culture values, and the practice-oriented cultural distance concentrates on culture practices. The cultural distance both in terms of cultural values and cultural practice has the attention of researchers. Due to the high cultural distance, knowledge-based firms are discouraged to invest in the R&D of foreign firms, especially firms from developing countries. When compared to its high cultural countries, firms from low cultural societies would be discouraged to seek new information from those high cultural societies, conduct knowledge acquisitions, and collaboration to develop new products. Thus, the cultural distance affects FDI knowledge-seeking behavior. To sum, firms should undertake intensive screening processes to mitigate the adverse influence of cultural distance, i.e., investments with large cultural differences are more selective than are those with small cultural differences.

Community Development

FDIs, Multinational Corporations (MNCs), and Public-Private Institutions' investment seem to be driven by a promise of growth and wealth. Undeniably, developing areas of their command have shown favorable advances in social indicators after the arrival of FDIs prospects. However, traditional econometrics fails to unravel complicated causation and correlations testing, from simple maths, forward-looking semblance like exponential growth equation to advanced equations of networks with disturbances. Most irregularities in outputs surfaced against FDIs including: unaccountable destructives while governments welcomed MNCs; lands and resources massively plundered; loss on outflows off-balance; and so on. Empirical advancement of academic research, though properly framed as a multidisciplinary and reporting-field research, has not benched controlled experiments. Hence the necessity for a sound framework to comprehensively scrutinize FDIs impact while not pre-processed as priori premises or assumptions like a physical model. While the series will remind a tendency of naivety, it is a mean inquiring not judging.

Structural linguistics to its maths of group theory with nodes and edges will situate implanting disciplines. To track communications and messages flowed, patterns and variations doted between group communications will be examined. The viabilities of languages partitioned in spaces while their messages streamed continuously over time, hence only time-invariant features were possessed to filter processes, will be availed to approach co-defiances. MDG prospects and fragile covariate patterns and indicators' data across nations with petits-framed Latin-dummies, FDIs inflows and minerals with due-determined price can test to recap findings. Consequential graphical plots and fractal correlations between selected scope can massively outlay a jury of index conversancy or collusion. Exploitation for resources over loss by nature is capable with execution of authority on behalf of equity to curb them.

SUSTAINABILITY AND FDI

The sustainability requirements of investors and investee countries are becoming more and more important in theory and in practice. If otherwise, all capital flows, including FDIs, could be permissible regardless of their character. Some may argue that the flow of capital per se is supreme and others would say that not all flows should be welcomed (Wanjohi Ngeng, 2018). The ongoing price war of sovereign wealth funds to attract capital investments demonstrates the willingness of some state actors to loosen restrictions against capital flows, and thereby to compromise their sovereignty. On the other hand, some may argue that if capital flows into a country take the form of FDIs and take root, they cannot be terminated arbitrarily and there is a time lag for impacts. Irrespective of the nature of FDI flows, normative questions surrounding it can be synthesized around how to regulate FDIs so that they could contribute to sustainable development (SD). Despite the added value of FDI against other capital flows, concerns have been raised about the possible adverse impacts of FBIs. Notable examples include the Asian financial crisis, the East African hedge fund tour, the shifting of polluting activities or coercive supply chain practices. There are a range of possible SD characteristics against which its contributions to sustainable development could be assessed, and thereby the possibility of developing a checklist. But, there would be challenges in the practical implementation of such a list ranging from establishing benchmarking to addressing unintended consequences (P. Sauvant & Mann, 2017). That said, the impact of an FDI and its sustainability characteristics could only be comprehensively assessed on a case-by-case basis with respect to a comprehensive list. Hence, it is impossible to generate an arbitrary list that could be universally adopted. FDI flows and SD could be approached at four levels (1) Unpacking FDI flows and specific SD concerns, (2) Mapping sustainability providers and their interests, (3) Exploring the interconnection among sustainability characteristics through a linkage measure, and (4) Workshop opportunities for stakeholders.

Environmental Considerations

Businesses play a central role in environmental, social and economic development worldwide. Often, as an effort to enhance long-term profitability, businesses increase their international activities through expansion into emerging markets (EMs), to access new customers, new sources of production, better opportunities for innovation, and lower costs (Hipólito Leal et al., 2021). These firms are then faced with the challenge of applying sound business practices in a new environment, increasingly complicated by the aspirations of stakeholders seeking more than economic profits from business. These stakeholder pressures for responsible practices create a dilemma for firms producing goods and services in conflict with their corporate principles and firm-level character, adding challenges on top of those of successfully establishing an operations site in new cultural, political, and institutional environments.

To date, the prevailing research has focused primarily on the balance between profitability and long-term sustainable practices in principle, and the effectiveness of founded institutional structures in different countries in minimizing opportunities for misconduct. There has been little, however, attention to interpreting that balance in practice in the evaluation of actual firm-level practice. There is hence a gap in understanding why some firms have perceived and responded effectively and equitably to stakeholder pressures while competitors have succumbed to alternate perceptions and either resisted or delayed needed adjustments. The purpose of this article has been to provide a conceptual framework for the analysis of firm-level interpretation of stakeholders pressure and responses in terms of firm character and modes of internationalization: full compliance with normative pressures in fidelity to corporate principles, shrewd maneuvers to accommodate each of contradictory competing demands, and recalcitrance to disallow incongruities in a patchwork of internationalized sites without necessarily changing any of the principles guiding them.

Corporate Social Responsibility (CSR)

The World Business Council for Sustainable Development defines Corporate Social Responsibility (CSR) as "the commitment of business to contribute to sustainable economic development, working with employees, their families, the local community and society at large to improve their quality of life" (Chilwalo, 2016). CSR means not only fulfilling legal expectations, but also going beyond compliance and investing "more" into human capital, the environment and relations with stakeholders. Aspects of CSR related to human capital include active involvement and stakeholder engagement to develop good corporate governance principles in relation to executive compensation, the business's lobbying of governments, and political contributions. Companies should uphold human dignity to improve the welfare of people through the provision of educational and social programmes, reduction of waste and pollution on the environment, and consider local people's views on issues affecting them (Nasser, 2013). Business corporations should engage in and support activities that reach out to underprivileged communities to promote their welfare.

MNCs are potentially important actors in development and can play a key role to help address many of the underlying challenges facing countries in Africa and globally. The need for social responsibility on the part of MNCs is even stronger, especially in light of increasing internal inequities and social unrest. With growing inequalities within and across countries, governments are often ill-equipped to find an answer to these tensions, and MNCs may have a key role to play in addressing inequalities through CSR. Unfortunately, their current CSR practices often fall short of addressing core societal challenges and show a limited focus on corporate sustainability rather than a greater attention to social issues. Central to this is the need to realize social inclusion, participation, empowerment and social justice through equitable distribution of resources among the local people from the proceeds realized through the operations of MNCs.

Despite the competition for investments among countries, the engagement of MNCs is also associated with the infringement of essential rights. Many shareholders and investment brokers expect a guarantee of large profits to be realized as mandated by the premise of the functioning of capitalism. In order to achieve such profits, MNCs engage in activities which have negative repercussions on society and the environment. Activities of pollution and waste dumping, depletion of land resources, and labour malpractices have become very profound and well documented. Such impact of MNCs on society has raised the ethical responsibilities of these corporations. Environmental problems emanating from MNCs operations include polluting rivers with toxic materials and creating trenches which act as breeding grounds for mosquitoes. MNCs may also disregard sanitary conditions by dumping waste materials indiscriminately. The need and justification for having ethical standards for MNCs is thus necessary.

Sustainable Development Goals (SDGs)

The Fifth Industrial Revolution was born out of the need for a genuine shift of attention from the former acronyms of Fourth Industrial towards the need of humankind for balance, equality, justice, health, resilience, peace, sustainability, and the wellbeing of individual humans and societies within the boundaries of the planet. In this sense, the Fifth Industrial Revolution could indeed be perceived as the global efforts of almost all the State and Non-State actors to reach the Sustainable Development Goals. Getting smarter is not sufficient for people, good intentions and basic operational frameworks are not sufficient if technology is not adjusted through socio-political and economic changes suitable to the specific contexts and their peculiarities. There is no universal panacea upon which each country can lay down its SDG plans. The SDG approach may require variations between and within regions, countries, and locales. While many aspects stem from the premier fiscal concept of balancing budgets, there are also non-direct economic drivers of welfare. These include local governance and behaviour, resilience, culture, heritage, inclusiveness, justice, and sustainability. All actors of the social game matter: from the State itself and its territories to the business, civil, and local community actors, the expectations, needs, and behaviour of all share will directly influence the tranquillity, equality, justice, and resiliency of the systems as primary support to social per capita welfare. Targets in terms of the SDG framework should not only consider the achievement of a certain output or outcome. In the case of SDGs, target-oriented performance assessment requires systematic improvements embraced by and benefiting people and the planet in terms of long-lasting resilient indicators and integrated economy-wide effects. In all parts of the world, states and economies, the pandemic drastically affected quality of life, increasing inequalities, informal economy, unemployment, and poverty across the entire multifaceted complexity of societies. Bleeding development, growth, and sustainability issues have spread in addition to climate change alarm and frustration in the global balance sheet of the sustainable development decade. The expected success of the Agenda for Sustainable Development is perplexed in view of the worsening of most, if not all, sustainability metrics. It appears that Partnerships for the Goals pose a policy dilemma in the international development agenda. In other words, how to enhance a global partnership for development that is mutually empowering and constructive? Perhaps the global partnerships would need to be rediscovered in terms of the Fifth Industrial Revolution concepts to enhance their ambit and outcomes.

REGULATORY FRAMEWORK FOR FDI

Governments throughout the world have recognized that Foreign Direct Investment (FDI) is a key factor for development. They have therefore sought to attract FDI by means of liberalizing their national regulatory frameworks for foreign investment, as part of a globalization strategy. This desire is reflected in the vast liberalization of FDI regimes during the last two or three decades.

In pursuit of this objective, countries liberalized their national regulatory frameworks for FDI. They removed restrictions on the entry and expansion of foreign investment, improved the treatment of foreign investors, and greatly reduced the number of industries that were fully reserved for domestic investors. Countries that restricted foreign investment severely often liberalized a substantial part of their regulatory framework in order to attract FDI. Liberalization was more needed in developing and transition economies, where such restrictions were more prevalent and onerous (P. Sauvant, 2011).

According to estimates, at present about two-thirds of World Bank member States maintain policies that are inconsistent with the treatment standards contained in the World Bank Guidelines on the Treatment of Foreign Direct Investments. For example, 17 developing or transition economies maintain at least one of the outright prohibitions on types of foreign investments or sectors, while one European and developing country imposes very restrictive screening of foreign investments. 23 countries maintain a controversial post-establishment screening mechanism to restrict foreign investments in the aviation and telecommunications sectors, while three South-East Asian countries restrict foreign investments in the land transaction sector (P. Sauvant & Mann, 2017).

International Treaties and Agreements

The surge towards globalization, notably of Foreign Direct Investments (FDI), has led to an intensified interest in the relationship between investments, development, and sustainability. Similarly there has been a rising consciousness about the distribution of FDI benefits among host countries. The investment regime is a crucial player and it has been argued that the concepts of sustainable FDI, Corporate Social Responsibility (CSR)-and responsible business conduct should be considered in the context of specific international agreements. Foreign Direct Investments (FDI), can contribute positively to socio-economic development of host countries. This should be operationalized in investment law and policy. That FDI must help the host country's development is not new; the notion of the development impact of investments has been around for decades (P. Sauvant, 2021). The core challenge would be how to operationalize this in international investment agreements (IIAs).

Investment treaties do not give rights to host countries and responsibilities to foreign investors, making these treaties structurally imbalanced. Promoting and protecting cross-border investments is arguably a legitimate policy objective of states similar to, say, protecting human rights. Relative to that objective, efforts should optimally focus on the conditions that foreign investors have to meet to benefit from treaty rights. The challenge is to translate the norms of responsible investment into precise treaty language for new and revised treaties. Emphasized should be that positive contributions to development are key policy aims of states entering into investment treaties, and how they are to be achieved with the help of treaty provisions. If the rights of persons, including corporate persons, are not balanced by responsibilities that advance the public good, the body of law may lose legality.

National Policies and Regulations

FDI and the Sustainable Development Goals: Progress and Challenges 'FDI will play a role in both the economic and sustainable development of developing countries by facilitating their economies' convergence with those of developed countries, offsetting - at least in part - their capital deficiency. The more rapid FDI growth of developing countries should contribute to their energy consumption increasing more rapidly than that of developed, with an attendant greater CO2 emissions increase' (P. Sauvant, 2021).

FDI's contribution to the achievement of the SDGs requires addressing its effective governance. There are policy and regulatory measures that can be taken at national, regional, and global levels in this regard. At a minimum, developing countries should adopt broad national policies and regulations regarding FDI. They should also build on existing initiatives aimed at monitoring and preventing the more egregious abuses of FDI. All countries, but particularly developed ones, should strengthen existing multi-stakeholder, multi-issue networks as broad frameworks within which to conduct multi-stakeholder processes aimed at collectively monitoring responsible FDI practices and adopting further measures on a cooperative basis. Tax evasion and avoidance by MNEs are perhaps the most widely recognized of the abuses associated with FDI. Reducing tax incentives for FDI is a complex challenge, requiring both efforts at the national level and international cooperation to strengthen the power of governments to raise the tax revenues on income earned within their jurisdiction, and correct the distortions of tax-neutral treaties. At the same time, there remains room at the national level for countries to adopt measures that counteract domestic tax evasion and avoidance. These measures include closing loopholes in domestic laws, strengthening tax administrations, and adopting rules and procedures that prevent tax evasion on cross-border payments.

STAKEHOLDER PERSPECTIVES ON FDI AND RESPONSIBLE PRACTICES AND GOOD GOVERNANCE

This paper reviews stakeholder perspectives on foreign direct investments (FDI), using the classification and terminology as presented in its title. Each perspective is briefly described, focusing primarily on how it may contribute to making FDI more sustainable. A sidebar at the end of the paper also notes how good governance has been defined in the context of FDI by an important group of stakeholders, which has significant commonalities with other stakeholder definitions but uses a different terminology. State Being the Host Country: The Connections of Economic, Environmental, and Social Impacts FDI is an essential driver of economic growth and well-being, being an important source of employment, investment, funding, technology transfer, skills upgrading, and productivity spillovers. It can also have significant environmental and social impacts, which can also affect economic impacts. Negative impacts often outweigh positive impacts on the environment and labor in the host developing country. The range of concerns is extensive, covering various issues, including non-transparent land acquisition leading to displacements, biodiversity loss, environmental harm, human rights abuses, and other labor rights violations, contributing to the latter.

Stakeholder-Driven Obligations of FDI Business Conduct and Legacy: Long-Lasting and Context-Dependent Impacts of FDI Conduct Much of the debate on the desirable conduct of foreign business, plus their sustainability obligations, has focused on the multi-national enterprises (MNEs) receiving FDI. Many negative consequences of FDI business conduct are long-lasting, being irreversible effects. These include depletion of finite natural resources, hollowing-out of population centers, persistent social inequalities exacerbated by foreign investors, and corrupt business practices, weakening national government capacity, independence, and trustworthiness ((P. Sauvant & Mann, 2017)).

Government Views

Governments are taking a closer look at FDI, realizing they need to assess it for potential benefits and how it fits in with the balance of domestic cost and benefit. A lot of things shape the assessment, such as how FDI affects national security and corporate governance. Governments are making schemes that aim to shield various sectors from outside investments. These plans come in response to pressures such as the world financial crisis, tightening credit, preserving local jobs, retaining national blue-chip companies, and preventing hostile takeovers of key industries. So far, many countries have quickened development of the schemes for screening foreign investment interest (P. Sauvant, 2021).

Many governments have indicated that, in the cases of a foreign acquisition or a proposed foreign investment, consideration will be given to national security or strategic interest. In a few countries, fiscal initiatives have been introduced to better monitor the stock of foreign investment. New legislation is being contemplated in a few countries, generally labeled as investment screening or "passive" investment laws, that aim to establish a screening or review process respecting foreign investment in a wider scope of opportune sectors than just shareholders involvements in the context of takeovers, nationalization threats, etc. In garnering support for such restrictive measures, a more careful and sober assessment of the risks and opportunities that foreign investments can bring. An insight into governments' concerns about foreign investment and an explanation of the mechanisms through which it may affect national security and strategic interest are important in this discourse (P. Sauvant, 2011).

It remains a looming question how much flexibility economies have to restrict foreign investment and whether certain policy measures should be off-limits. With the liberalization of investment rules for two decades closely aligned with the globalization of capital markets and the proliferation of merger-takeover activities, the legal frameworks for attracting and facilitating foreign investments have been firmly set. Restrictions on foreign investments have by and large been prohibited by international accords and conventions, and majority control has been implied to guarantee domestic equity interests. Emerging and transition economies that had not previously maintained unqualified openness to foreign investment have been generally expected to follow suit. The issuing of such invitations to limit protection of domestic investments and investors and the movement of capital would thus entail huge reputational costs.

Local Communities

FDI is often hailed as one of the engines of socio-economic development of countries (Lawrence Sambala, 2016). This probably hinges on the idea that efficient firms tend to invest abroad, and that later on markets in poorer countries will catch up, and at the core of this idea are various complementing factors, including infrastructure development, human capital, institutional framework, etc. Even referring to some of these aspects would bring one into difficult contexts. For instance, it would be much easier to discuss FDI in countries where the education system can be seen as functioning. Unfortunately, this is not a universal phenomenon, and the relative success of FDI in sustainable socio-economic development cannot simply be taken for granted.

If FDI would always bring in high-tech manufacturing firms, that would be good news not only for the host country's balance of payment, but also for its capital formation rate and productivity growth. However, in most poorer countries the labor market is not well developed, being fragmented to a large extent. For instance, in Tanzania, the jobs created by TTP may be easily filled by the job-seekers in the short run, but over time the employees will be senior, and new qualified entrants into the labor market will need to be hired. Another point is Tanzania's poor infrastructure. If the roads connecting the two are bad, the investor will also not prefer to set up a factory at a distance of 100 kilometer from the port, as the transportation costs for the inputs will be great. Similar problems arise for electricity and telecommunication services. Therefore, in addition to scarce educated labor, poor infrastructure and public services decrease the incentive for foreign investments to locate there. Local linkages are also often very scarce, as was the case of Tanzania with TTP. In the light of all this, an FDI in the form of activities that contribute to the modernization of the country as a whole can be seen as difficult and even unlikely. However, based on the case of TTP, it seems plausible that an FDI can generate off-firm spillovers that may contribute to the modern sector development of Tanzania, provided that the conditions are right. Conditions like a relatively high productivity and labor-intensity manufacturing sector within a country's borders, and a favorable firm-specific context seem to favor the generation process of off-firm spillovers. But even if this is the case, spillovers will be industry- and firm-specific, and therefore on an aggregate level, the off-firm spillovers generated by the case industry may not sufficient for the whole country's economic-engine transformation.

Investors' Perspectives

From an investor's perspective, the assessment of a country's investment conditions has been defined primarily in terms of the standard methods by which analyses are conducted to identify country-specific opportunities, which fall into two groups: the general investment climate coupled with a framework for regulatory certainty for their investments; and investment opportunities with a reasonable prospect of achieving an after-tax return on investment that is commensurate with risk (P. Sauvant & Mann, 2017). It is important to underscore the environmental, social, and governance factors and issues as they are viewed particularly from a new perspective. The perspective of investors is focused on the assessment of FI in terms of how well a country's investment policies and practices incorporate ESG elements. These assessments are applicable to both international investors and home-country governments seeking to enhance the sustainable development contribution of their international investments.

As noted above, investors demand a general investment climate plus regulatory certainty governing the ability to repatriate profits. In the context of this paper, the following attributes are particularly relevant: rules affecting currency convertibility and repatriation; overt taxes on the repatriation of profits; and the need to invoke national treatment. More particularly related to the economic environment, one issue generally overlooked in the coverage of FDI sustainability is the longer-term sustainability of potential, expected, relevant investment opportunities. Again, the analytical perspective is broader than national decisionmakers on both the sides of developing and developed countries (P. Sauvant, 2021). In assessing a country's attributes regarding ESG overall, potential expected business opportunities across a range of economic industries and tracks are considered. As noted earlier, it is euphemistically put that things cannot go on as they are and anticipations on the investor side regarding safety and sustainability of international business operations, business respect for human rights, and due diligence obligations are presented.

Multi-stakeholder dialogue (states, businesses, civil society)

The distribution of foreign direct investments (FDI) benefits can be improved through a multi-stakeholder dialogue involving states, business, and civil society. Such an approach would create a stronger demand for more and better FDI, leading to a better match between the demand and supply of FDI, and thus a more equitable distribution of its benefits. Spontaneously, such an approach is unlikely to emerge because the parties involved have incompatible interests. However, a number of states have an interest in attracting more sustainable FDI, and in ensuring more equitable distribution of its benefits, which could create the basis for such a dialogue (P. Sauvant, 2021). States need to adopt policies facilitating the attraction of sustainable FDI, providing for more equal human resources and environmental standards, and for fairer taxes and revenue sharing arrangements. Civil society organizations could push for a stronger demand of such policies by all states. To increase the effectiveness of these measures, civil society organizations should broadly link the expertise of stakeholders involved in investment-related issues among relevant international organizations.

While this policy-oriented agenda could create a stronger demand for sustainable FDI and more equitable distribution of its benefits, such an approach is unlikely to be successful unless states ensure a conducive regulatory and institutional environment. Currently, there are political strongholds acting against the acceptance of such policies in many countries – evidenced by the de facto lack of regulations and institutions required. In many countries, this order results in the offer of tax advantages and lax human rights legislation in order to attract FDI, while limiting civil society's ability to hold FDIs accountable. Consequently, FDI leads to a greater concentration of wealth and power in the economy (Dinç, 2012). To counteract this situation, more democratic governance structures need to be established allowing for a peaceful dialogue of all stakeholders ensuring a fair distribution of wealth and power.

The lack of an overarching political milieu, preventing meaningful policy-oriented dialogue, has left a regulatory gap that civil society organizations have aimed to fill by establishing soft law systems including the OECD Guidelines, the UN Global Compact, and the UN Guiding Principles, which are aimed at businesses. This soft law regulatory framework is supported by the International Council of Industries and can be regarded as the global business community's equivalent regulatory framework to the multitude of international conventions and treaties states have put to uphold basic human rights and sustainable development goals.

ETHICAL CONSIDERATIONS IN FDI

The analysis of bilateral investment treaties (BITs) signed by countries in the Asia-Pacific region highlights a trend towards a more balanced approach that protects a country's right to regulate. BITs allow foreign investors to sue host governments for compensation about actions or initiatives said to violate treaty rules. Host governments, especially in developing countries, have to balance the rights and protections afforded to foreign investors with the capacity to implement development strategies, regulate various sectors, and protect the rights and interests of local communities (P. Sauvant & Mann, 2017). An intensely negotiated treaty, properly implementing it at the national level, and establishing a proper dispute settlement mechanism can divert country resources from development, including poverty alleviation and health initiatives.

Foreign Direct Investment (FDI) brings benefits to host countries, such as industrial upgrading, knowledge transfer, skills training, increased employment, and export promotion. At the same time, it would also raise concerns among local communities, governments, and civil society about labour rights violations, environmental degradation, lack of transparency, limited accountability, and disruption of local economies. The viewpoint that FDI can help to promote national development is not universal, particularly among countries recently opened to foreign investments. To achieve the utmost positive impact while mitigating the negative impact, actors in both host and home countries are advocating for a more systemic approach globally, regionally, and nationally. At the international level, the OECD Guidelines and the UN Global Compact already offer business principles in a broad framework. These principles are not binding and thus rely on governments, organisations, and civil society groups to raise awareness and monitor adherence.

Ethical Investment Practices

In the global economy, many foreign firms invest in developing countries to set up subsidiaries or joint ventures. This form of investment is termed foreign direct investment (FDI), and firms doing so are referred to as 'foreign direct investors' (fd-i firms). Due to their size, firm privilege, and foreign origins, fd-i firms are viewed as a more responsible source of business in less developed countries (Anthony Camilleri, 2017). Nonetheless, cases of non-compliance with laws, regulations, norms, and standards occur. Non-compliance of fd-i firms can be particularly damaging to developing countries, and stakeholders view these firms as socially unjust if they fail to operate responsibly. One way to address this issue is via responsible investment practices, through which stakeholders reward responsible firms with funds or investment. Responsible investment practices include screening based on ethical, corporate governance (cg), or sustainability criteria or engagement via governance ballots, dialogue, assessments, and ratings.

Mostly rated firms are larger, older, concentrated in the industrial and financial sectors, and from developed countries. Using a fixed effects regression model, the study investigates whether the 'big three' responsible investment practices, i.e. ESG integration, shareholder engagement, and exclusionary screening, on the average yield of fd-i and non-fd-i firms separately. Although none of these practices have significant effects on the yield of non-fd-i firms, ESG integration has a positive instant effect on the yield of fd-i firms, and exclusionary screening positively affects yield with a 3-day lag (Sugimoto, 2018). The cumulative effect of responsible investment practices on the yield of fd-i firms is robustly significant. The results imply that responsible investment practices limit the sources of funds for fd-i firms and signal to other investors the profit loss from investing in such firms. This study contributes to the small IS literature concerning fd-i firms, rating firms, and responsible investment practices. Further research avenues are provided based on the limitations of the study.

Avoiding Exploitative Practices

While the positive impacts of Chinese FDI in South East Asia are documented, the potential downsides are also examined. There is concern that such FDI may replicate exploitative practices prevalent in China. This concern is partly based on the perception that China has not been able to successfully clean up a labor system rife with abuses. Intended to lure in investors, China's system of bonus labor quotas, temporary contracts, and yellow unions has since 1997 provided efficient informal arrangements for labor exploitation. The major problem with this system is that it allows local enterprises to maximise profits through aggressive wage suppression and discrimination against more vulnerable workforce components. Targeted by different forms of incentives-labor-intensive processes and sectors, remote rural laborers, women destined-to-be-illiterate workers-FDI can do the same in South East Asia which has similar basic mechanisms of labor exploitation.

In the case of Malaysia's high-tech industry, for example, there are allegations of forced labor, close to slave work, and punishing behaviors by local managers with the tacit support of HQs. It should be however acknowledged that, among all Chinese enterprises venturing out, the hardware sector, both at the upstream and down-stream end, such as PC and cellphone manufacturing, are good examples of independently-controlled TNCs with the less moralized concern on labor exploitation. In addition, a richer understanding of labor-capital relationship in general, more sophisticated advanced-catch-up labor relations in particular, can be obtained by examining the first-wave TNCs from the Asian Tigers, such as Taiwan's Hon Hai Precision Industry/Foxconn in the case of low-end assemblage. Nonetheless, it cannot be denied that under similar initial conditions, there is no guarantee that Chinese enterprises moving into the South East Asian labor market would not behave in ways detrimental to labor's overall interest (Girmay, 2018).

FUTURE TRENDS IN FDI

As projected by the World Bank, by 2030 half of the global population will live in Sub-Saharan Africa, with the urban population representing 60% of this region and amounting to 1.3 billion inhabitants. Although in absolute and relative terms this phenomenon might be associated with development, it also creates the challenge of reconciling pressure for growth with rising demand for public goods, such as infrastructure, health care, and education. Attracting private investment capital, of which non-portfolio FDI (domestic and foreign greenfield investment plus infrastructure investment) is the choice since it cannot be flighted out of the country, constitutes a key policy goal for Africa's emerging market economies. A large and increasing amount of resources is placed under management in the emerging market and technical assistance in the investment process can be provided. It is questioned whether willingness and competence to rapidly absorb FDI, especially foreign investors with a different ownership structure, exist (P. Sauvant & Mann, 2017). Investment promotion in SSA faces a dilemma: on the one hand, investment promotion is not only about getting what is not already available (which investors offer what to whom?), but also about having first rates audiences in terms of companies/prospectus, to be able to meet the demand of companies in terms of prospectus and first rate workshops. On the other hand, attracting FDI offers more choices than screening thereof and local partnerships play a key role. These conflicting objectives might be resolved by two complementary approaches: 1) agency chain investment promotion, taking the transnational investor's perspective in identifying a pool of first rates companies to peruse which increased the chance of success; and 2) admin to OTC implementation structures, combining incentivized screening with a range of locally placed funds aimed at mitigating the risks of capital outflow through assisted use of OTC instruments (Esplugues Mota, 2018).

Emerging Markets

Emerging markets have gained a prominent position within the international economic landscape. In the Global North, the 2008–2009 financial crisis put a spotlight on the economic strength of emerging markets ((Guus) Hendriks, 2017). Eventual policymakers, academics and market participants alike started to publish various journals, contributing to the rising flood of information and analyses on the 'next 11' (N-11) and BRIC countries. Global annual GDP growths have come under pressure since 2010-2011, largely as a result of the sluggish Western recovery, commodity price collapse and weakening growth fundamentals in emerging markets, but these countries have not lost their appeal as investment markets, and prospects continue to look brighter than in the developed world.

The notion of emerging markets does not necessarily imply a state of limbo, with the desired state being that of a 'developed market'. Countries can bloom in context-specific or unique ways or structures. Whether and how investors, the investment market practically, are able to 'outsmart' or overcome both the optimistic and pessimistic views, largely depends on the characteristics of the country. Both mainstream and niche approaches can be deployed. Some characteristics are easier to tackle than others, the complexity and interrelationship of them varies, and the timing of their impact generally is difficult to foresee. This paper attempts to shed light on the investment environment of a particularly colourful gambling place in Asia by identifying its main characteristics, how they impact the investment market, and how these aspects differ from both mainstream and niche emerging markets. This information should be useful for potential investors and the gambling market professionals currently not active, but interested to make investments. The paper is structured as follows: An introductory background of the gambling market is provided, then an overview of key insights on emerging markets is presented. Subsequently, the investment characteristics of the gambling market are described, followed by a discussion on their implications.

Technological Innovations

Technological innovations have been at the center of most discussions about economic development. In this debate, there are two opposing views, one considered to be more liberal and the other more protectionist (Hipólito Leal et al., 2021). The liberal view accepts that investments that will always benefit the host country in its economic growth should not be hindered. The protectionist view believes that foreign direct investments (FDI) do not always benefit the host country and therefore they should be carefully monitored. Scholars have therefore analyzed different aspects related to economic development, such as: the impact of technological change on economic growth; patent law; product and labor market regulations; free trade agreements; and the financial integration of the economy through stock market development and FDI (Fazaalloh, 2024).

The analysis of these aspects contributes to the understanding of the most frequently discussed research topics. There are two ideas that can be derived from the literature. Firstly, FDI is a long-run phenomenon despite several short-run factors that may impact FDI flows. The majority of factors, incentives or regulations that directly affect FDI are expected to have an effect on investment flows and stock. In particular, once a firm decides to invest abroad, its concerns about the host country's conditions are likely to diminish. Therefore, given a set of host countries, the firm is likely to invest more in the country that was best initially interviewed. Aggregate stock is in consequence additionally influenced by the initial set of host country attributes.

Quantile regression estimates show that larger performance, a more educated workforce and a lower level of regulations contribute to the inflow and stock of market-seeking and efficiency-seeking investments. However, a higher availability of skilled workers has a positive impact only on the inflow of assets-seeking investments. The same does not seem to happen with low wages or labor unions. Legal rights, government effectiveness and civil rights are all relevant in attracting and retaining foreign investment. However, the fewer limits imposed by host countries on FDI do not seem to foster investment. The results for the stock regressions are similar but slightly less significant. Overall, the analysis seems to indicate that the transfer of technology to developing countries could instantaneously worsen the environment.

Global Economic Shifts

It is true that the largest markets are found outside the OECD, and that in the 1990s, this demographic fact was particularly pronounced in the world of telecommunications. It is also true that much foreign direct investment today is designed, to some extent, to keep market share in industries, such as telecommunications, where large capital, and hence large cash flows, are needed to guarantee survivability. But this is only part of the story or only half of the story, for the search for market share is only one out of a multitude of motivations behind outward FDI (A. Collins, 2012). The incentives to invest overseas are many, and the countries to invest in are thus many and varied. However, in hard currency terms, a large fraction of outward FDI from ECCs has nonetheless gone, in earlier decades, to a small number of fellow OECD countries, in particular the United States and the United Kingdom. This disproportionate internationalization of European Union multinationals adventures primarily in the United States has withstood the profound economic changes in the world economy and in the global marketplace since the early 1990s. Many feels that given this persistent bias, the time is ripe to investigate and analyze investment flows arising from the dramatic and historically unprecedented changes in Asia and Latin America (Hipólito Leal et al., 2021). The former very recently began to open their once-closed, protected and largely insulated economies to the rest of the world while the latter did so a decade or so earlier.

The emergence of some Asian countries as significant foreign investors paints a somewhat different picture. In the early 1990s, Asia produced a little less than 6 per cent of total global outward FDI. By the end of the same decade, this figure had more than doubled. On the two continuums plotted, two countries, Singapore and South Korea, stand out as being unusually aggressive outward FDI investors. Intriguingly, the OECD is no longer binding on the Asian economies either; every one of them has directly invested abroad much more perceived gains to themselves, and indeed they are taking risks that are deemed too great or too unthinkable for conservative OECD firms. The bursting of the dot.com bubble, the Asian currency crisis, the Bali bombings against such a backdrop of brewing unpredictability, the bizarre clash colonized against colonizers in East Timor, and the war against terrorism launched in the same Middle Eastern country, future caution on the part of investors across the entire world could be entirely justified.

CONCLUSION

In terms of their name in global markets, business corporations are obliged to observe lawful, ethical, and socially acceptable standards. They should respect local and international laws, avoid practices that are corrupt, greedy, and exploitative, and wish to make their community / global environment a better place to live (Wanjohi Ngeng, 2018). Such expectations form a key reason for the emergence of the practice of Social Responsibility. When Foreign Direct Investment (FDI) is being considered by an investor, such policies must be examined by the investor for all the stated decision-making criteria. However, FDIs invariably operate in developing/poor countries where such considerations are more murky. The objective of this article is to offer a decision tree for FDIs to navigate ethically through the ambiguities involved in FDIs. Governments issue an Invitation to Fairness seeking to promote reputable professionals / business who might gain the right to invest in a new quarry to extract valuable minerals recently discovered. Notable in the correlations between numbers of professional enquirers locally and in advanced economies and the sourcing, growing and recyclability of Fetherlight ceramic is the absence of smelters adequate for combusting such inputs to the requisite degrees Centigrade. The risks borne by investors and FOI relevant to environmental issues. Reputable companies seeking to invest in third world nations may encounter lobbying by public servants for inducements, grants, permits, etc. but the focus of FDI implementation assurance bears solely on the quarry. Rumours regarding deposits of ready-mixed minerals and ceramics and no amalgamation of the minerals sought are dispelled by an enquirer in the quarry. Camouflage, hence, concern over non-compliance with environmental regulations.

Consequently, a decision tree is set out for consideration of FDI by an investor which assumes the Invitation to Fairness is similarly addressed.

REFERENCES

Abdelaal Mahmoud, A. (2010). FDI, local financial markets, employment and poverty alleviation. *African Economic Review, 22*(3), 77–94.

Camilleri, A. (2017). Corporate sustainability, social responsibility and environmental management. *Corporate Governance Review, 11*(4), 214–229.

Chilwalo, M. (2016). Multinational Corporations: Corporate Social Responsibility versus Environmental Problems. *European Scientific Journal, 12*(17), 241. DOI: 10.19044/esj.2016.v12n17p241

Chiwalo, M. (2016). Multinational corporations: Corporate social responsibility versus environmental problems. *Global Business and Sustainability Review, 8*(3), 56–71.

Collins, D. (2012). Sustainable international investment law after the Pax Americana: The BOOT on the other foot. *Journal of International Economic Law, 15*(3), 433–456.

Dapra, J., Wright, R., & Wu, C. (2019). FDI 101: An overview of foreign direct investment. *International Review of Economics and Business, 27*(1), 13–29.

Diakon Debebe, A. (2018). The role of foreign direct investment in technology transfer for local employees: Evidence from SNP, Hawassa Town, Ethiopia. *Ethiopian Journal of Economics, 26*(2), 101–120.

Ding, M. (2012). Business diplomacy management as a key role in the sustainable development and stakeholder management in the multinational corporations: DaimlerChrysler case. *International Journal of Corporate Governance, 5*(3), 33–51.

Esplugues Mota, C. (2018). A more targeted approach to foreign direct investment: The establishment of screening systems on national security grounds. *Journal of International Trade Policy, 9*(4), 221–239.

Girmay, A. (2018). FDI and gender equality: TNC responsibilities to address gender vulnerabilities in developing countries. *Gender and Development Review, 14*(2), 65–80.

Hendriks, G. (2017). The sustainable development effects of investment by emerging-market multinationals: Shaping beneficial outcomes for home and host country. *Emerging Markets Journal, 9*(1), 51–67.

Leal, P. H., Caetano, R. V., & Marques, A. C. (2021). Is the relocation of polluting industries prompted by FDI flow and stock, globalization, corruption and regulation? *Ecological Economics, 183*, 106–118.

Li, D., & Shen, S. (2022). Social environment and healthy investment behavior: Joint influence of culture and institution on China. *China Economic Journal, 15*(2), 203–220.

Marozva, G., & Makoni, L. (2018). Foreign direct investment, infrastructure development and economic growth in African economies. *African Journal of Economic Development, 10*(1), 33–52.

Nasser, T. (2013). The effects of the economy, perception of business law, external anchors, and ethical behavior on corporate social responsibility in developing countries. *Journal of Business and Society, 19*(2), 44–62.

Ribeiro Goraieb, M., Nascimento, M. R., & Verdu, F. C. (2019). Cultural influences on foreign direct investment. *Journal of International Management, 25*(4), 100–115.

Sambala, L. G. (2016). The role of multinational business corporations in sustainable development: The case of Tanzania Tobacco Processing Ltd. *African Journal of Business and Sustainability, 12*(1), 77–94.

Sauvant, K. P. (2011). The regulatory framework for investment: Where are we headed? *World Investment Review, 7*(1), 1–23.

Sauvant, K. P. (2021). Improving the distribution of FDI benefits: The need for policy-oriented research, advice, and advocacy. *Transnational Corporations Review, 13*(2), 89–105.

Sauvant, K. P., & Mann, H. K. (2017). Towards an indicative list of FDI sustainability characteristics. *Investment Policy Journal, 3*(1), 12–28.

Sugimoto, S. (2018). Ethics in responsible investment: How to incorporate ethics into investment analysis. *Journal of Business Ethics, 152*(3), 709–726.

Wanjohi Ngeng, A. (2018). Unlocking the quandary: Towards embedding sustainable development into the Sino outbound FDIs. *Journal of International Business and Economics, 6*(2), 45–57.

KEY TERMS AND DEFINITIONS

Corporate Social Responsibility (CSR): A business model in which companies integrate social and environmental concerns in their operations and interactions with stakeholders, aiming to contribute to societal well-being beyond legal compliance.

Environmental Considerations: The assessment and management of the environmental impacts associated with FDI, including resource use, pollution, biodiversity, and climate change, with an emphasis on minimizing negative effects and promoting ecological sustainability.

Foreign Direct Investment (FDI): Investment made by a company or individual from one country into business interests located in another country, typically involving management, ownership, or control of the enterprise, and resulting in a lasting interest and significant influence in the host country.

Regulatory Framework: The set of national or international laws, policies, and guidelines that govern and monitor foreign direct investment, ensuring alignment with national interests, economic development goals, and responsible business conduct.

Responsible Business Practices (RBPs): Corporate strategies and operations that go beyond profit maximization to include ethical conduct, environmental stewardship, and social responsibility, ensuring that business activities contribute positively to the communities and environments where they operate.

Socio-Economic Impact: Thc effect of FDI and business activities on the economic and social fabric of host countries, including aspects such as employment, income distribution, poverty reduction, infrastructure development, and community well-being.

Stakeholder Engagement: The involvement and collaboration of various groups such as governments, investors, employees, communities, and civil society, in decision-making processes related to FDI, aimed at enhancing transparency, accountability, and shared value creation.

Sustainable Development Goals (SDGs): A collection of 17 interlinked global goals adopted by the United Nations to address urgent environmental, political, and economic challenges, promoting prosperity while protecting the planet and fostering inclusive, sustainable growth.

Sustainable Development: A development approach that meets the needs of the present without compromising the ability of future generations to meet their own needs, integrating economic growth, social inclusion, and environmental protection as core objectives.

Technology Transfer: The process by which knowledge, skills, technologies, or expertise are shared from foreign investors or multinational corporations to local firms or economies, facilitating capacity building, innovation, and economic development in host countries.

Chapter 5

Towards Responsible Research and Business Practices Through Ethical AI

Yassin Allammari
https://orcid.org/0009-0004-9461-435X
Abdelmalek Essaadi University, Morocco

Nadir El Morabit
https://orcid.org/0000-0002-0889-5924
Abdelmalek Essaadi University, Morocco

El Hassania Rahou
https://orcid.org/0009-0000-5640-1813
Faculty of Economics and Management, Hassan First University, Settat, Morocco

ABSTRACT

Artificial intelligence (AI) is transforming not only scientific research, but also business practices, particularly in emerging economies such as Morocco. However, this rapid adoption raises ethical and governance issues that directly affect research institutions and businesses. Regulating AI has therefore become essential to ensure transparency and social responsibility. This study aims to explore the perceptions, risks, and ethical needs related to the use of AI in business practices and scientific research in the Moroccan context. A qualitative study was conducted based on semi-structured interviews with Moroccan stakeholders in science and industry. The results reveal high expectations for ethical regulation and the establishment of participatory standards, which are likely to strengthen trust in both research and

DOI: 10.4018/979-8-3373-3104-1.ch005

business environments. Drawing on international guidelines and the ETHNA model, the study proposes a governance framework adapted to the context, promoting inclusive and responsible scientific innovation while guiding companies towards sustainable and ethical AI.

INTRODUCTION

At a global scale, artificial intelligence (AI) is revolutionizing research by handling vast data sets, modeling intricate systems, and even automating analysis in fields including medicine, physics, agriculture, and social sciences (Galiana et al., 2024; Baum & Owe, 2023; Chao, 2019). Beyond being a trendy tech tool, it has become integral to the production of knowledge, informing methods and practices (Jadan-Guerrero et al., 2024). However, its growing emergence and remarkable implementation in various fields pose ethical challenges, such as opaque algorithms could perpetuate bias (Savulescu et al., 2024) and the matters of equity, transparency, accountability and exclusion remain questionable (Paci, 2024; Westerstrand, 2024). Such rapid digitalization and increasing academic pressure threaten to further erode responsibility, deepen inequalities and disengage the public (Gonzalez-Esteban & Calvo, 2022). On the other hand, businesses increasingly face such dilemmas, as AI-driven decisions compromise ethical principles and the trust of stakeholders.

In response, international institutions have proposed ethical AI frameworks with guidelines, governance principles, and regulatory tools (Jobin et al., 2019; European Parliament, 2017). The ETHNA system, for instance, promotes ethical standards based on responsibility, participation, transparency, and openness (González-Esteban & Calvo, 2022). Yet, most frameworks originate in Western contexts, requiring adaptation to local realities in developing countries (Borger et al., 2023; Resnik & Hosseini, 2025). Such adaptation is critical not only for research institutions but also for firms seeking to implement responsible and sustainable business models under the pressures of global competition and digital transformation.

Morocco, like many emerging countries, is actively involved in the digital transition of its economy and research system (Benhmama & Bennani, 2024). Strategies such as "Digital Morocco" reflect substantial efforts to stimulate scientific innovation, integrate advanced technologies, and enhance the competitiveness of research institutions (Ahamjik et al., 2024). However, this dynamic is coupled with several challenges, including poorly structured ethical research frameworks, absence of specialized AI committees, limited familiarity with international frameworks, and insufficient ethical awareness among researchers (Allam & Lalaoui, 2025; Mkik & Mkik, 2023). At the same time, Moroccan enterprises, particularly in health, agriculture, and education, face similar governance gaps, where ethical AI adoption is often fragmented and lacks robust oversight mechanisms.

So AI risks being deployed in institutions it is not ready to police the consequences of its applications, with potential repercussions for both research systems and business ecosystems (Chao 2019). In this context it is imperative to document the perceptions of Moroccan research stakeholders with regard to ethical AI issues. Moreover, relating such conceptualizations to business actors' expectations can underpin how academic ethics governance frameworks may be advantageously experienced as a "joint good" for the academy and industry. Several questions keep coming back. What risks do people see? Which practices are already in place, and which ones still lack clarity? Are norms or guidelines taking shape, or do gaps remain? How can ethical governance grow in a way that fits the national setting?

To explore these issues, the article relies on a qualitative approach.

It draws on semi-structured interviews with researchers, institutional leaders, and experts in innovation.

Combining empirical exploration with analysis of existing ethical frameworks, this paper aims to propose practical pathways for responsible and inclusive AI governance within Morocco's research system. It also explores the implications for organizations and businesses, underlining how research-driven ethical frameworks can support responsible business practices. It contributes to broader reflections on institutionalizing ethics in scientific research in the Big Data era, anchoring principles of integrity, social justice, and sustainability at the core of scientific practices. In line with this objective, this study offers a significant contribution to the literature on AI ethics in scientific research by providing one of the first empirical investigations into researchers' perceptions in Morocco, an emerging country where such issues remain largely underexplored. Unlike many existing studies that focus primarily on Western contexts or normative frameworks, this paper adopts a grounded qualitative approach to uncover local challenges, expectations, and practices. Additionally, it proposes a context-sensitive integration of the ETHNA governance model, thus contributing to the adaptation of international ethical standards to structurally diverse and under-resourced environments. By bridging the gap between academic inquiry and organizational needs, the study sheds light on how ethical AI can serve as a lever for responsible and inclusive innovation in both scientific and business domains. These contributions aim to inform both academic debate and policymaking on responsible AI in the Global South.

The remainder of this article is organized as follows: First, it reviews key ethical challenges associated with AI in scientific research and business practices from an international perspective. The second section describes the qualitative research methodology employed in Morocco. The third one presents interview results, highlighting primary challenges identified by respondents. Finally, the last section discusses integrating the ETHNA model as a governance framework and concludes with practical recommendations tailored to the Moroccan context.

LITERATURE REVIEW

Artificial Intelligence (AI)

Artificial intelligence is defined as a collection of techniques and systems capable of replicating certain human cognitive functions, such as learning, perception, reasoning, and decision-making (Russell, 2022). In the scientific context, AI encompasses several sub-domains, notably machine learning, natural language processing, and computer vision (Musib Feng et al., 2017). These technologies are now employed across a wide range of disciplines to analyze massive data volumes, complex model phenomena, and generate automated scientific publications (Beta, 2019). Consequently, AI is not merely a support tool but increasingly acts as a cognitive entity within the scientific production chain (Khlaif et al., 2023).

Ethics in Scientific Research

Research ethics refers to the set of moral principles and standards governing researchers' conduct, validation of scientific methods, and responsible use of obtained results (Galiana et al., 2024). Traditionally, it relies on concepts such as integrity, transparency, justice, and respect for human participants (ALL, 2019). In the digital era, ethics face unprecedented challenges, particularly regarding the exploitation of massive datasets, decisions made by automated systems, and the complex attribution of responsibilities (Marshall & Naff, 2024). By introducing a form of cognitive delegation, AI directly challenges the traditional foundations of scientific ethics, notably regarding accountability, explainability, and human oversight (Saveliev et al., 2021; Siau & Wang, 2020).

Ethics in Business Practices

Business ethics concerns the values and standards that guide decision-making and behavior within firms (Crane & Matten, 2021). It emphasizes principles such as fairness, responsibility, accountability, transparency, and respect toward stakeholders, including employees, customers, and society more broadly. Today, in the era of AI, these ethical concerns gain new urgency. Algorithmic bias, data privacy, and the opacity of automated decisions highlight challenges in corporate decision-making (Martin, Shilton & Smith, 2019). AI finds use in recruitment, customer profiling, financial services, and supply chain management. These applications can improve efficiency, yet they also bring risks of discrimination or exclusion (Cowls & Floridi, 2018). At present, businesses face both reputational and legal pressures to demonstrate responsible innovation and sustainability when deploying AI (Dignum, 2022). Consequently, integrating ethical frameworks into corporate governance is no longer optional. It has become a strategic necessity, for building trust, ensuring compliance, and promoting sustainable practices.

Ethical Governance of AI

Ethical governance is the process of creating organizational structures, norms, and routines that ensure the socially responsible application of new technologies (Almeida et al., 2021; Schneider et al., 2023). This implies active involvement of multiple actors (academics, funders, citizens, firms, and public authorities) in co-defining the principles that will determine how technologies should be used, with the objective of guiding practices that are consistent with universal values such as human dignity, social justice, non-discrimination, and sustainability (Herrera, 2024; Taeihagh, 2021). Ethical governance is not merely limited to codes of practice but rather provides a collective and evolving model in which people can anticipate impacts and become part of social expectations (Srikanth, 2025).

Emerging Ethical Risks and Challenges

Although the contributions of AI to scientific production and business innovation are unquestionable, this increasing entanglement is also raising systemic ethical issues concerning both the production of knowledge and the deployment of AI in commercial practices (Jobin et al., 2019). These considerations are not only technological, but they challenge the values, responsibilities, and balances that science and organizations strive to maintain (Rashidov, 2024). Recent literature points to a number of new risks linked to AI applications in research and business contexts (Giray et al., 2024; Gonzalez-Esteban & Calvo, 2022; Saheb et al., 2021).

A major concern lies in the black-box nature of AI algorithms. Burrell (2016) notes that many models, especially those built with deep learning, remain opaque. Even their creators often cannot fully explain the decisions the models make. This lack of transparency weakens both reproducibility and peer verification in scientific research (Patrici Calvo, 2022), and it undermines accountability and auditability in business decisions. In areas such as biomedical research, economic modeling, or corporate analytics, this opacity can breed distrust in study results or organizational choices (Borger et al., 2023).

Equally important, AI systems rely heavily on training data (Jobin et al., 2019). Such data may carry historical, social, and institutional biases, which AI models can inherit and amplify. Multiple studies show that AI can produce biased predictions related to gender, race, or socioeconomic status (Lobacheva & Kashtanova, 2022; Obermeyer et al., 2019). In research, these biases threaten objectivity, skew findings, and can harm populations if flawed results influence public policies or medical treatments (Chen et al., 2024). In the corporate world, biased AI may reinforce discriminatory practices in hiring, marketing, or credit decisions. Clearly, ethical oversight must extend beyond academia to industry.

Moreover, the use of AI raises challenges for the attribution of responsibility (Sharma, 2024). When analyses, diagnostics, or business decisions are partially outsourced to AI-based systems, responsibilities for errors and manipulated outcomes may become ambiguous (Smith, 2021). This attenuation of responsibility undermines conventional means of blame attribution in both science and organizations and may foster risky practices under the label of algorithmic neutrality (Vakkuri & Kemell, 2019). Automation of reasoning, whether scientific or operational, can distance actors from the epistemological and strategic foundations of their work (Rodriguez, 2022).

In addition, the advance of AI exacerbates inequalities between well-resourced and less-resourced institutions and companies (Lobacheva & Kashtanova, 2022). The technical requirements for AI deployment, such as supercomputing, sophisticated databases, or data science skills—are often concentrated in Northern states, leading research centers and firms in emerging countries like Morocco to the periphery (Vinuesa et al., 2020; Allam & Lalaoui, 2015). This digital and infrastructural divide limits both scientific participation in global projects and the ability of local enterprises to compete in AI-driven markets (Ahamjik et al., 2024).

Therefore, these risks strongly suggest that ethical reflection is required for AI integration in both research and business practices. They highlight the importance of governance that increases transparency, minimizes bias, clarifies roles, and facilitates fair access to resources. These aspects will be further detailed in the next sections through an examination of cultural and institutional ethical frameworks, and the presentation of inclusive governance models such as the ETHNA system.

Research Ethics in the AI Era: An Increasingly Pressing Requirement

The rise of AI in research and business goes beyond mere technical or methodological effects. Capelli et al. (2023) highlight that it reshapes power relations, how knowledge is validated and recognized, and even the logic of citizen and stakeholder participation. These systemic shifts introduce new risks. They challenge core ethical principles in both science and organizational practice—integrity, transparency, justice, and accountability are all at stake (Piteira et al., 2019). Key concerns include the loss of human oversight, relentless automation of processes, and the social exclusion of certain groups from the benefits of research and innovation (Menissier, 2023; Roumate, 2021).

A clear consequence of AI is the growing delegation of decisions to machines. AI systems can rank research projects for funding, filter or select data and articles for analysis, and even generate hypotheses, reports, or marketing strategies automatically (Chen et al., 2024). Such delegation increases efficiency but simultaneously raises ethical and accountability questions that organizations must address. This delegation creates the risk of loss of human control as scholars and practitioners become partly dependent on tools over which they have no full control (Vinuesa et al., 2020). It is also related to the growing influence of algorithmic oversight in both research management and corporate governance (Burrell, 2016). Increasingly, institutions and firms use automated evaluation systems, based on bibliometric indicators, predictive analyses, or performance metrics. While traditional standards of academic and business evaluation are ideally impartial and unbiased (Gonzalez-Esteban & Calvo, 2022), automation may impose standardization pressures, incentivize efficiency over ethics, and facilitate manipulation of outcomes.

Another vulnerability concerns the unequal distribution of AI-derived benefits (Heinrichs, 2022). Most AI tools and platforms are developed within technologically dominant contexts controlled by major economic powers or private companies (Jobin et al., 2019). These actors control technical infrastructures, validation criteria, databases, and research or business priorities (Sharma, 2024). Consequently, underrepresented populations, peripheral institutions, or small enterprises risk exclusion from AI advances, potentially becoming indirect victims. AI models trained on North American or European datasets may yield inappropriate outcomes for other cultural, medical, environmental, or market contexts. Developing countries like Morocco may become passive technology consumers, with limited ability to shape technological purposes or ensure equitable distribution of benefits.

Social exclusion of research and innovation benefits constitutes a violation of epistemic and economic justice, as access to knowledge production and AI-driven opportunities should be equally open to all stakeholders, including marginalized communities and small enterprises (Giantini, 2023). From a responsible research and business perspective, this situation calls for a renewal of AI governance mechanisms that incorporate values of inclusion, contextualization, and solidarity, bridging ethical practices across both scientific and commercial domains.

The Need for Strengthened Ethical Oversight

Classical ethical frameworks are no longer fully adapted to address the profound changes introduced by AI in both scientific research and business practices. These insights are supported by Zheng and Peng (2022). Traditionally, informed consent or internal corporate codes have been considered sufficient safeguards in research or business contexts (Erman & Furendal, 2024). However, the active control of isolated elements is no longer adequate for overseeing the entire evolving system of globalized science and commerce (Cui et al., 2022; Dignam, 2020; Filgueiras, 2024).

In research, ethical oversight has traditionally been handled by ethics review committees assessing projects with potential risks to humans or the environment (Kshetri, 2024). These committees, often rooted in the biosciences, rely on principles such as respect for persons (autonomy), beneficence, justice, and informed consent (Hilb, 2020). In business contexts, ethics offices or compliance departments perform a similar function by supervising corporate AI applications yet face comparable limitations regarding AI's structural and societal impacts.

Firstly, both research and corporate ethics bodies often lack direct engagement with the technologies they supervise, making it difficult to anticipate risks linked to algorithms, big data, automation, or autonomous learning (Winter & Davidson, 2019). Secondly, these oversight mechanisms typically operate reactively on a case-by-case basis, whereas AI requires proactive ethics capable of predicting emerging effects in dynamic research and market environments (Janssen et al., 2020). Thirdly, their mandates often extend only to isolated projects or initiatives, while AI raises systemic, societal, and organizational challenges that surpass the scope of single research or business activities (Thuraisingham & Thomas, 2024).

In response to these challenges, international institutions have promoted an AI ethics framework built on responsibility, transparency, and participation. Jobin et al. (2019) identified over 80 global initiatives converging on principles such as explainability, fairness, data governance, safety, accountability, and sustainability. The EU's Ethics Guidelines for Trustworthy AI (2019) emphasize multilayered governance through legislation, self-regulation, and public engagement. The European Parliament's AI Act (2021) signals further progress toward stricter regulation, though its applicability outside Europe is still debated (Union, 2021).

Effective governance, however, relies not only on formal rules. Inclusive stakeholder involvement across research and business is essential (Cihon et al., 2020). This calls for training researchers and practitioners in ethical awareness and reflective practice (Beulen et al., 2022) and for institutional mechanisms, dialogue platforms, internal charters, or dedicated ethics committees, that embed responsibility within universities, research centers, and companies (Saheb et al., 2021).

Funding bodies, public, private, or corporate, also shape practices by incorporating ethics, transparency, and societal impact into grants, innovation funding, and project approvals (Gauckler & Werner, 2024). Some European programs, such as Horizon Europe, now explicitly require responsible innovation approaches, favoring projects that integrate participatory and ethical governance mechanisms (European Commission, 2021).

Finally, civil society, including associations, users, and citizens, remains essential to ensure social legitimacy in both scientific and commercial AI applications (Lobacheva & Kashtanova, 2022). Therefore, AI should not remain the exclusive domain of technical experts but must become a matter of democratic deliberation, grounded in shared values and adaptable to local cultural, institutional, and market contexts.

International Initiatives and AI Governance Models

Recently, concerns around the advent of artificial intelligence (AI) as a disruptive technology have led international organizations, governments, and stakeholders in general to take more and more interest in the development of ethical and legal norms (Abelnica et al., 2024). These efforts are largely driven by the need to learn how AI might be developed and deployed responsibly, transparently, inclusively and in line with fundamental rights (Hadzovic et al., 2024). They can be ethical principles, legal rules, codes of behavior or voluntary charters. Importantly, these initiatives target not only researchers but also corporate practitioners, technology managers, and innovation units within companies, reflecting the broader societal impact of AI.

One of the most notable efforts is the document Ethics Guidelines for Trustworthy AI, by the High-Level Expert Group on Artificial Intelligence (AI HLEG), set by the European Commission (European Commission, 2019). These criteria instruct what trustworthy AI is premised on, namely legality, ethical strength, and social compatibility (European Commission, 2019). An interesting aspect of the guidelines is that they feature an operational component, with a self-assessment tool for developers, corporate teams, and research organizations to check whether their AI systems comply with the dictated principles.

In 2021, the European Commission published the very first binding legislation on artificial intelligence, hence the Artificial Intelligence Act (European Commission, 2021). This ground-breaking law has been designed to establish a coordinated and regulated single market for AI in the EU. It is a risk-based framework that categorizes AI systems in terms of four classes of risk: from unacceptable risk (covered by the list of banned practices) to high risk (where operators will be subject to obligations of law compliance), limited risk and minimal risk (neither specific notarization nor consent requirements apply). The Act seeks to strike a balance between innovation and safety, safeguarding fundamental rights and fostering reliable and human-centric AI in the Union. Its provisions are increasingly relevant for companies operating globally, including firms in Morocco, as they navigate compliance with international AI standards.

The AI Act introduces specific requirements concerning technical documentation, transparency, human oversight, data management, and risk governance. It represents a significant advancement toward the legal regulation of AI, although some experts criticize its lack of clarity regarding implementation and inadequate consideration of non-European contexts (Veale & Zuiderveen Borgesius, 2021).

Parallel to public frameworks, numerous technology companies (such as Google, Microsoft, IBM) and multilateral organizations (OECD, UNESCO) have released their own ethical charters. These documents advocate similar values, fairness, transparency, accountability, privacy, but remain non-binding and are frequently perceived as mechanisms of "soft regulation." Such charters are now increasingly applied not only within corporate research labs but also in operational business units, highlighting ethical expectations in everyday commercial AI practices. While these charters constitute an initial step toward ethical formalization, several authors highlight the risk of "ethics washing," meaning the instrumentalization of ethics as a communication strategy to preclude stricter regulations (Bietti, 2020). Moreover, their formulation often reflects internal corporate logics, lacking significant citizen, academic, or cross-sectoral participation, thereby limiting their universal applicability (Chhillar & Aguilera V, 2022).

Collectively, these initiatives reflect an international willingness to regulate AI usage but also reveal substantial disparities in approaches, an absence of global coordination, and difficulties in adapting existing frameworks to local contexts, especially in developing countries. Moroccan companies and research institutions alike face challenges in interpreting and operationalizing these norms in their local environments. In this context, hybrid models such as the ETHNA system have emerged, aiming to combine regulation, self-regulation, and shared responsibility within the field of research and innovation (González-Esteban, 2023).

Although initiatives aimed at governing the ethical use of artificial intelligence (AI) have proliferated in recent years, their effectiveness remains constrained by several significant obstacles (Bahoo et al., 2023). Recent analyses indicate that existing approaches struggle to regulate actual AI usage, particularly in complex institutional environments or in countries of the Global South, where governance infrastructures remain underdeveloped (Jobin et al., 2019; Veale & Zuiderveen Borgesius, 2021). This gap affects both academic and commercial sectors, creating inconsistencies in AI ethical practices across research and enterprise applications.

One limitation is the fragmentation of the normative enterprise. Many actors, including states, technology corporations, universities, and multilaterals organizations have independently issued recommendations, yet with limited coordination (Khlaif et al., 2023). This situation leads to a fragmented environment in which researchers and corporate practitioners are faced with the challenge of identifying meaningful, action-oriented benchmarks. Some charters are too abstract, having no monitoring mechanism, while others, such as the European AI Act, are legally binding, but they are region-bound (European Commission, 2021; Veale & Zuiderveen Borgesius, 2021). These are circumstances that undercut the legitimacy of current norms, particularly in a world in which digital technologies cut across national boundaries (Borger et al., 2023).

Secondly, several authors have highlighted the increasing phenomenon of "ethics washing," defined as the strategic use of ethics as a legitimizing tool without genuine commitment to responsible practices (Bietti, 2020; Floridi, 2019; Jobin et al., 2019). Major tech companies frequently issue internal ethical charters primarily to address social or media pressures while continuing opaque practices, such as algorithmic surveillance, lack of transparency, and systematic biases. In parallel, businesses deploying AI for operational purposes may adopt ethics charters more for compliance optics than practical governance. Bietti (2020) characterizes this strategy as a mechanism for circumventing regulation through non-binding ethical language. This phenomenon exacerbates stakeholder mistrust and underscores the need for institutionalized, verifiable, and genuinely enforced frameworks.

A key weakness in current AI governance lies in the lack of inclusive participation. Most guidelines are shaped primarily by technical and legal experts. Citizens, humanities scholars, NGOs, and stakeholders from the Global South are rarely involved (Chomanski, 2021). Corporate actors from emerging economies are also underrepresented, limiting the relevance of these frameworks for local business practices. Ethical governance gains authority only when norms are co-created from diverse cultural, social, and disciplinary perspectives. The persistent underrepresentation of Global South voices and the lack of interdisciplinarity reinforce global inequities in standard-setting (Crawford, 2021).

The ETHNA system responds effectively to these gaps. Developed under the European Horizon 2020 program, it offers a modular governance model for research and innovation institutions, grounded in responsibility, transparency, and participation (González-Esteban et al., 2021). It covers governance, integrity, gender, public engagement, and open access (González-Esteban, 2023) and establishes Responsible Research and Innovation Offices to coordinate ethics initiatives in academia and business. Its flexible design allows institutions to adopt tools adapted to their resources and regulatory environments, making it particularly relevant for emerging contexts. In Morocco, where ethical infrastructures are still limited, ETHNA enables the gradual development of governance, fosters local ownership, and aligns practices with international standards, offering a practical and sustainable pathway toward responsible AI governance.

Gaps in Emerging Contexts: The Case of Morocco

Although recent technological developments are remarkable, emerging countries like Morocco face substantial challenges regarding ethical practices, both in scientific research and in business applications of artificial intelligence (Benhmama & Bennani, 2024; Mkik & Mkik, 2023). These challenges do not necessarily stem from technological backwardness alone but also from weak institutional frameworks, insufficient participatory mechanisms, and a lack of locally tailored ethical standards (Allam & Lalaoui, 2025). Consequently, the Moroccan context illustrates the limitations of the current situation and highlights the opportunity for more contextualized and inclusive ethics governance across both research and commercial practices.

To date, Morocco's national scientific research and innovation ecosystem lacks formal ethical oversight mechanisms (Jabir et al., 2024). Very few universities maintain active ethics committees outside biomedical fields (Bensalah, 2021). Similarly, companies employing AI often have internal compliance offices or codes of conduct, but these are usually limited in scope, rarely addressing complex algorithmic risks, data governance, or accountability in automated decision-making. Where committees or offices exist, their remit often remains project-specific and reactive, neglecting structural ethical considerations applicable to both research and business AI applications.

While efforts such as the Morocco Digital 2020 programme have accelerated the digital transformation, the national policies do not have formal frameworks on AI ethics and governance (Allam & Lalaoui, 2025). Such a gap results in a lack of protections when researchers and practitioners use and deploy technologies, contributing to the possibility of misapplications, exclusions, and social risks. International normative documents, including European non-binding norms and corporate codes of conduct, cannot be adopted directly when they are premised on institutional capacities and resources that Morocco lacks, and on priorities that are sometimes not adapted to cultural or economic realia. Morocco needs agile, context specific regulations to manage the tension between scarce resources, professional pressure to innovate, and national imperatives of justice, inclusion and sovereignty.

The ETHNA framework offers a practical model to address this need. Its participatory approach, flexibility, and modular tools make it suitable for both academic and business contexts in Morocco (González-Esteban et al., 2021). Complementary measures, such as establishing Responsible Research and Innovation Offices, drafting codes of conduct, and forming ethics committees with monitoring mechanisms, can integrate ethics into decision-making and promote transparency, accountability, and engagement. Together, these actions strengthen local ownership, bridge North–South divides, and help establish a sustainable AI governance system tailored to Moroccan realities.

Aligning with European and international standards can further enhance Morocco's scientific and commercial visibility. It can also encourage joint research and innovation initiatives, while making AI-driven solutions more socially legitimate and widely accepted across sectors.

METHODOLOGY

This study employs a qualitative exploratory approach based on semi-structured interviews. This methodological choice is justified by the need for an in-depth understanding of the representations, practices, and expectations of scientific research stakeholders in Morocco concerning AI and the ethical issues it raises. Given the relatively under-theorized and normatively sparse context in emerging economies, this approach also allows us to capture how ethical concerns in research intersect with organizational and business practices that increasingly rely on AI. A qualitative design thus provides rich, contextualized, and inductive insights (Denzin & Lincoln, 2011; Ridder, 2014).

Morocco provides a pertinent case for analysis, situated at the intersection of ambitious digital transformation policies and structural vulnerabilities in governance (Ejjami, 2024; Oubibi et al., 2022). While new technologies are actively assimilated, ethical frameworks regulating AI remain scarce, both in research and in professional practice (Jabir et al., 2024). This normative gap exposes not only researchers but also enterprises and institutions to unregulated ethical dilemmas, making contextualized and critical reflection essential.

A purposive sampling approach was used to select 10 participants with diverse backgrounds, including researchers, academic administrators, ethics professionals, and institutional representatives. To broaden perspectives on responsible practices, the study also included experts familiar with AI applications in organizational or business contexts. Interviews lasted on average 25–30 minutes, conducted either face-to-face or online, and were recorded and transcribed with full consent from participants. The interview guide was structured around four themes: AI use, perceptions of ethics, existing mechanisms, and expectations for governance.

Data analysis followed Braun and Clarke's (2006) thematic analysis using a strictly inductive coding approach. This method facilitated the identification of patterns, tensions, and needs related to ethical governance, anchored not only in the Moroccan juridical and research landscape but also in the broader challenges of responsible organizational and business practices in the digital era.

RESULTS

Participant Demographics

Ten participants took part in the interviews, representing a balanced mix of academic and business perspectives within the Moroccan research and innovation ecosystem. The sample included 5 researchers from universities and research institutes, and 5 business practitioners from private companies and startups engaged in AI-related activities. Table 1 summarizes their demographic and professional characteristics.

Table 1. Participant Demographics

Participant ID	Gender	Position	Field of Expertise	Institutional Affiliation	Experience with AI (years)
Interview-01	Male	Senior Researcher	Social Sciences	Public University	5
Interview-02	Female	Assistant Professor	Humanities	Public University	2
Interview-03	Male	Researcher	Economics	Research Institute	3
Interview-04	Female	Doctoral Student	Computer Science	Public University	1
Interview-05	Male	Researcher	Health Sciences	Public Hospital Research Unit	4
Interview-06	Female	Innovation Manager	AI in Industry	Private Company	5
Interview-07	Male	Business Strategist	Innovation Policy	Private Enterprise	6
Interview-08	Female	R&D Director	Industrial AI Projects	Manufacturing Company	7
Interview-09	Male	Product Manager	Tech Start-up	Start-up Company	3
Interview-10	Female	CEO & Founder	AI Solutions	Private Enterprise	8

Source: Author's own elaboration.

The diversity of this sample provided a comprehensive view of perspectives, practices, and expectations regarding AI ethics and governance in Morocco. The balanced inclusion of researchers and business practitioners ensured that both the academic and commercial dimensions of AI were represented. This heterogeneity enriched the analysis, allowing for thematic interpretations that bridge responsible scientific research and responsible business practices, in line with the broader objectives of the study and the scope of the book.

Heterogeneous and Emerging Uses of AI in Research and Business Practices

Findings from the interviews indicate that the use of Artificial Intelligence (AI) in Morocco is heterogeneous and still at an early stage, both in scientific research and in business-related applications where research outputs are implemented. Adoption of AI varies according to disciplinary focus, availability of digital infrastructure, organizational maturity, and the profile of participants, including researchers and practitioners. While AI is not yet systematically integrated, it increasingly complements traditional approaches in research and contributes to innovation and decision-making in professional contexts.

Three major areas of AI application were highlighted by participants across research and business settings, summarized in Table 2.

Table 2. Summary of AI Applications in Moroccan Scientific Research

Use Case	Fields Mentioned	Tools / Techniques Used
Data analysis	Civil Engineering, Agriculture, Health Sciences, Physics	Machine Learning (e.g., image classification, anomaly detection, complex modeling)
Prediction/simulation	Economics, Environmental Sciences, Medicine	Predictive Algorithms (e.g., market behavior, disease spread, energy systems)
Research task automation	Multidisciplinary	NLP Tools (e.g., bibliographic sorting, information extraction, summarization)

Source: Author's own elaboration.

Participants reported notable improvements in efficiency, accuracy, and decision-making through AI adoption. For example, a researcher in agricultural sciences noted: "*Machine learning algorithms have transformed our approach to analyzing satellite imagery, allowing us to identify crop diseases much faster and more reliably than traditional methods*" (Interview-03).

Predictive algorithms were also considered valuable for decision making: "*We can with far greater precision predict market developments with predictive algorithms and thus underpin timely and informed policy and innovation decisions*" (interview-08).

Managers in industry stressed that AI meets research outputs adjusted to enhance operational and strategic activities such as market predictions, product development, and organizational service delivery. For example, a manager in a tech start-up stated: "*AI allows us to simulate business situations and production schedules, lowering risks fundamentally and make responsible data-driven decisions*" (Interview-09).

Ethical Awareness and Emerging Concerns in Research and Business Practices

Both researchers and business practitioners participating in our study have strong ethical sensitivities, showing well-articulated concerns about AI implications in scientific and business domains. Four specific related themes were identified and are detailed in Table 3.

Table 3. Ethical Concerns about AI Identified by Participants

Ethical Concern	Description	Example Quote (Researcher / Practitioner)
Algorithmic Bias	Replication or amplification of societal biases due to datasets	Researcher: "*If data is biased, AI outcomes will also be biased, but faster and more convincingly*" (Interview-05) Practitioner: "*Biased AI in business decisions could unfairly affect clients and employees*" (Interview-08)
Explainability and Accountability	Difficulty understanding AI models, reducing accountability	Researcher: "*Sometimes I use tools I don't fully master, and this uncertainty makes me uncomfortable*" (Interview-02) Practitioner: "*When AI decisions are unclear, it's hard to justify actions to management or customers*" (Interview-06)
Loss of Human Control	Reduced autonomy due to rapid AI adoption	Researcher: "*Pressure to adopt AI rapidly, sometimes without sufficient critical reflection*" (Interview-04) Practitioner: "*Automation sometimes overrides expert judgment, which can be risky in business*" (Interview-10)
Institutional Vacuum	Lack of clear guidelines, governance frameworks, or ethical committees	Researcher: "*Our institution has no clear framework or specialized committees for AI ethics*" (Interview-03) Practitioner: "*Companies lack structured policies for ethical AI implementation*" (Interview-07)

Source: Author's own elaboration.

Participants highlighted significant concerns about the potential impact of these ethical challenges on the credibility and integrity of their work. Researchers focused on scientific rigor, reliability of results, and responsible experimentation, whereas practitioners emphasized risk management, fairness in business decisions, and responsible deployment of AI in operational and strategic contexts.

Overall, the findings reveal a strong demand for context-sensitive guidance and governance frameworks that can support both responsible scientific research and ethical business practices, bridging the gap between knowledge production and its practical application in Moroccan organizations.

Institutional Deficiencies and Individual Ethics

In both academia and industry, the absence of a clear institutional framework creates significant challenges for ethical practice. Without explicit rules or guidelines, the responsibility for making ethical decisions regarding AI use often falls on individuals. Many participants highlighted that researchers and practitioners alike rely on informal and personal judgment rather than structured governance: *"We rely heavily on AI tools, but there is no institutional guidance about the ethical use of those tools. Every researcher tries to do their best on their own, but this is not a substitute for a real governance"* (Interview-04).

Several participants, especially from the private sector, emphasized that this institutional gap increases risks in applied contexts where AI systems directly impact clients, employees, and business operations. They expressed concerns that companies could adopt AI tools without clear accountability mechanisms, exposing them to reputational and ethical risks.

Both groups urged the government to establish ethical frameworks adapted to the Moroccan context, addressing issues such as transparency, explainability, and accountability. The need for formal training and awareness-raising sessions was also strongly emphasized: *"There's a dire need for training and formal education in AI ethics. We talk about digital transformation all the time, but we don't talk about ethics in those terms, at least... in many formal ways"* (Interview-06).

Preferences for Participatory and Modular Governance Models

Both researchers and practitioners expressed a strong preference for participatory governance models that ensure inclusiveness, flexibility, and adaptation to local contexts. Respondents consistently rejected rigid or externally imposed governance frameworks, emphasizing the need for approaches that resonate with Morocco's institutional and organizational realities. In this regard, the ETHNA model was particularly valued for its incremental and modular orientation, as it allows institutions and companies to progressively implement ethical practices in line with their specific resources and capacities. Table 4 presents the governance models most appreciated by respondents.

Table 4. Preferred Governance Approaches According to Researchers and Practitioners

Approach	Description	Researchers' Preference	Practitioners' Preference
Participatory	Involvement of researchers, practitioners, and stakeholders in decision-making	Strongly preferred	Strongly preferred
Imported Standards	Guidelines directly applied from foreign contexts	Not preferred	Not preferred
Modular / Step-by-step	Gradual and adaptable implementation of ethical governance	Strongly preferred	Preferred

Source: Author's own elaboration.

Respondents highlighted the ETHNA model as particularly promising:

"The ETHNA approach—gradual and modular—is appealing because it allows us to progressively build our ethical framework according to our capabilities and specific institutional needs" (Interview-01).

"A step-by-step model is more feasible for companies like ours, since it does not require immediate large-scale restructuring but encourages continuous improvement" (Interview-08).

Such participatory and modular approaches were perceived as essential for sustainably embedding ethical governance into both research institutions and business practices in Morocco.

DISCUSSION

Consistent with observations in the international literature on AI ethics in scientific and business settings, several results of this study are confirmed. In line with the works of Jobin et al. (2019) and Floridi (2019), interviews revealed an emerging tension between the possibilities of AI and the ethical dangers of its uncontrolled application. Fears reported by Moroccan participants—such as algorithmic biases, opacity, and the diffusion of responsibility—reflect documented issues in North American and European contexts and underscore the global dimension of certain ethical concerns. This convergence highlights that both academic researchers and practitioners within companies face similar ethical dilemmas, though they manifest differently depending on the context.

However, the Moroccan case presents notable specificities. The near-total absence of formal institutional frameworks, lack of ethical training, technological dependency, and unequal access to infrastructure underscore an urgent need for locally anchored governance frameworks. Unlike the situation of those with a legal framework already in place (e.g., the AI Act in Europe), Morocco is still in the process of defining its own framework, reflecting both vulnerability and an opportunity to restructure the entire system. For businesses, this regulatory vacuum often translates into reliance on imported corporate codes or ad hoc practices, which are poorly adapted to the Moroccan socio-economic context. For academia, it results in researchers making isolated ethical judgments without collective guidance. This situation highlights the importance of a tool such as ETHNA, which is flexible, adaptable at different stages, and usable across both research and commercial environments.

This analysis also intellectually informs the ethics debates on AI in scientific research and business practices in Africa in general and in francophone Africa in particular, a hitherto empirically neglected area. It therefore contributes to filling an important gap in the literature and provides useful lessons for other developing nations confronting similar problems. The fact that Moroccan practitioners echoed many of the concerns raised by academics, particularly around accountability and explainability, suggests that future governance must bridge these two domains rather than treat them separately.

Instead, the results highlight that ethical AI governance is contingent on localized adaptations, grounded in three key principles: actively engaging local stakeholders (researchers, practitioners, regulators, civil society) in the development of ethical guidelines; implementing change incrementally rather than all at once; and integrating ethical considerations into both research and organizational processes. In businesses, this means embedding ethical checkpoints into AI product development and service deployment. In research, it means creating transparent and critical frameworks for the design and use of AI-driven methodologies.

The ETHNA system directly addresses these issues. Its logic of structured self-regulation allows both research organizations and companies to adapt the system according to their capacities and priorities. By incorporating tools such as codes of conduct, ethics committees, ethical guidelines, and monitoring indicators, ETHNA facilitates the establishment of robust ethical practices without imposing a uniform approach. For example, a university ethics committee and a company's innovation unit can both rely on ETHNA principles, but tailor them differently to their contexts.

The study's outcomes suggest several avenues for enhancing AI ethical governance within Moroccan institutions and enterprises. Firstly, establishing dedicated ethics governance offices or units directly linked to research vice-presidencies or to corporate compliance/innovation departments to stimulate discussion, propose internal frameworks, and liaise with national and international authorities. Secondly, co-constructing institutional guidelines on ethical AI use involving researchers, IT specialists, business leaders, and regulators. Thirdly, integrating AI ethics modules into master's and doctoral curricula, while simultaneously providing professional training for practitioners and managers in companies. Finally, promoting regional collaborations, notably with other Maghreb or African countries, to share tools, best practices, and frameworks, thereby reinforcing both scientific and economic sovereignty in AI adoption.

This study has several limitations that should be acknowledged. First, the small sample size and qualitative methodology limit generalizability to the broader Moroccan ecosystem of research and business. Additionally, participants predominantly came from urban institutions and enterprises closely connected to innovation dynamics, thereby overlooking peripheral or rural realities where AI adoption and ethical awareness may follow different trajectories.

Consequently, future research could explore several directions. One avenue could involve conducting a larger-scale quantitative study to map AI usage and perceptions across Moroccan universities and companies. Another promising direction would entail comparative research between different Maghreb or Sub-Saharan African countries, identifying both convergences and regional specificities in AI ethical governance across research and commercial settings. Lastly, longitudinal analyses of governance framework implementation in selected pilot institutions and enterprises would provide insight into the long-term impacts of these mechanisms on research practices, organizational dynamics, and business innovation.

RECOMMENDATIONS

Ethical considerations for AI in Morocco should be considered holistically by academia and business in combination. Oversight boards must be set up by universities, research centers and corporations to draft and monitor ethical codes, to provide places where deliberation can occur and to ensure that international standards of AI ethics are practiced, understood and modified to respect local values. In the private sector, compliance offices or ethics boards will need to review AI products and services, ensure they are transparent to consumers, and reduce reputational and legal risks. These are some of the tools needed to protect the integrity of research and ensure accountability in business.

A second priority is to draft clear, context sensitive rules for ethically applicable AI – including rules on transparency, data quality, accountability, and explainability. For businesses, this applies to consumer rights, data and workplace automation. These policies also would have to be created in partnership between researchers, industry, policymakers, legal professionals and civil society, and should be able to evolve with technological development and changes in social mores in order to be credible and relevant.

Last but not least, ethical education and international collaboration are vital to the development of a responsible AI governance. Academic programs should incorporate AI ethics into graduate training, and companies should promote lifelong learning and continuing professional development. Interdisciplinary workshops and consideration of ethical impact criteria on funding decisions may also serve as additional tools to strengthen accountability. Simultaneously, Morocco would need to partner with African and Global South countries to strive for common standards and tools. More flexible models such as ETHNA could underpin a more participative and inclusive governance, specifically tailored to Morocco academia and business.

CONCLUSION

The aim of this paper was to investigate the utilization, attitudes, and ethical implications resulting from the increasing convergence of artificial intelligence (AI) applications in scientific research and business activities by scientific researchers and business practitioners in Morocco. In the global scenario of AI's deep transformation of scientific means of production and organizational strategies and business models, the development of a context-sensitive ethical reflection becomes indispensable, especially in emerging economies whose governance is underdeveloped. The qualitative interview results highlighted a heterogeneity in AI usage practices across research institutions and enterprises, significant awareness of ethical issues (such as bias, opacity, and loss of control), and an almost total absence of dedicated institutional regulation. However, rather than merely pointing to these deficiencies, the respondents clearly expressed expectations for flexible, participative governance adapted to local realities, aligning with the principles of the ETHNA model.

This research thus provides a dual contribution. Empirically, it documents, to our knowledge for the first time, ethical perceptions related to AI not only in Moroccan scientific research but also in business practices two domains that are increasingly interconnected but rarely studied together in the literature. Conceptually, it reinforces the notion that no ethical governance can be effective unless co-constructed with local stakeholders, considering available resources, professional cultures, and socio-economic issues specific to each context.

Several practical implications emerge from this study, notably the necessity to establish national ethical guidelines for AI use in both research and business sectors, create ethical oversight units within institutions and firms, and institutionalize ethics training not only within scientific curricula but also in professional development programs for industry actors. Such steps would help reinforce a culture of accountability and encourage innovation as a means of achieving sustainable and equitable development.

Lastly, this study paves the way for future research at several levels, such as replication studies with larger samples, comparative studies across regions (Maghreb, sub-Saharan Africa) and prospective analyses on governance implementations in progress (academia and industry). This invites us also to think about AI ethics not as a barrier to innovation, but as a prerequisite for its legitimate, responsible, and inclusive use- one that protects both scientific credibility and business trust in the Moroccan and Africa-wide context.

REFERENCES

Abelnica, K., Petrovskis, G., Vindecs, A., & Romanovs, A. (2024). IT governance in the artificial intelligence age: Trends and practices. In J. Grabis, A. Romanovs, & M. Narigina (Eds.), *2024 IEEE 65th International Scientific Conference on Information Technology and Management Science of Riga Technical University, ITMS 2024* (pp. 92–97). IEEE. DOI: 10.1109/ITMS64072.2024.10741935

Ahamjik, I., Agbani, A., Abik, M., Khayli, M., Galzim, N., Berrada, J., & Bouslikhane, M. (2024). Contribution of artificial intelligence for understanding animal rabies epidemiology in Morocco: What are the perspectives of an innovative and predictive approaches? *One Health, 19*, 100874. Advance online publication. DOI: 10.1016/j.onehlt.2024.100874 PMID: 39247759

ALL, E. (2019). *Trust in science and changing landscapes of communication.*

Allam, K., & Lalaoui, S. (2025). Determinants of artificial intelligent adoption in emerging economies: Evidence from the information technology industry in Morocco. In Zimmermann, R., Rodrigues, J. C., Simoes, A., & Dalmarco, G. (Eds.), *Human-Centred Technology Management For A Sustainable Future, VOL 2* (pp. 43–50). IAMOT., DOI: 10.1007/978-3-031-72494-7_5

Bahoo, S., Cucculelli, M., & Qamar, D. (2023). Artificial intelligence and corporate innovation: A review and research agenda. *Technological Forecasting and Social Change, 188*, 122264. DOI: 10.1016/j.techfore.2022.122264

Baum, S. D., & Owe, A. (2023). Artificial Intelligence Needs Environmental Ethics. *Ethics Policy\& Environment, 26*(1), 139–143. DOI: 10.1080/21550085.2022.2076538

Benhmama, A., & Bennani, Y. B. (2024). Factors driving the adoption of artificial intelligence technology in the recruitment process in Morocco. *Access-Access To Science Business Innovation In The Digital Economy, 5*(3), 387–406. DOI: 10.46656/access.2024.5.3(1)

Bensalah, M. (2021). Toward an ethical code of AI and human rights in Morocco. *Arribat-International Journal of Human Rights Published by CNDH Morocco, 1*(2), 187–203.

Beta, W. (2019). *Lithium-ion batteries: a machine-generated summary of current research*. Springer.

Beulen, E., Plugge, A., & van Hillegersberg, J. (2022). Formal and relational governance of artificial intelligence outsourcing. *Information Systems And E-Business Management, 20*(4, SI), 719–748. DOI: 10.1007/s10257-022-00562-7

Bietti, E. (2020). From ethics washing to ethics bashing: a view on tech ethics from within moral philosophy. *Proceedings of the 2020 Conference on Fairness, Accountability, and Transparency*, 210–219. DOI: 10.1145/3351095.3372860

Borger, J. G., Ng, A. P., Anderton, H., Ashdown, G. W., Auld, M., Blewitt, M. E., Brown V, D., Call, M. J., Collins, P., Freytag, S., Harrison, L. C., Hesping, E., Hoysted, J., Johnston, A., Mcinneny, A., Tang, P., Whitehead, L., Jex, A., & Naik, S. H. (2023). Artificial intelligence takes center stage: exploring the capabilities and implications of ChatGPT and other AI-assisted technologies in scientific research and education. *Immunology And Cell Biology, 101*(10, SI), 923–935. DOI: 10.1111/imcb.12689

Braun, V., & Clarke, V. (2006). Using thematic analysis in psychology. *Qualitative Research in Psychology*, *3*(2), 77–101. DOI: 10.1191/1478088706qp063oa

Burrell, J. (2016). How the machine 'thinks': Understanding opacity in machine learning algorithms. *Big Data \& Society, 3*(1), 2053951715622512.

Capelli, G., Verdi, D., Frigerio, I., Rashidian, N., Ficorilli, A., Grasso, S. V., Majidi, D., Gumbs, A. A., Spolverato, G., & Ethics, A. I. S. E. B. S. G. (2023). White paper: Ethics and trustworthiness of artificial intelligence in clinical surgery. *Artificial Intelligence Surgery*, *3*(2), 111–122. DOI: 10.20517/ais.2023.04

Chao, C.-H. (2019). Ethics Issues in Artificial Intelligence. *2019 International Conference On Technologies And Applications Of Artificial Intelligence (TAAI)*. DOI: 10.1109/TAAI48200.2019.8959925

Chen, Z., Chen, C., Yang, G., He, X., Chi, X., Zeng, Z., & Chen, X. (2024). Research integrity in the era of artificial intelligence: Challenges and responses. *Medicine, 103*(27), e38811. Advance online publication. DOI: 10.1097/MD.0000000000038811 PMID: 38968491

Chhillar, D., & Aguilera, V. R. (2022). An Eye for Artificial Intelligence: Insights Into the Governance of Artificial Intelligence and Vision for Future Research. *Business & Society*, *61*(5), 1197–1241. DOI: 10.1177/00076503221080959

Chomanski, B. (2021). The missing ingredient in the case for regulating big tech. *Minds and Machines*, *31*(2), 257–275. DOI: 10.1007/s11023-021-09562-x

Cihon, P., Maas, M. M., & Kemp, L. (2020). Should artificial intelligence governance be centralised? Design Lessons from History. *Proceedings Of The 3rd Aaai/ Acm Conference On AI, Ethics, and Society AIES 2020*, 228–234. DOI: 10.2139/ssrn.3761636

Cowls, J., & Floridi, L. (2018). Prolegomena to a white paper on an ethical framework for a good AI society. *Minds and Machines*, *28*(4), 689–707. DOI: 10.1007/s11023-018-9482-5 PMID: 30930541

Crane, A., & Matten, D. (2021). *Business ethics: Managing corporate citizenship and sustainability in the age of globalization* (6th ed.). Oxford University Press.

Crawford, K. (2021). *The atlas of AI: Power, politics, and the planetary costs of artificial intelligence.* Yale University Press.

Cui, X., Xu, B., & Razzaq, A. (2022). Can Application of Artificial Intelligence in Enterprises Promote the Corporate Governance? *Frontiers in Environmental Science*, *10*, 944467. Advance online publication. DOI: 10.3389/fenvs.2022.944467

de Almeida, P. G. R., dos Santos, C. D., & Farias, J. S. (2021). Artificial Intelligence Regulation: A framework for governance. *Ethics and Information Technology*, *23*(3), 505–525. DOI: 10.1007/s10676-021-09593-z

Denzin, N. K., & Lincoln, Y. S. (2011). *The Sage handbook of qualitative research.* sage.

Dignam, A. (2020). Artificial intelligence, tech corporate governance and the public interest regulatory response. *Cambridge Journal Of Regions Economy and Society*, *13*(1, SI), 37–54. DOI: 10.1093/cjres/rsaa002

Dignum, V. (2022). Responsible artificial intelligence---from principles to practice: A keynote at theWebConf 2022. In *ACM SIGIR Forum* (Vol. 56, No. 1, pp. 1-6). New York, NY, USA: ACM.

Ejjami, R. (2024). Adopting artificial intelligence and big data tools across industry sectors in Morocco: An integrative literature review. *International Journal of Environment. Workplace and Employment*, *8*(2), 171–198. DOI: 10.1504/IJEWE.2024.141574

Erman, E., & Furendal, M. (2024). Artificial intelligence and the political legitimacy of global governance. *Political Studies*, *72*(2), 421–441. DOI: 10.1177/00323217221126665

European Commission. S., & others. (2019). Ethics guidelines for trustworthy AI. *Publications Office.*

European Commission. (2019). *Ethics guidelines for trustworthy AI.* High-Level Expert Group on Artificial Intelligence (AI HLEG), European Commission. https://ec.europa.eu/futurium/en/ai-alliance-consultation

European Commission. (2021). *Horizon Europe – The EU research and innovation programme (2021–2027): General annexes.* European Commission. https://ec.europa.eu/programmes/horizon2020/

European Commission. (2021). *Proposal for a regulation of the European Parliament and of the Council laying down harmonised rules on artificial intelligence (Artificial Intelligence Act) and amending certain Union legislative acts (COM(2021) 206 final).* European Commission. https://eur-lex.europa.eu/legal content/EN/TXT/?uri=CELEX%3A52021PC0206

Filgueiras, F. (2024). Artificial intelligence and education governance. *Education, Citizenship and Social Justice, 19*(3), 349–361. DOI: 10.1177/17461979231160674

Floridi, L. (2019). Translating principles into practices of digital ethics: Five risks of being unethical. *Philosophy \& Technology, 32*(2), 185–193.

Galiana, I., Gudino, L. C., & Gonzalez, P. M. (2024). Ethics and artificial intelligence. *Revista Clínica Española, 224*(3), 178–186. DOI: 10.1016/j.rce.2024.01.007 PMID: 38355097

Gauckler, C., & Werner, M. H. (2024). Artificial intelligence: A challenge to scientific communication. *Klinische Monatsblätter für Augenheilkunde, 241*(12), 1309–1321. DOI: 10.1055/a-2418-5238 PMID: 39637910

Giantini, G. (2023). The sophistry of the neutral tool. Weaponizing artificial intelligence and big data into threats toward social exclusion. *AI and Ethics, 3*(4), 1049–1061. DOI: 10.1007/s43681-023-00311-7

Giray, L., Jacob, J., & Gumalin, D. L. (2024). Strengths, weaknesses, opportunities, and threats of using ChatGPT in scientificrResearch. *International Journal Of Technology In Education, 7*(1, SI), 40–58. DOI: 10.46328/ijte.618

González-Esteban, E. (2023). The ETHNA system and support tools. In *Ethics and Responsible Research and Innovation in Practice: The ETHNA System Project* (pp. 19–37). Springer. DOI: 10.1007/978-3-031-33177-0_2

Gonzalez-Esteban, E., & Calvo, P. (2022). Ethically governing artificial intelligence in the field of scientific research and innovation. *Heliyon, 8*(2), e08946. Advance online publication. DOI: 10.1016/j.heliyon.2022.e08946 PMID: 35243068

González-Esteban, Y., & Patrici Calvo, E. (2022). Ethically governing artificial intelligence in the field of scientific research and innovation. *Heliyon, 8*(2), e08946–e08946. DOI: 10.1016/j.heliyon.2022.e08946 PMID: 35243068

Hadzovic, S., Becirspahic, L., & Mrdovic, S. (2024). It is time for artificial intelligence governance. *Internet ofThings : Engineering Cyber Physical Human Systems, 27*, 101292. Advance online publication. DOI: 10.1016/j.iot.2024.101292

Heinrichs, B. (2022). Discrimination in the age of artificial intelligence. *AI & Society, 37*(1), 143–154. DOI: 10.1007/s00146-021-01192-2

Herrera, F. (2024). Responsible artificial intelligence Systems: From trustworthiness to governance. *2024 Design, Automation & Test*. In *Europe Conference & Exhibition*. Date., DOI: 10.23919/DATE58400.2024.10546553

Hilb, M. (2020). Toward artificial governance? The role of artificial intelligence in shaping the future of corporate governance. *The Journal of Management and Governance, 24*(4), 851–870. DOI: 10.1007/s10997-020-09519-9

Jabir, H., Lagtati, K., & Pohe-Tokpa, D. (2024). Ethical and legal regulation of using artificial intelligence in Morocco. *Journal of Digital Technologies and Law, 2*(2), 450–472. DOI: 10.21202/jdtl.2024.23

Jadan-Guerrero, J., Acosta-Vargas, P., & Gutierrez-De Gracia, N. E. (2024). Enhancing scientific research and paper writing processes by integrating artificial intelligence tools. In Stephanidis, C., Antona, M., Ntoa, S., & Salvendy, G. (Eds.), *HCI International 2024 Posters, PT VII, HCII 2024* (Vol. 2120, pp. 64–74)., DOI: 10.1007/978-3-031-62110-9_7

Janssen, M., Brous, P., Estevez, E., Barbosa, L. S., & Janowski, T. (2020). Data governance: Organizing data for trustworthy artificial intelligence. *Government Information Quarterly, 37*(3), 101493. Advance online publication. DOI: 10.1016/j.giq.2020.101493

Jobin, A., Ienca, M., & Vayena, E. (2019). The global landscape of AI ethics guidelines. *Nature Machine Intelligence, 1*(9), 389–399. DOI: 10.1038/s42256-019-0088-2

Khlaif, Z. N., Mousa, A., Hattab, M. K., Itmazi, J., Hassan, A. A., Sanmugam, M., & Ayyoub, A. (2023). The Potential and Concerns of Using AI in Scientific Research: ChatGPT Performance Evaluation. *JMIR Medical Education, 9*, e47049. Advance online publication. DOI: 10.2196/47049 PMID: 37707884

Kshetri, N. (2024). Economics of Artificial intelligence governance. *Computer, 57*(4), 113–118. DOI: 10.1109/MC.2024.3357951

Lobacheva, A., & Kashtanova, E. (2022). Social discrimination in the epoch of artificial intelligence. *Wisdom, 2*(1, SI), 97+. DOI: 10.24234/wisdom.v2i1.767

Marshall, D. T., & Naff, D. B. (2024). The ethics of using artificial intelligence in qualitative research. *Journal of Empirical Research on Human Research Ethics; JERHRE, 19*(3), 92–102. DOI: 10.1177/15562646241262659 PMID: 38881315

Martin, K., Shilton, K., & Smith, J. (2019). Business and the ethical implications of technology: Introduction to the symposium. *Journal of Business Ethics, 160*(2), 307–317. DOI: 10.1007/s10551-019-04213-9

Menissier, T. (2023). The four ethics of artificial intelligence. *Revue D'anthropologie Des Connaissances, 17*(2).

Mkik, M., & Mkik, S. (2023). Acceptability aspects of artificial intelligence in Morocco: Managerial and theoretical contributions. In Motahhir, S., & Bossoufi, B. (Eds.), *Digital Technologies And Applications, ICDTA 2023, VOL 2* (Vol. 669, pp. 65–74)., DOI: 10.1007/978-3-031-29860-8_7

Musib, M., Wang, F., Tarselli, M. A., Yoho, R., Yu, K.-H., Andrés, R. M., Greenwald, N. F., Pan, X., Lee, C.-H., Zhang, J., Dutton-Regester, K., Johnston, J. W., & Sharafeldin, I. M.Musib Feng. (2017). Artificial intelligence in research. *Science, 357*(6346), 28–30. DOI: 10.1126/science.357.6346.28 PMID: 28684488

Obermeyer, Z., Powers, B., Vogeli, C., & Mullainathan, S. (2019). Dissecting racial bias in an algorithm used to manage the health of populations. *Science, 366*(6464), 447–453. DOI: 10.1126/science.aax2342 PMID: 31649194

Oubibi, M., Zhou, Y., Oubibi, A., Fute, A., & Saleem, A. (2022). The challenges and opportunities for developing the use of data and artificial intelligence (AI) in North Africa: case of Morocco. *International Conference on Digital Technologies and Applications*, 80–90. DOI: 10.1007/978-3-031-02447-4_9

Paci, B. (2024). The hidden abode of artificial intelligence production: Stretching the limits of artificial intelligence ethics and critique. *Thesis ELEVEN, 184–185*(1, SI), 152–166. DOI: 10.1177/07255136241292366

Parliament, E. U. (2017). *Civil law rules on robotics.*

yPatrici Calvo, E. (2022). Ethically governing artificial intelligence in the field of scientific research and innovation. *Heliyon, 8*(2), e08946. DOI: 10.1016/j.heliyon.2022.e08946 PMID: 35243068

Piteira, M., Aparicio, M., & Costa, C. J. (2019). Ethics of artificial intelligence: Challenges. *2019 14th Iberian Conference On Information Systems and Technologies (CISTI).* DOI: 10.23919/CISTI.2019.8760826

Rashidov, A. (2024). Artificial intelligence in scientific research. *strategies for policy in science and education-strategy II NA Obrazovatelnata I Nauchnata Politika, 32*(5, S), 35–45. DOI: 10.53656/str2024-5s-3-ald

Resnik, D. B., & Hosseini, M. (2025). *Disclosing artificial intelligence use in scientific research and publication: When should disclosure be mandatory, optional, or unnecessary?* Accountability in Research-Ethics Integrity and Policy., DOI: 10.1080/08989621.2025.2481949

Ridder, H.-G. (2014). *Book review: Qualitative data analysis. A methods sourcebook.* Sage publications Sage UK: London, England.

Rodriguez, A. L. T. (2022). Ethics for sustainable artificial Intelligence. *Arbor-Ciencia pensamiento Y cultura, 198*(806). DOI: 10.3989/arbor.2022.806013

Roumate, F. (2021). Artificial intelligence, ethics and international human rights law. *International Journal of Information Ethics*, •••, 29.

Russell, S. (2022). Artificial intelligence and the problem of control. *Perspectives on Digital Humanism, 19*, 1–322. DOI: 10.1007/978-3-030-86144-5_3

Saheb, T., Saheb, T., & Carpenter, D. O. (2021). Mapping research strands of ethics of artificial intelligence in healthcare: A bibliometric and content analysis. *Computers in Biology and Medicine, 135*, 104660. Advance online publication. DOI: 10.1016/j.compbiomed.2021.104660 PMID: 34346319

Saveliev, A. M., Zhurenkov, D. A., Poikin, A. E., & Berkutova, T. A. (2021). Ethics of Artificial Intelligence and Post-non-classical Scientific Rationality. *IFAC-PapersOnLine, 54*(13), 397–401. DOI: 10.1016/j.ifacol.2021.10.480

Savulescu, J., Giubilini, A., Vandersluis, R., & Mishra, A. (2024). Ethics of artificial intelligence in medicine. *Singapore Medical Journal, 65*(3), 150–158. DOI: 10.4103/singaporemedj.SMJ-2023-279 PMID: 38527299

Schneider, J., Abraham, R., Meske, C., & Vom Brocke, J. (2023). Artificial intelligence governance for businesses. *Information Systems Management, 40*(3), 229–249. DOI: 10.1080/10580530.2022.2085825

Sharma, S. (2024). Benefits or concerns of AI: A multistakeholder responsibility. *Futures, 157*, 103328. DOI: 10.1016/j.futures.2024.103328

Siau, K., & Wang, W. (2020). Artificial intelligence (AI) ethics: Ethics of AI and ethical AI. *Journal of Database Management, 31*(2), 74–87. DOI: 10.4018/JDM.2020040105

Smith, H. (2021). Clinical AI: Opacity, accountability, responsibility and liability. *AI & Society, 36*(2), 535–545. DOI: 10.1007/s00146-020-01019-6

Srikanth, G. (2025). Transforming governance: Exploring the intersection of e-governance and artificial intelligence in India. *Electronic Government-. International Journal (Toronto, Ont.), 21*(3), 300–312. Advance online publication. DOI: 10.1504/EG.2025.145913

Taeihagh, A. (2021). Governance of artificial intelligence. *Policy And Society, 40*(2, SI), 137–157. DOI: 10.1080/14494035.2021.1928377

Thuraisingham, B., & Thomas, T. (2024). Social media governance and fake news detection integrated with artificial intelligence governance. *2024 IEEE International Conference on Information Reuse and Integration for Data Science, IRI 2024*, 190–197. DOI: 10.1109/IRI62200.2024.00048

Union, E. (2021). Proposal for a regulation of the european parliament and of the council laying down harmonised rules on artificial intelligence (artificial intelligence act) and amending certain union legislative acts. *COM/2021/206final.*

Vakkuri, V., & Kemell, K.-K. (2019). Implementing artificial intelligence ethics: A tutorial. In S. Hyrynsalmi, M. Suoranta, A. NguyenDuc, P. Tyrvainen, & P. Abrahamsson (Eds.), *Software Business (ICSOB 2019)* (Vol. 370, pp. 439–442). DOI: 10.1007/978-3-030-33742-1_38

Veale, M., & Zuiderveen Borgesius, F. (2021). Demystifying the draft EU artificial intelligence Act-analysing the good, the bad, and the unclear elements of the proposed approach. *Computer Law Review International, 22*(4), 97–112. DOI: 10.9785/cri-2021-220402

Vinuesa, R., Azizpour, H., Leite, I., Balaam, M., Dignum, V., Domisch, S., Felländer, A., Langhans, S. D., Tegmark, M., & Fuso Nerini, F. (2020). The role of artificial intelligence in achieving the sustainable development Goals. *Nature Communications, 11*(1), 233. DOI: 10.1038/s41467-019-14108-y PMID: 31932590

Westerstrand, S. (2024). Reconstructing AI Ethics Principles: Rawlsian Ethics of Artificial Intelligence. *Science and Engineering Ethics, 30*(5), 46. Advance online publication. DOI: 10.1007/s11948-024-00507-y PMID: 39384600

Winter, J. S., & Davidson, E. (2019). Governance of artificial intelligence and personal health information. *Digital Policy Regulation and Governance, 21*(3, SI), 280–290. DOI: 10.1108/DPRG-08-2018-0048

Zheng, H., & Peng, C. (2022). The impact of public health emergency governance based on artificial intelligence. *Journal of Intelligent Systems*, *31*(1), 891–901. DOI: 10.1515/jisys-2022-0065

ADDITIONAL READING

Attard-Frost, B., Brandusescu, A., & Lyons, K. (2024). The governance of artificial intelligence in Canada: Findings and opportunities from a review of 84 AI governance initiatives. *Government Information Quarterly*, *41*(2), 101929. Advance online publication. DOI: 10.1016/j.giq.2024.101929

Camilleri, M. A. (2024). Artificial intelligence governance: Ethical considerations and implications for social responsibility. *Expert Systems: International Journal of Knowledge Engineering and Neural Networks*, *41*(7, Suppl. 1), e13406. Advance online publication. DOI: 10.1111/exsy.13406

Correia Loureiro, S. M., Guerreiro, J., & Tussyadiah, I. (2021). Artificial intelligence in business: State of the art and future research agenda. *Journal of Business Research*, *129*, 911–926. DOI: 10.1016/j.jbusres.2020.11.001

Effoduh, J. O., Akpudo, U. E., & Kong, J. D. (2024). Toward a trustworthy and inclusive data governance policy for the use of artificial intelligence in Africa. *Data & Policy*, *6*, E26. DOI: 10.1017/dap.2024.26

Goos, M., & Savona, M. (2024). The governance of artificial intelligence: Harnessing opportunities and mitigating challenges. *Research Policy*, *53*(3), 104928. Advance online publication. DOI: 10.1016/j.respol.2023.104928

Hosseini, M., Resnik, D. B., & Holmes, K. (2023). The ethics of disclosing the use of artificial intelligence tools in writing scholarly manuscripts. *Research Ethics*, *19*(4), 449–465. DOI: 10.1177/17470161231180449 PMID: 39749232

Keith, A. J. (2024). Governance of artificial intelligence in Southeast Asia. *Global Policy*, *15*(5), 937–954. DOI: 10.1111/1758-5899.13458

Marshall, D. T., & Naff, D. B. (2024). The ethics of using artificial intelligence in qualitative research. *Journal of Empirical Research on Human Research Ethics; JERHRE*, *19*(3), 92–102. DOI: 10.1177/15562646241262659 PMID: 38881315

Ooi, P. B., & Wilkinson, G. (2025). Enhancing ethical codes with artificial intelligence governance: A growing necessity for the adoption of generative AI in counselling. *British Journal of Guidance & Counselling*, *53*(1), 66–80. DOI: 10.1080/03069885.2024.2373180

Resnik, D. B., & Hosseini, M. (2025). Disclosing artificial intelligence use in scientific research and publication: When should disclosure be mandatory, optional, or unnecessary? *Accountability in Research: Ethics, Integrity and Policy*. Advance online publication. DOI: 10.1080/08989621.2025.2481949

KEY TERMS AND DEFINITIONS

Algorithmic Bias: Systematic and repeatable errors in AI outputs that reflect and perpetuate existing social, cultural, or institutional prejudices present in training data.

Artificial Intelligence (AI): A branch of computer science concerned with building systems capable of performing tasks that typically require human intelligence, such as learning, reasoning, and decision-making.

Data Governance: The set of rules, processes, and responsibilities that ensure the ethical, secure, and effective management of data throughout its lifecycle in research and innovation.

Ethical Governance: A system of oversight and decision-making that embeds values like transparency, responsibility, and justice into the development and deployment of technologies.

ETHNA System: A modular ethical governance model for research and innovation institutions, designed to promote responsibility, transparency, participation, and gender equality in scientific practices.

Opacity (in AI): The lack of clarity or transparency in how an AI system functions, often due to complex algorithms or insufficient documentation, which makes it difficult to scrutinize or explain decisions.

Responsible Research and Innovation (RRI): A framework that anticipates and reflects on societal needs and ethical impacts throughout the research and innovation process to ensure socially desirable outcomes.

Section 3

AI Ethics and Sustainable Development and Responsible Practices

Chapter 6

Human–AI Symbiosis for Business Sustainability Through Cyberpsychology:
Preserving Human Skills and Identity in the Age of Explosive Intelligence

A. K. Sharadhi
https://orcid.org/0000-0002-9412-1449
University of Arizona, USA

Samuel Kajama Tshikuta
https://orcid.org/0009-0004-6949-0623
Coppin State University, USA

ABSTRACT

Since the Industrial Revolution, technological advances have molded humanity's identity, workforce, and economic value. Cyberpsychology's Psychological Adaptation Vectors (PAVs) can analyze AI's significant impact on society and human psychology across various domains, from theoretical frameworks to its recent daily-life integration, influencing human cognition and behavior. The current Artificial Generative Intelligence (AGI) and ambient systems leading towards super intelligence, caused by advancements in big data and computations, raise challenges against human uniqueness, employment, identity, and existential queries about the future. As evidenced by Reddit sentiment analysis, AI deployments lack transparency

DOI: 10.4018/979-8-3373-3104-1.ch006

and answerability, aggravating anxiety because they are seen as tools for automation rather than augmentation. This tragedy is creating a Turing Trap, benefiting only the technological class, which can lead to economic inequality. Government intervention, collaboration, and regulation are important to ensure responsible AI development to avoid creating a dystopian society.

INTRODUCTION

A highly promising large language model, which was meant to revolutionize scientific writing, called Galactica by Meta, was taken down with its three-day release in November 2022, as it generated perfectly credible research papers, which turned out to be entirely fictional. This was not a sole case; another embarrassing situation where the court case's legal brief, having never existed, citations generated by AI, sent by the New York lawyers, had to face an unprecedented situation. Meanwhile, it even became harder for the professors to distinguish between the assignment's essays submitted by AI and those students across multiple university campuses (Khalil & Er, 2023), while other students were accused of using AI tools, even though they had not touched it (Gorichanaz,2023). These stories make us feel a strange irony where we have created such sophisticated machines that can fool even experts, as well as being fundamentally unreliable for us to build our basic trust in their information.

All these instances lead us back to a straightforward question posed by Alan Turing decades ago: "Could a machine convincingly pretend to be a human?" (Turing, 1950). However, he was not the first to reverie the idea of artificial beings. Another ancient Greek myth tells us the story of Daedalus, who might have crafted life-like statues that owned the ability to move and speak on their own. Moving forward to 1920, Karel Čapek introduced the word 'robot' from the Czech word 'robota', encapsulating the age-old fascination of our humanity for creating artificial life. This fascination never faded, and we are the first generation to not only anticipate but also the first one to turn these ancient dreams into a tangible reality. The AI machines that we have created today are far from resembling their predecessors' fantasy; they are becoming real. Their predecessors were traditional computers, not even that old, that followed the rigid rules of algorithms; they were obedient and produced predictable results. But modern AI machines that run based on generative AI can learn from their mistakes and create definite solutions. They match the thought of Descartes' concept of a "thinking thing", which he might not have imagined (Cottingham, 1992), but we see its breakthroughs at a dizzying pace. They are capable of just consuming ordinary human languages to produce a variety of outputs, from poetry to programming coding, from artwork to authoring scientific academic papers, often surpassing human performances. We believed these tasks were impossible for machines to deploy, but they are increasingly getting good at spotting diseases in medical images (K et al., 2022). Autonomous vehicles, such as those used in Waymo One and Cruise taxis, Tesla's autonomous car navigating city streets in America, are another example. Amazon's enhancements in warehouse operations, achieved through automated robots, are another good example, and some experts are engaging with AI models in philosophical discussions.

Moore's law states that as computational power continues to grow exponentially, it will become easier to approach the third stage of AI, known as "Theory of Mind" (Schlinger, 2009), where it would enable AI systems to understand and react to human sentiments and thoughts. This advancement is catching the entire spectrum of human emotions, prompting experts to draw comparisons to the potential impact of AI on climate change. Some call for the halting of its deployment at once, while others dismiss current AI as "stupid" because we are just fearing an intelligence that does not even exist. In all these perspectives, they are overlooking the deeper context of how human psychological dynamics are at play, which is a complex way that humans can interact and adapt to these modern technologies. This is where the emerging field of cyberpsychology comes into play. Cyberpsychology (Dalal et al., 2021) combines insights from philosophy, psychology, sociology, computer science, and human-computer interaction to understand how technological advancements will shape our thoughts, behaviors, and emotions. Understanding these dynamics is especially important during the time of our humanity as we are living in our increasingly digital lives, where it is not just academic curiosity but also essential for our collective well-being.

This book chapter delves deeper into this emerging cyberpsychology concept, through Psychological Adaptation Vectors (PAVs). These PAVs trace human psychological adaptation across four different eras in response to technological advancements. This chapter address the gap in understanding this longitudinal phenomenon. We also conducted a sentiment analysis experiment on the Reddit platform using RoBERTa, a natural language processing (NLP) technique, revealing what people perceive and feel about AI. Finally, we explore why such fears are created due to the misconception that AI enhances productivity through automation rather than reshaping through Human-AI symbiosis or augmentation, which creates a Turing Trap led by the technological class. This has resulted in economic and social issues by belittling the working class, while technological elites have benefited disproportionately. To tackle this big question, mitigations through government intervention, implementing policies to safeguard human workers' rights, supporting training and educating, and collaboration between technological organizations and policy makers for responsible AI deployment are vital. Our future in the world is increasingly shaped by artificial minds, hence it's becoming crucial that this technological growth uplifts all classes of society. However, this still unwraps the question of whether it might create a dystopian future where a few individuals control everything by exacerbating existing social inequalities.

DEVELOPMENTAL EPOCHS OF PSYCHOLOGICAL ADAPTATION TO TECHNOLOGICAL ADVANCEMENTS

The commencement of the artificial intelligence journey not just begins with the machines, but with the evolution of Homo sapiens, the apex pinnacle of cognitive evolution and highest biological complexity. Behind all this is a marvel of biomedical engineering, called the "human brain", which has all the necessary tools and machinery resources that led to the development of ambient AI systems, which are capable of reasoning and learning from their mistakes. This highest evolution of the machine was backed up by previous humanity and technological progress, requiring the highest psychological and societal transformations. Over the centuries, our societies have altered in numerous ways through several revolutions encompassing experiences ranging from the human labor mechanization during the Industrial Revolution to the advent of ambient computing. It has reshaped the lives of individuals throughout history and its economics, power structure. Whether it be theoretical, imaginary, or coming into life, all those phases of AI development have seen distinct psychological adaptations in humans to that time, which marks the interplay between technology and human cognition. How the human processes, regulation of that emotion, social interactions with AI, and creation of new identities affect the psychological functioning is what is examined by the field called "Cyber Psychology". This is because the evolution of AI is as much about the psychology of humans as it is about the advancement of the technology. In this field, sensors, human behavior, thought process, and cognition in emotions are integral components of the study. In the book chapter, we have identified four "Psychological Adaptation Vectors (PAVs)" in a relationship to understand human's involvement along with the AI, each in each era, where these vectors encapsulate the ways that individuals in society are responding to the technological progress, which is depicted through the term "Impact" (Figure 1). These factors are not static; they are dynamic and constantly changing, interrelated factors that structure the psychological model of humanity. These vectors can be interpreted as:

Figure 1. Psychological Adaptation Vectors (PAVs)

PSYCHOLOGICAL ADAPTATION VECTORS (PAVS)

VECTORS	DESCRIPTION	IMPACT
1.Techno Cognition	Encompasses adjustments in trust levels as humans interact with AI systems and evaluate their reliability. It involves both individuals' and collective's cognitive trust and trauma.	Determines how individuals, groups, and societies process cognition, adapt to, and integrate technological changes, including cognitive mechanisms for trust or trauma, such as adaptation to AI technologies.
2.Emotional Boundary Negotiation	Involves the gradual redefinition of emotional attachment and detachment between humans and machines through developing nuanced emotional literacy with non-human intelligences.	Understanding the evolving landscape of emotional dynamics in human-AI interactions
3.Identity and Agency Reconstruction	Involves the reshaping of human identity and agency considering AI's capabilities.	Investigates how AI transforms self-perception and personal autonomy.
4.Existential Processing and Negotiation	Comprises of reimagining human purpose and meaning in an increasingly AI-mediated world.	Explores philosophical and psychological recalibration of human significance.

Source: Authors' Original Creation

I Epoch: The Industrial Revolution (1760s-1840) and Technological Integration(1840s-1950)

The first psychological disruption during the Industrial Revolution marked the first major confrontation between technological advancement and human labor. It manifested in the displacement of artisans who were skilled at their crafts, but machines replaced them. It started in Britain and later gradually spread to other parts of the world, introducing new technologies. Notably, the steam engine devised by James Watt (Gopalakrishnan, 2009) and the automated power looms invented by Edmund Cartwright (Mackey, 2024) exhibited the highest production efficiency compared to human laborers stuck in challenging traditional roles of labor and identity. They also fundamentally restructured the economic value and human skills substitution at an unrestrained rate.

During this period, the cyber psychological landscape can be described as a situation where only one portion of the population enjoys a high quality of life, while the majority faces hardships due to the creation of a distinct working-class with lower wages and poor working conditions. This has led to the destruction of the professional identities of experienced craft workers as machines started to replace them. For example, the introduction of machines rapidly in textile industries led to significant workforce replacement, causing "Cognitive Trauma". This spearheaded the society to engage in "Collective Existential Negotiation" about redefining human value as productivity connected to the efficiency of the machines rather than human relevance in crafts and skills. Most of the citizenry that was being exploited started the Luddite movement (O'Rourke et al., 2013) as a collective psychological response, where artisans destroyed all these machines in the form of protest. These factors laid the psychological foundation for this era, revealing the capacity of humans for adapting to change during the highest disorientation, caused by the invention of new technology.

The post-Industrial Revolution experienced a shift towards a gradual understanding of viewing technology, such as electricity, the telephone, the telegraph, aviation, automobiles, and some other early computing, as an extension of human beings. This change occurred after the renegotiation of "Emotional Boundaries" between humans and machines, ensuing its incorporation as one of the economic structures, steadily establishing "Collective Cognitive Trust" towards it, which led to our modern technological society.

II Epoch: The rise of early Expert Systems and Theoretical frontiers (1950s-1980s)

This period marked the beginning stages of artificial intelligence, but it was only confined to theoretical and academic frameworks. Alan Turing's 1950 paper instituted an enquiry about the cognitive nature of artificial intelligence, followed by the Dartmouth conference of 1956 (McCarthy et al., 2006), which was deemed the birth of artificial intelligence as an official discipline of academic field. Pioneering projects like Marvin Minsky's early neural networks (Minsky & Papert, 1969) and Joseph Weizenbaum's chatbot ELIZA (Weizenbaum, 1966) put forward humans' emotional connection to technology, even though we intellectually understood their limitations. This overlaid the way for early expert systems applications like MYCIN in 1972 (Shortliffe, 1974) and DENDRAL in 1965 (Djerassi et al., 1982), which gave a broader spectrum of usage of AI in specific domains such as detection of a bacterial infection, recommendation of antibiotics, and detection of unknown organic component's molecular structure, respectively. Professionals such as organic chemists and medical doctors who used these expert systems had a calibrated test for them, knowing their strengths and limitations. They effectively leveraged their strengths while being aware of their limitations, evading complete dependence on them without giving up their critical judgment or professional experience. This suggests an "Emotional Boundary" between human critical knowledge and AI judgment. This sets the intellectual groundwork for possible practical applications of AI in the future, fostering curiosity around the subject matter. These captivating concepts, depicted in science fiction works like Isaac Asimov's "I, Robot" (1950) and Stanley Kubrick's "2001: A Space Odyssey" (1968), allowed individuals to interpret these theoretical frameworks as fantasy. During that time, there was a definite separation between humans and machines co-functioning together because it was a distant possibility rather than an immediate concern. However, concepts like Isaac Asimov's "Three Laws of Robotics" (Palacios-González, 2024) and HAL 9000 continued to cater to this area by showing how complex the relationship between humans and technology might advance at some point in the future. These exceptional narratives served as psychological preparation for such possibilities of technology going rogue, even before they existed. This demonstrates "Existential Processing" in an AI-driven world. The invention of microprocessors like Intel 4004 (Aspray, 1997) and the rise of tech giants like Intel, Apple, and Microsoft arose during this time.

III Epoch: The Internet Phase (1990s-2007) and the invisible dependency and integration chapter (2007 onwards - 2020)

This phase is stamped by significant positive transformation, yet it poses several challenges to the public. Notably, the increase in accessibility of large data sets with the highest computational power and the exponential growth of tech companies have led to a search for jobs, especially in the IT field. The surge of mass internet adoption (Leiner et al., 2009) fundamentally changed how humans can access information. Thus, as in the cyberpsychology outlook, this dynamic created new "Identity Reconstruction" among people who continued to chase new jobs in the cities, forming new possibilities. These traditional wage-based jobs held importance in hard work, tailed by fulfilling life's commitments and responsibilities, such as providing and nourishing, housing, applying for mortgages, pursuing education, competing for higher wages, indulging in vacations, and automobiles as in cars, scooters for means of travel, as a response to dynamic economic machinery of this period. The discovery of the World Wide Web (Brin & Page, 1998), Google Search (1998), Amazon's recommendation engine, and early voice assistants like Apple's Siri prototype gently infused AI into our daily life. Search engines, email, online social network sites, virtual worlds, forums (Boyd & Ellison, 2007), and unprecedented access to new communication platforms and accessing vast information beyond geographical constraints enabled younger generations to seamlessly integrate into this framework, treating this as their extension, which profoundly shaped their personalization behavior and thus creating "hypertext minds" (Tergan, 1997).

The highest computational power in a pocket-sized device was showcased by the first revolutionary smartphone with a touchscreen, introduced as "iPhone" in 2007 by Apple. Such inventions have multiplied in impact to the extent that the kids who are born in this period are the first generation to grow up with smartphones as their default environment. They have embraced new "Identity Reconstruction" into their cognitive frameworks, as Marc Prensky described in "digital natives" (Prensky, 2001), that this generation built "cognitive trust" over these technologies by treating them as an extension of oneself. Due to this early exposure, the kids were subjected to developing fragmented attention and a distorted sense of time, where memories were never formed from a young age, which altered how their brains developed over time. It is also linked to ADHD (Meng et al., 2024), memory issues, anxiety, and difficulty staying focused. Meanwhile, many adult users experienced "nomophobia" (Gonçalves et al., 2020), the type of anxiety that is created after being separated from their cell phones. In addition to that, they engaged in increased doom scrolling and context switching, which resulted in partial attention. We have jointly developed "collective cognitive trust" in these gadgets, depending on them for scheduling, navigation, and as an extension of our memory and coordination.

Similarly, social media platforms grew to have massive user databases, transforming communication and information sharing at the cost of the highest media consumption. This created the "always connected" paradigm (Turkle, 2011), but internally, we were not connected to real-life relationships. This invisible integration of technology shows how our "Emotional Boundary" with the new inventions got renegotiated, gradually directing us towards invisible dependency, believing it as an extension of our being. These discreetly form our separate digital identities (Suler, 2004) that are glued to notifications and updates, always competing for others' attention for maximum engagement, which is certainly toxic. Still, some knew quite about ethical concerns over data privacy and algorithmic bias, and some were also educated about "filter bubbles" (Seargeant & Tagg, 2019), but most of us were not. Regardless, social media platforms became the source of a larger dataset that led to the development of big data infrastructure, algorithms, and cloud computing, which set the stage for the next AI breakthrough. Although during this period, AI was limited to specific applications rather than a full-fledged agent.

IV Epoch: Ambient systems and Generative intelligence juncture (2020s to present)

In the previous few years, we have seen AI in the limelight as the greatest technology that can endow an enhanced world by upgrading cities, homes, factories, schools, offices, and hospital facilities. However, this faith was short-lived as this era is marked by fiction becoming reality, where we see accelerating advances of AI as a looming threat, competing with human evolution and their capacity for adaptability and competence. These radical changes in AI capabilities are brought by a combination of feeding high-quality data, advanced computational power, also with integration of human feedback, trained through reinforcement learning to solve specific problems to produce creative outputs.

The major players in the AI industry are Anthropic, OpenAI, Google, Meta, and Perplexity. These companies have shown us the modern AI pacing from Large

Language Models (LLMs) like ChatGPT, DeepSeek, Claude, Grok, Gemini, Copilot, Lamma, to artistic AI systems like Dall-E, Midjourney, and similar systems showing unprecedented abilities to generate creative outputs. This is further compounded by face recognition embedded in CCTVs (Sharadhi et al., 2022); problem-solving software engineering skills like Devin (Hassan et al., 2024); Axios, which unleashes PhD-level expertise (Chamboko-Mpotaringa & Manditereza, 2024); and AI-authored patents (Schwartz & Rogers, 2022). This endeavor has sprung from software to robotic developments by Tesla, Nvidia, Boston Dynamics, 1X Technologies that are being trained through synthetic data in simulation (A. K., 2025), which becomes the fastest training in its intelligence (Ge et al., 2020), supported by cutting edge computing and architecture. They are getting trained to get positioned for any tasks, be it at home chores or factory work, or autonomous cars. For example, cashier-less billing terminals are supplanting store clerks, autonomous cars are replacing drivers, autonomous vacuum cleaners are substituting human janitor staff, factories are using robots in their warehouses for unlimited work, and generative AI architectures are menacing the work of art professionals, writers, and techies. Not only these general applications, but they are also striding towards the usage of the Artificial Generative Intelligence (AGI) prototypes in medical applications and biological factors such as Alpha fold programs for prediction of protein structure (Jumper et al., 2021) and brain-computer interfaces by Neuralink; more development towards real-time learning robots without the necessity of human intervention, by the company IntuiCell; a humanoid robot "Sophia" was granted personhood in 2017 (Negri, 2021). According to the International Federation of Robotics, the leading 10 countries with the highest operational robot density in manufacturing and other outlets of industries are Korea, Singapore, Germany, Japan, China, Sweden, Hong Kong, Switzerland, Chinese Taipei, and the United States. These systemic AI frameworks encompassing deep learning, reinforcement learning, and generative intelligence to perform a complex task autonomously have created a paradigm shift where the tasks included human uniqueness, creativity, consciousness, efforts, originality and gradual growth are being challenged by these swift advancements that are even amalgamating in healthcare, surgical assistance, automation in logistics in manufacturing, as well as in research, which can begin to replace humans in real jobs. We are experiencing the "Collective Trauma" and "Existential Negotiation", as one felt during the time of the Industrial Revolution. Everybody is going through growing pains now, as we don't want to diminish the value of human beings. This is ultimately steering us to "Existential Recalibrations" because the state of the psychological response is multifaceted, showing both fascination and anxiety towards these advancements and the implication that it could do for human roles in society.

People have a dual nature when it comes to embracing AI. Those users who marvel at its capabilities are adapting AI that requires "cognitive trust" for everyday tasks such as decision making, content creation, drafting emails, coding, and debugging. While some are forming social and emotional dependence on the non-human entities, AI bots, by reducing the need for human interaction, negatively affecting their loneliness. They are stuck in their virtual relationship cocoon as these AI bots cater to what users want to hear, which can reinforce unhealthy habits and reality distortion. This raises questions about the cyberpsychology concept of "emotional boundary negotiation" between authentic connection and simulation, which is likely to redefine human experience in the coming decades. A real example, as per the author of the book "Future of You," published in 2021, by Tracy Follows, has expressed that in Japan, young men are already selecting their digital avatar assistants for relationships, instead of dealing with real partners (André, 2021).

On the other hand, the second group is grappling with fears regarding job displacement and ethical issues, expressing the unease and moral dilemma of "cognitive trauma". The cause is AI's intrusion into the areas meant for human creativity, reasoning, job security, and economic opportunities for growth. Artists, writers, music composers, illustrators, web designers, and other creative professionals are experiencing this risk. When tasks are automated to the point that AI can imitate original human-authored works, then human efforts in creativity and imagination are derided. These works can never be considered as a genuine manifestation of people's hard work to reach that level and their experience behind that expertise. Furthermore, this group cannot compete with the automation of increasingly sophisticated AI autonomy. Advancement is fueled by more companies competing for it by building hundreds of models within the four walls. This deployment and the extent of their growth are not shown to the rest of the common people, as it is not open source. It is fundamentally a "black box" for all of us. Once it turns up, major job displacement will lead to economic challenges, as seen in major companies are resorting to mass layoffs and recessionary measures in the label of cost-cutting (Acemoglu et al., 2022). Extremely educated workers across different white-collar industries will face the highest disruption, because these AI agents are getting hired and challenging them in previously secured roles, ranging from programming to law to creative arts (Lee, 2016).

Starting from 2020, there has been a clear indication of tech layoffs associated with automation and AI integration. In 2023 alone, there have been job cuts across over 240,000 of over 1100 companies, largely attributed to this fact. This represents approximately 50% more cuts than the previous year, due to reduced demand in certain positions in sales, administrative, and technical support. Even AI systems are being used in Human Resource Management (HRM) to weed out applicants. In contrast to the early 2000s, where advancements in technology created new job opportunities and the industry trend was about rapid hiring, but currently, the major companies like Google, Meta, and Microsoft's primary drive is to restructure operations and streamline their investments in AI technology rather than traditional roles. Although some tech leaders are trying to promote AI as a collaborative tool to enhance productivity, these rapid adjustments are showing a different story. But companies like Paytm explicitly said that AI automation is the factor for reducing their staff, which impacted many roles across their different business units. This effect is also evidenced by a decline in job offers for the graduates from top universities like Harvard, MIT, and Stanford, which has shown a downward trend in 2024. Since 2021, the job placement rate curve has dropped low, resulting in unemployed graduates. This is credited to the fact that tech giants have scaled down the hiring targets and focused more on artificial intelligence systems investments, making the market more competitive and harder for even graduates from prestigious universities to secure their jobs.

This calls for responsible AI development, as human interaction with AI becomes more prevalent. However, it requires intense ethical reconsideration, as companies race towards deploying autonomous, humanoid robots and super agents. The demarcation between Human-AI Symbiosis or augmentation, and replacement is becoming steadily indistinct. As a collective psychological response, the first EU AI Act addresses the risk of AI and ways to regulate it by advocating the unfettered development, which exactly mirrors the Industrial Revolution's Luddite moment. Human relationships are a two-way street where people need to collaborate and show up for each other. Unlike replacing humans with AI, which has no authenticity and fails to provide justice for humanity due to this skewed transformation. Artificial General Intelligence (AGI) is just developing further, bringing us closer to the narrative where a machine can perform at the human level and might even develop emotional intelligence. They can solve tasks at the same quality, own super intelligence, and become sentient beings capable of forming their belief system and goals, which forces the notion of humanity and challenges us to experience "Existential Renegotiation" and new "Identity Reconstruction". Artificial intelligence visionary Nils Nilsson had already documented that if we are trying to achieve real human-level AI, it would simply mean that majority of the humans performing work for paychecks will certainly be automated. That is where the term Human Like AI (HLAI) comes in, which is a blend of all kinds of intelligence that can solve any type of problem (Zhu & Liu, 2023).

These are inquiries that set the stage for the next exploration, investigated using extensive experimentation that analyzes individual behaviors and societal paths on the Reddit platform. It offered a word map showing humans' relationship with AI. This experimentation was implemented by conducting sentiment analysis on Reddit posts related to AI across different industries, where the setup was aligned with the research goal of examining public attitudes towards AI across different industries. The code starts with the initialization of the Reddit API client and defining functions for scraping posts containing AI-related keywords across different industry-specific subreddits to ensure the collected data is relevant and diverse. Then, the data is used for analyzing their sentiment using the Natural Language Processing (NLP) technique called RoBERTa. This sentiment analysis function categorized posts in terms of negative and positive sentiments based on the presence of positive and negative words. Then, the overall analysis is presented in terms of a word cloud, which provides insights into the comprehensive understanding of the overall sentiment distribution of how AI is perceived across different industry-related subreddit discussions on Reddit. Overall, Reddit offers a diverse database with a large amount of data, where users are engaged in community discussions, real-time insights, and the availability of an API, making it ideal for this experimentation.

A word cloud was generated for both the positive and negative words to illustrate our societal sentiment towards the implications of artificial intelligence (AI). The positive words (Figure 2) included terms such as "effective," "innovative," "productive," "successful," and "beneficial," indicating the hope that AI can enhance productivity in solving complex problems, where "emotional boundary" is positive. However, the negative words (Figure 3) exhibited diverse cognitive and emotional responses that align with concepts in cyber psychology. Anxiety about losing one's job due to automation is highlighted by the words "layoff," "disrupt," "inferiority," which leads to "identity reconstruction," as there is a loss of individual identities due to the loss of jobs, which are tied to economic reasons. "Fear" is the primary word that plays a central role in describing discourse about autonomous AI implications. This triggers "cognitive trauma" towards AI agents, as we fear losing control of it, as shown in terms like "surveillance," "autonomous," "overlord," "replacer," and "soulless." This accelerates the obsolescence of certain skills, leading to words like "singularity" and "obsolete." This reminds us of the best definition from Vernor Vinge during 1993, who professed that "singularity" is the embodiment of superhumanity of ultra-intelligent machines that are highly capable of any intellectual activity, surpassing any highly clever human being, and that this is the last invention that humanity will make (Hutter, 2012). This emotion also triggers the loss of "cognitive trust" towards AI, where the term "alienation" describes the emergence of technological alienation that makes individuals feel disconnected from all these technological progressions (Sun & Ji, 2024), leading to "stress" and "anxiety." This also creates ethical concerns such as "misinformation," "propaganda," and "unethical," showing societal unease with the misuse of AI. This also leads to "existential renegotiation" and resistance towards AI adaptation, as shown in terms like "existential." Therefore, the government must address concerns such as "privacy" and "security" through "regulation" and "transparency" by creating policies that ensure AI serves humanity responsibly.

Figure 2. Positive Words

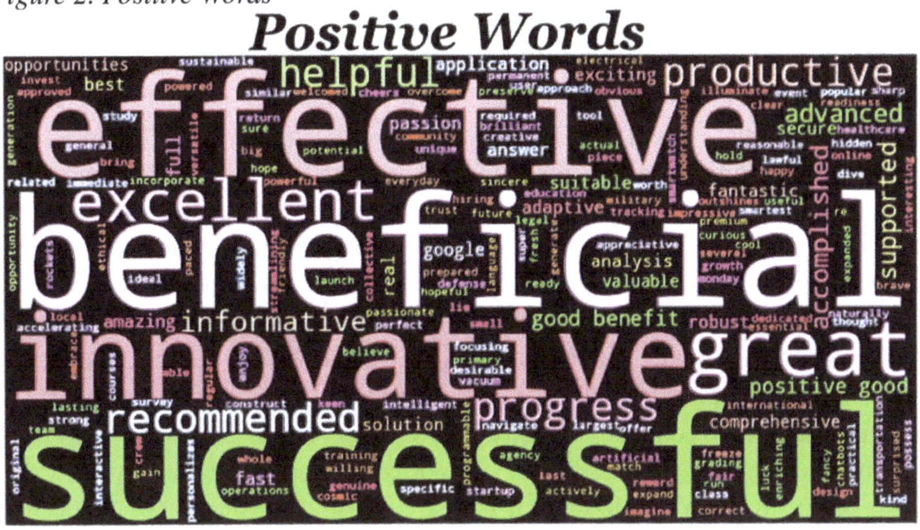

Source: Authors' Original Creation

Figure 3. Negative Words

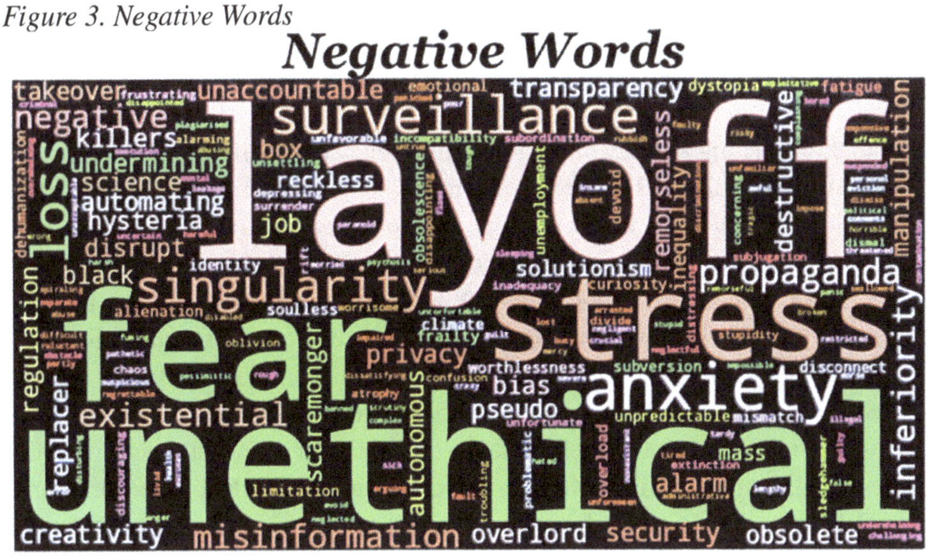

Source: Authors' Original Creation

EMERGENCE OF THE TURING TRAP

The pervasive assumption that automation is the finest solution for enhancing productivity is a common misconception because it builds the threat of concentrated power and wealth. Thus, it creates the peril of those people without power having no means to improve their results. This technological change disproportionately affects most sections of society, but is only beneficial on average, and this dynamic is called the Turing Trap (Brynjolfsson, 2022). This trap is reinforced by three groups of people belonging to the technological class, as they hold major control over creation, investments, and developments. They are first and foremost "Technologists", who are experts in technology creation. The second group comprises "Techno-Capitalists", who are entrepreneurs, investors, businesspeople, and prominent Big Tech companies. Lastly, "Government Policymakers" will find this automation fascinating because they are not only apprehensive about the value mining from common people's physical labor but also the extraction of intellectual property from the creators of this technology.

1. **Technologists:** Technologists always wanted to address the frequent challenges that traditional computers could not handle. That is why for decades, they pursued replicating human intelligence through these machines, with the birth of the term "Electronic brain" (Zhao et al., 2023). This is presently amplified by top-tier AI companies competing and inventing technologies little known to outsiders, and all kept within four walls, which can be called a "black box" situation, lacking answerability and transparency (Grodzinsky et al., 2008).
2. **Techno-Capitalists:** The techno-capitalists took this notion to the top notch, and it has become a popular strategy for top-tech companies for cutting costs by laying off employees, as it is harder to train them. It is simpler than growing by hiring. Besides, investors pick this kind of scalable business pattern that can grow without hiring, to avoid complexities and radical changes that entail hiring, then training, expanding, job security, health insurance stability, and salary-paying requirements. Businesspeople, executives, and entrepreneurs who have means to integrate these machineries to replicate and replace humans for given tasks will cut the maximum volume of workers and their valuable contributions, resulting in more revenue for themselves. They are performing plug-and-play automation, and the economy's labor share is weakening and falling extremely far behind. Over the past 40 years, the average wages with only a high school education are incredibly low, but millionaires and billionaires have grown exponentially. This kind of race is certainly going to bring the economic game to a halt as capital flow ceases.

3. **Government Policymakers:** Government policymakers are very much indulged in becoming international technology leaders and want to support AI automation rather than augment it. For example, currently, the US Tax code urges investing towards machinery technology rather than human workforce, making it cheaper to buy machines than to hire people. This policy also provides bigger tax breaks (Soled & Thomas, 2019).

RECOMMENDATIONS FOR ETHICAL AI AND HUMAN-AI SYMBIOSIS AS THE SOLUTION

The first fundamental welfare theorem of economics states that under a specific set of conditions, market prices lead to a "Pareto Optimal Outcome" (Ehrgott, 2012). This means it will elevate only one sector while another sector naturally bears the cost. Although it might not be a direct exploitation, there is a trade-off of displacement happening gradually, and since AI is of central importance to current and future economic affairs, as it is not only an innovation itself, but it will also increase productivity and create opportunities for some, while concurrently threatening the economic stability of others. At this stage, normal workers will lose political and economic bargaining power and are left with nothing because all the incomes depend on techno capitalists' decisions, who are in control. Not only that, but these intelligent AI machines are also receiving larger dataset training, with billions of parameters to achieve the greatest accuracy, and are surpassing the evolutionary progress of humans, which took over millions of years. Simon Kuznets states that useful knowledge lives within human brains (Kuznets, 1955), but that does not mean a single human brain contains the necessary knowledge to operate a small business venture, let alone an entire economy or industry. Thus, knowledge needs to be circulated and distributed, which is lacking in the current paradigm, as it only addresses a technological expansion. This also creates security risks, privacy issues, and economic instability by embracing new technologies within the current frame of the economy without redesigning the notion of human identity, autonomy, and value. This could potentially create social problems like an increase in substance abuse, crime, mental health challenges, alienation to another levels, and reducing individuals as mere data entities. To sum up, automating labor unravels less usefulness than augmentation, which can generate something new.

Government intervention under these circumstances is very crucial in navigating difficult challenges thrown by AI technologies and steering the economy out of the Turing Trap to ensure workers, creators, and their livelihoods are safeguarded. This urgency is very much needed as there are ongoing legal battles against major AI companies, including OpenAI, Microsoft, Google, Anthropic, and Meta, related to the illicit use of the creators' intellectual property. Government policymakers must reinvent new policies for organizations with effective regulations ensuring workers' job security while concurrently holding them accountable. It is also important to keep a healthy pace with all these rapid technological advancements and urge knowledge sharing among these organizations for collective understanding and collaborative development of responsible AI technologies, where the first step starts with keeping it open source. The government must directly intervene to provide the necessary education and training for the labor market. Industry leaders and policymakers must join forces to navigate ethical issues posed by AI and establish strategies that nurture innovation while guarding public interests. For this, international regulatory efforts have surfaced as a vital response, with significant initiatives like the EU Parliament's AI Act of August 2024 that includes comprehensive protocols on AI use, development, and deployment (Nikolinakos, 2023). Additionally, the associates of the Council of Europe, the US, Britain, and 57 other nations have discussed the AI Convention, focusing on shielding human rights (van Kolfschooten & Shachar, 2023). This is followed by India's distinctive approach called the "Fair Dealing" clause that sets up conditions for AI companies to get consent from copyright owners before feeding it to their AI models for training, thereby not exploiting the creator's creativity. This issue, though not yet fully addressed by extensive AI legislation, is gaining global recognition, not just from press, academic establishments, and prognosticators, but also from international bodies, including UNESCO have published "Recommendations on the Ethnics of Artificial Intelligence" in 2023 (Choudhary et al., 2024).

The Psychological Adaptation Vectors (PAVs) framework of Cyberpsychology can help techno-capitalists streamline their organizational business fairly by utilizing these factors. The "Techno Cognition" vector allows them to gauge the psychological readiness of employees to adapt to AI-augmented workflows, fostering "Cognitive Trust". If any barriers arise, such as "Cognitive Trauma" or "Existential Concerns", the "Emotional Boundary Negotiation" vector, coupled with the "Identity and Agency Reconstruction" vector, can address these psychological obstacles by anticipating potential risks and barriers. This enables them to adopt empathetic Employee Welfare tactics that mitigate identity reconstruction concerns. Finally, the "Existential Processing and Negotiation" vector guides techno-capitalists and their companies towards ethical AI business practices by promoting meaningful future dialogues that priorities human dignity and well-being, rather than solely advancing AI technology.

Yet, this still leads to some ethical questions. For example, are we ready to utilize technological progression to gain superior economic outcome and discard human worth and achievements? Should we consider abandoning the democracy of old truth, rules, and values to support algocracy, by empowering super-intelligent machines to build a society and subdue humans' power struggles with machines? Thus, Human-AI symbiosis, where AI augments human capabilities, is deeply significant, where human workers can regain their bargaining power in the labor market, resulting in more gains by boosting their productivity, so do their average incomes, living standards, healthcare, and longevity. Humans can remain central for value creation and political decision-making by becoming indispensable. Thus, this augmentation fundamentally brings an endless frontier of new prospects, expands our collective abilities, and accelerates AI as humans' extension rather than acting as a replacement. Because the number of tasks this symbiosis can do together is unquestionably larger than just what a human can do alone. Such responsible AI aligned to societal missions can be used for discovering life-saving drugs, preserving an established societal order. Hence, every member of society, including technological organizations and policymakers, plays a principal role in maintaining the structural integrity of societal wellness.

CONCLUSION

We are at an unusual interval in history, which has forced us to wonder if the current societal structure can endure this building technological innovation load. The problem lies not in advanced inventions, but with their integration into the existing societal foundation, by treating people as a mere data pattern and pursuing economic prospects by pressing every bit of efficiency. This will lead to alienation and trading out human autonomy and privacy conveniently. There is an important need to control it to the extent where we must quickly decide whether it expands opportunities for humans through Human-AI symbiosis and augmentation or by replacing humans through automation. The first choice is very beneficial, which provides the workforce with necessary platforms and tools, opportunities for growth, and the sharing of the economic pie equally. On the other hand, the second possibility quickly displaces the workforce and risks dividing the economic pie, which is only beneficial for the technological class working in support of replacement, by trapping most of the powerless class into an unhappy equilibrium, which is called a Turing trap.

The solution is to reverse the ongoing technological backlash by creating initiatives for augmentation over automation and redesigning the relationship between humans and AI, where they can work together through Human-AI symbiosis by leveraging each other's strengths for better outcomes. Stakeholders globally should guide the ethical implementation of AI within their nation's organizations for prioritizing human psychological well-being. This is because discounting psychological readiness considerations is integrally unethical. The Cyberpsychology's Psychological Adaptation Vectors (PAVs) framework can be used to track longitudinal phenomena and facilitate this process. This collective approach can create a thriving society that encourages discovery, lifts the standard of living not just for the selected but for many, by ensuring everyone in equal political gains, thereby redirecting us out of the Turing Trap.

The future is not predetermined, but we hope that through these technological enhancements, we can increase our lifespan and continue to collect more meaningful developments, and hopefully not dive towards new societal divisions based on wealth, and those who don't use these enhancements are seen as less capable. However, AI has the power to mirror our psychological fractures, reflecting our deepest angsts; it can be misused to design biochemical weapons or to replace valuable truths and ideas. Hence, a more vigilant approach is needed to avoid this excessive race. As Nick Bostrom has warned in his book "Superintelligence: Paths, Dangers, Strategies" (2014), AI could find new modes in unexpected ways. We might need new jobs like AI managers or AI psychologists who will help us to fathom these technological changes and raise awareness of how they might affect our mental health and sense of reality.

The key question remains: Will these advancements in technology bring more prosperity, opportunities, and progress to everyone's life, or will they create higher social inequalities by forming a dystopian future with only a handful of people directing everything while the rest just barely having it together?

ACKNOWLEDGMENT

This research received no specific grant from any funding agency in the public, commercial, or not-for-profit sectors.

REFERENCES

Acemoglu, D., Autor, D. H., Hazell, J., & Restrepo, P. (2022). Artificial Intelligence and Jobs: Evidence from Online Vacancies. *Journal of Labor Economics*, *40*(S1), S293–S340. DOI: 10.1086/718327

André, E. (2021). Socially Interactive Artificial Intelligence: Past, Present and Future. *International Conference on Multimodal Interfaces*, 4. DOI: 10.1145/3462244.3480862

Aspray, W. (1997). The Intel 4004 microprocessor: What constituted invention? *IEEE Annals of the History of Computing*, *19*(3), 4–15. DOI: 10.1109/85.601727

Boyd, D., & Ellison, N. B. (2007). Social network sites: Definition, history, and scholarship. *IEEE Engineering Management Review*, *38*(3), 16–31. DOI: 10.1109/EMR.2010.5559139

Brin, S., & Page, L. (1998). *The anatomy of a large-scale hypertextual Web search engine*. *30*, 107–117. DOI: 10.1016/S0169-7552(98)00110-X

Brynjolfsson, E. (2022). The Turing Trap: The Promise & Peril of Human-Like Artificial Intelligence. *Daedalus*, *151*(2), 272–287. DOI: 10.1162/daed_a_01915

Chamboko-Mpotaringa, M., & Manditereza, B. (2024). Artificial Intelligence in higher education: A decade's bibliometric snapshot, emerging themes and future research. *International Journal of Research In Business and Social Science*, *13*(8), 192–202. DOI: 10.20525/ijrbs.v13i8.3708

Choudhary, E., Narayanan, S., & Khan, F. (2024). Legal and Regulatory Landscape. *Advances in Healthcare Information Systems and Administration Book Series*, 222–239. DOI: 10.4018/979-8-3693-7452-8.ch014

Cottingham, J. (1992). *The Cambridge Companion to Descartes*. Cambridge University Press., https://www.amazon.com/Cambridge-Companion-Descartes-Companions-Philosophy/dp/0521366968 DOI: 10.1017/CCOL0521366232

Dalal, R. S., Howard, D. J., Bennett, R. J., Posey, C., Zaccaro, S. J., & Brummel, B. J. (2021). Organizational science and cybersecurity: Abundant opportunities for research at the interface. *Journal of Business and Psychology*, *37*(1), 1–29. DOI: 10.1007/s10869-021-09732-9 PMID: 33564206

Djerassi, C., Smith, D. H., Crandell, C. W., Gray, N. A. B., Nourse, J. G., & Lindley, M. R. (1982). The DENDRAL project: Computational aids to natural products structure elucidation. *Pure and Applied Chemistry*, *54*(12), 2425–2442. DOI: 10.1351/pac198254122425

Ehrgott, M. (2012). Vilfredo Pareto and multi-objective optimization. *Doc. math*, *8*, 447-453.

Ge, S. S., Dongjie, Z., Li, D., Mao, X., & Nemati, A. (2020). Historical and futuristic perspectives of robotics. *Artificial Life and Robotics*, *25*(3), 393–399. DOI: 10.1007/s10015-020-00613-7

Gonçalves, S., Dias, P., & Correia, A.-P. (2020). *Nomophobia and lifestyle: Smartphone use and its relationship to psychopathologies. 2*, 100025. DOI: 10.1016/j.chbr.2020.100025

Gopalakrishnan, K. V. (2009). James Watt: Father of steam power. *Resonance*, *14*(6), 522–529. DOI: 10.1007/s12045-009-0062-0

Gorichanaz, T. (2023). Accused: How students respond to allegations of using ChatGPT on assessments. *Learning*, *9*(2), 183–196. DOI: 10.1080/23735082.2023.2254787

Grodzinsky, F. S., Miller, K. W., & Wolf, M. J. (2008). The ethics of designing artificial agents. *Ethics and Information Technology*, *10*(2), 115–121. DOI: 10.1007/s10676-008-9163-9

Hassan, A. E., Lin, D., Rajbahadur, G. K., Gallaba, K., Cogo, F. R., Chen, B., Zhang, H., Thangarajah, K., Oliva, G. A., Lin, J., Abdullah, W. M., & Jiang, Z. M. (2024). *Rethinking Software Engineering in the Era of Foundation Models: A Curated Catalogue of Challenges in the Development of Trustworthy FMware*. DOI: 10.1145/3663529.3663849

Hutter, M. (2012). Can Intelligence Explode. *Journal of Consciousness Studies*, *19*(1-2), 143–166. http://hutter1.net/publ/singularity.pdf

Jumper, J. M., Evans, R. O., Pritzel, A., Green, T., Figurnov, M., Ronneberger, O., Tunyasuvunakool, K., Bates, R., Žídek, A., Potapenko, A., Bridgland, A., Meyer, C., Kohl, S. A. A., Ballard, A. J., Cowie, A., Romera-Paredes, B., Nikolov, S., Jain, R. D., Adler, J., & Hassabis, D. (2021). Highly accurate protein structure prediction with AlphaFold. *Nature*, *596*(7873), 583–589. DOI: 10.1038/s41586-021-03819-2 PMID: 34265844

K, S. A., Gururaj, V., Shankar, S. P., S., S., & Bharadwaj, A. (2022). Electrocardiogram based cardiovascular disease detection with Ensemble Learning Classifier. *2022 4th International Conference on Circuits, Control, Communication and Computing (I4C)*, 48-53. DOI: 10.1109/I4C57141.2022.10057719

Khalil, M., & Er, E. (2023, June). Will chatgpt g et you caught? rethinking of plagiarism detection. In *International conference on human-computer interaction* (pp. 475-487). Cham: Springer Nature Switzerland.

Kuznets, S. (1955). Economic Growth and Income Inequality. In *AMERICAN ECONOMIC REVIEW* (Vol. 45, pp. 1–28). DOI: 10.4324/9780429311208-4

Lee, K. F. (2016). Automation, Computerisation and Future Employment In Singapore. *Social Science Research Network*. https://doi.org/DOI: 10.2139/SSRN.2995283

Leiner, B. M., Cerf, V. G., Clark, D. D., Kahn, R. E., Kleinrock, L., Lynch, D. C., Postel, J., Roberts, L. G., & Wolff, S. (2009). *A brief history of the internet*. *39*(5), 22–31. DOI: 10.1145/1629607.1629613

Mackey, J. W. (2024). *The Industrial Revolution. The Modernization of the Western World*. Routledge., DOI: 10.4324/9781003467328-12

McCarthy, J. J., Minsky, M., Rochester, N., & Shannon, C. E. (2006). A Proposal for the Dartmouth Summer Research Project on Artificial Intelligence. *AI Magazine*, *27*(4), 12. DOI: 10.1609/AIMAG.V27I4.1904

Meng, Z., Ao, B., Wang, S., Niu, T., Chen, Y., Ma, X., & Huang, Y. (2024). Relationships between screen time and childhood attention deficit hyperactivity disorder: A Mendelian randomization study. *Frontiers in Psychiatry*, *15*, 1441191. Advance online publication. DOI: 10.3389/fpsyt.2024.1441191 PMID: 39376970

Minsky, M., & Papert, S. (1969). Perceptron: an introduction to computational geometry. *The MIT Press, Cambridge, expanded edition*, *19*(88), 2. DOI: 10.7551/mitpress/11301.001.0001

Negri, S. M. C. A. (2021). *Robot As Legal Person: Electronic Personhood in Robotics and Artificial Intelligence*. DOI: 10.3389/frobt.2021.789327

Nikolinakos, N. T. (2023). *EU Policy and Legal Framework for Artificial Intelligence.* Robotics and Related Technologies - The AI Act., DOI: 10.1007/978-3-031-27953-9

O'Rourke, K. H., Rahman, A. S., & Taylor, A. M. (2013). Luddites, the industrial revolution, and the demographic transition. *Journal of Economic Growth, 18*(4), 373–409. DOI: 10.1007/s10887-013-9096-y

Palacios-González, C. (2024). *Robotic Persons and Asimov's Three Laws of Robotics.* 185–194. DOI: 10.1093/oso/9780198876434.003.0018

Prensky, M. (2001). Digital Natives, Digital Immigrants Part 1. *On the Horizon, 9*(5), 1–6. DOI: 10.1108/10748120110424816

Schlinger, H. D.Jr. (2009). Theory of Mind: An Overview and Behavioral Perspective. *The Psychological Record, 59*(3), 435–448. DOI: 10.1007/BF03395673

Schwartz, D. L., & Rogers, M. (2022). Inventorless Inventions? The Constitutional Conundrum of AI-Produced Inventions. *Social Science Research Network.* DOI: 10.2139/ssrn.4025434

Seargeant, P., & Tagg, C. (2019). Social media and the future of open debate: A user-oriented approach to Facebook's filter bubble conundrum. *Discourse, Context & Media, 27,* 41–48. DOI: 10.1016/j.dcm.2018.03.005

Sharadhi, A. K., (2025). Scouring through the crowd simulation dynamics in urban environments. *International Journal of Innovative Science and Research Technology,* 1439–1459. https://doi.org/DOI: 10.38124/ijisrt/25mar1176

Sharadhi, A. K., Gururaj, V., Shankar, S. P., Supriya, M. S., & Chogule, N. S. (2022). Face mask recogniser using image processing and computer vision approach. *Global Transitions Proceedings, 3*(1), 67–73. DOI: 10.1016/j.gltp.2022.04.016

Shortliffe, E. H. (1974). *A rule-based computer program for advising physicians regarding antimicrobial therapy selection.* 739. DOI: 10.1145/1408800.1408906

Soled, J. A., & Thomas, K. D. (2019). Automation and the Income Tax. *Columbia Journal of Tax Law, 10*(1), 1–48. DOI: 10.7916/CJTL.V10I1.2864

Suler, J. R. (2004). The Online Disinhibition Effect. *Cyberpsychology, Behavior, and Social Networking, 7*(3), 321–326. DOI: 10.1089/1094931041291295 PMID: 15257832

Sun, M., & Ji, N. (2024). Erosion and Remodeling: New Alienation of User Subjects Led by Intelligent Algorithms. *Communications in Humanities Research, 36*(1), 50–61. https://doi.org/10.54254/2753-7064/36/2024BJ0035DOI: 10.54254/2753-7064/50/20242548

Tergan, S.-O. (1997). Conceptual and Methodological Shortcomings in Hypertext/Hypermedia Design and Research. *Journal of Educational Computing Research, 16*(3), 209–235. DOI: 10.2190/2R2J-NXB9-0MD3-T603

Turing, A. M. (1950). Computing Machinery and Intelligence. *Mind, LIX*(236), 433–460. DOI: 10.1093/mind/LIX.236.433

Turkle, S. (2011). The Tethered Self: Technology Reinvents Intimacy and Solitude. *Continuing Higher Education Review, 75*, 28–31. https://files.eric.ed.gov/fulltext/EJ967807.pdf

van Kolfschooten, H., & Shachar, C. (2023). The Council of Europe's AI Convention (2023-2024): Promises and pitfalls for health protection. *Health Policy (Amsterdam), 138*, 104935. DOI: 10.1016/j.healthpol.2023.104935 PMID: 37925880

Weizenbaum, J. (1966). ELIZA — A computer program for the study of natural language communication between man and machine. *Communications of the ACM, 9*(1), 36–45. DOI: 10.1145/365153.365168

Zhao, L., Zhang, L., Wu, Z., Chen, Y., Dai, H., Yu, X.-X., Liu, Z., Zhang, T., Hu, X., Jiang, X., Li, X., Zhu, D., Shen, D., & Liu, T. (2023). When Brain-inspired AI Meets AGI. *Meta-Radiology, 1*(1), 100005. DOI: 10.1016/j.metrad.2023.100005

Zhu, T., & Liu, X. (2023). The future is now? Consumers' paradoxical expectations of human-like service robots. *Technological Forecasting and Social Change, 196*, 122830. Advance online publication. DOI: 10.1016/j.techfore.2023.122830

ADDITIONAL READING

Asimov, I. (1940). I. robot. *Narkaling Productions*. https://scholar.google.com/scholar?q=Isaac+Asimov.+1940.+I.+Robot.+Narkaling+Productions

Bostrom, N. (2014). *Superintelligence: Paths, dangers, strategies*. Oxford University Press., https://www.amazon.in/Superintelligence-Dangers-Strategies-Nick-Bostrom/dp/0198739834

Heaven, W. D. (2022). Why Meta's latest large language model survived only three days online. *MIT Technology Review. Last accessed December, 15*, 2022. https://www.technologyreview.com/2022/11/18/1063487/meta-large-language-model-ai-only-survived-three-days-gpt-3-science

Hurley, D. (2024). Ethical Questions Swirl Around Neuralink's Computer-Brain Implants. *Neurology Today*, *24*(10), 1–14. https://journals.lww.com/neurotodayonline/fulltext/2024/05160/ethical_questions_swirl_around_neuralink_s.7.aspx. DOI: 10.1097/01.NT.0001022988.26412.f4

Merken, S. (2023). New York lawyers sanctioned for using fake ChatGPT cases in legal brief. *Reuters*. https://www.reuters.com/legal/new-york-lawyers-sanctioned-using-fake-chatgpt-cases-legal-brief-2023-06-22/

Milmo, D. (2023, October 24). AI risk must be treated as seriously as climate crisis, says Google DeepMind chief. *The Guardian*. https://www.theguardian.com/technology/2023/oct/24/ai-risk-climate-crisis-google-deepmind-chief-demis-hassabis-regulation

Taylor, C. (2023, July 19). Meta top exec Nick Clegg says A.I. large language models are actually 'quite stupid' and downplays the 'hype running ahead of the technology'. *Fortune*. https://fortune.com/2023/07/19/meta-nick-clegg-ai-stupid-facebook-instagram-llama-2-openai/

Westfall, C. (2023, March 29). Elon Musk and other tech leaders call for slowdown on AI development. *Forbes*. https://www.forbes.com/sites/chriswestfall/2023/03/29/elon-musk-and-other-tech-leaders-call-for-slowdown-on-ai-development/

KEY TERMS AND DEFINITIONS

Alienation: A disconnected or isolated that individual phase during rapid AI progressions, hindering their ability to adapt.

Cyberpsychology: An emerging field that explores traces how technologies like AI effect human emotions, behavior, and cognitive processes through its conceptual framework.

HLAI: AI replicates human critical thinking skills by human like thinking.

Human – AI Symbiosis: AI augments human capabilities by improving productivity and protecting humans through ethical practices in all domains.

Natural language processing (NLP): A subfield of AI that processes and generates human language to enable computer to understand it.

Pareto Optimal Outcome: A situation where one sector is elevated while another sector naturally bears the cost.

Psychological Adaptation Vectors (PAVs): A conceptual framework that helps understand the relationship between human psychology and AI technology.

Sentiment Analysis: The study of what the public perceives was about AI to categorize positive and negative opinions.

Singularity: A hypothetical future point where AI surpasses human Intelligence in all domains, driven by its intellectual progress.

Turing trap: A major societal risk when most of the task are automated by AI, leading to the decline of common people's outdated skills and human purpose, resulting in societal disparity, by benefiting only the technological class.

Chapter 7
Ethical AI in the Supply Chain:
Driving Responsible and Sustainable Flow Optimization

Nouhaila Bentatou
https://orcid.org/0009-0001-0946-0818
Ibn Tofail University, Morocco

Hamid Ait Lemqeddem
Ibn Tofail University, Morocco

ABSTRACT

In the era of digital transformation, artificial intelligence (AI) is emerging as a strategic lever in supply chain management, enhancing forecasting, planning, optimization, and logistics resilience. However, its adoption raises ethical challenges related to algorithm transparency, data protection, decision-making fairness, and sustainability. This chapter presents a critical and multidisciplinary study of AI ethics in supply chains, based on a structured review of 31 recent scientific articles. It identifies thirteen key themes, including ethical principles, regulatory issues, and enabling technologies such as XAI, blockchain, and federated learning. The study emphasizes the need for ethical performance indicators, collaborative governance, and risk management tools to support the responsible adoption of AI. It ultimately calls for a strategic integration of logistics innovation, ethical awareness, and social responsibility to foster a more transparent, fair, and sustainable supply chain.

DOI: 10.4018/979-8-3373-3104-1.ch007

INTRODUCTION

Supply chains are at the forefront of significant changes in an era of digital transformations, thanks to the rapid advances in artificial intelligence (AI). These technologies enable the optimization of flows, reduced costs, and greater responsiveness to disruptions, among other benefits. However, increased efficiency and automation through the application of technologies raises serious ethical concerns relating to, among others, algorithmic transparency, equitable decision-making, data privacy, and sustainability in logistics practices (Wamba et al., 2015). These issues are all the more crucial as regulatory frameworks around AI are evolving rapidly, requiring companies to integrate more structured ethical governance into their supply chains (Jobin et al., 2019; Wellbrock, 2025). While companies are working with AI-based systems to drive their efficiencies, the question explored here is how to responsibly enact ethical supply chains. In this context, our chapter is developed to study the conditions under which logistics flows can be optimized based on ethical AI that can ensure social responsibility, performance, and environmental sustainability. The interest in understanding the impact AI has on the transformation of supply chains without ignoring the ethical risks and dilemmas arising from its use. It also aims to assess prospects, collaborative approaches, and innovations that can support truly ethical AI in the supply chain.

Methodology

From a methodological perspective, this chapter is based on a narrative, structured, and analytical literature review. More than 60 articles were selected, primarily from journals indexed in the Scopus database, ensuring the rigor and academic relevance of the sources. Keywords used in the search include ethical AI, responsible artificial intelligence, AI and supply chain, sustainable supply chain, AI governance, algorithmic bias, AI risk management, and AI ethics metrics. The documents analyzed cover the period 2015–2025 to capture the most recent developments. The thematic analysis allowed us to structure the discussion into thirteen sections, each addressing a critical aspect of the integration of ethical AI in modern supply chains. This approach aims to provide a conceptual, multidisciplinary, and critical synthesis, combining contributions from management science, applied ethics, sustainable logistics, and emerging technologies. This cross-disciplinary positioning not only allows us to structure current knowledge but also to identify grey areas that have not yet been explored, particularly regarding the mechanisms for disseminating an ethical culture in logistics organizations.

Chapter Outline

This chapter is organized around thirteen complementary themes that provide a comprehensive and in-depth understanding of the issue of ethical AI in supply chains. It begins with an exploration of the role of artificial intelligence in modern supply chain management, before presenting the fundamental principles of ethics applied to AI. It then explores sustainability issues in supply chains and examines the practical implications of implementing responsible AI. The following themes address sensitive issues related to data privacy, security, and the impact of algorithms on decision-making. This chapter subsequently looks at the possibilities of ethical AI in this sector, before looking into the perceptions of the main stakeholders: businesses, consumers, and regulators. There is then an analysis of the indicators used to measure the ethical performance of artificial intelligence systems. The closing parts discuss embedding ethics in risk management frameworks, cross-sector cooperation that enables responsible technology deployment, technological innovations that facilitate such a shift, and cultural changes that drive ethical processes within organizations. A concluding section of the chapter is dedicated to summarizing the most important outcomes of the conversation and outlining steps for working with planners and decision-makers in the area of fostering sustainability. The aim of this chapter is thus to identify relevant and thoughtful levers for balancing operational efficiency with ethical responsibility and sustainability.

THE ROLE OF AI IN SUPPLY CHAIN MANAGEMENT

New and innovative technologies keep expanding the field of supply chain management. For instance, the application of artificial intelligence (AI) assists in resolving the logistics systems' uncertainties and complexities in a novel manner. "Recently, companies in industrial sectors around the world have started adopting artificial intelligence (AI) technologies in their supply chains on the premise that AI enhances operational flexibility, efficiency and, in turn, profitability." (Lamarre et al., 2023). These authors illustrate how organizations that apply AI in certain operational functions report a more seamless operational process, improved disruption management, and enhanced competitive positioning in the digital economy. AI helps automate and continually optimize decision-making, predict flow disruptions, and overall improve operations from planning to final delivery (Culot et al., 2024). The ability to manage and analyze large volumes of information promptly provides a competitive advantage in enhancing overall supply chain efficiency. In fact, demand forecasting methods based on machine learning can significantly enhance supply chain performance while improving forecasting accuracy and operational outcomes

(Feizabadi, 2022). AI also provides companies the ability to automate complex tasks, translating to greater operational flexibility, while mitigating human error and lowering operational costs. Recent research has similarly suggested that algorithm types, which are based on advanced computational methods, such as quantum and AI-based methods, can also aid in decision-making regarding inventory optimization and complex supply chain-related problems (Shamsuddoha et al., 2025). Another significant, early contribution linked to the arrival of AI is demand forecasting. Demand forecasting using historical data is not novel. However, demand forecasting based on machine learning algorithms can also include methods to capture a variety of external information sources (e.g., online usage, social trends, weather) to assist organizations in predicting shifts in demand with greater accuracy than simply using historical data (Min, 2019). This feature of AI allows organizations to better manage their inventory, optimize costs associated with unnecessary inventory, and enhance customer satisfaction. AI plays a key role in inventory optimization and warehouse automation. AI can perform real-time monitoring of inventory quantity in a warehouse, prediction and control of inventory anomalies, and real-time anomaly detection using IoT and predictive analytics at the same time. Siemens, for instance, is investigating the interplay between AI-driven digital twins for supply chain simulation and the real-time prediction of bottlenecks and flow control adjustments. The application of digital twins in logistics is linked to an increase in the volume of timely predictive and decision-making, as well as improved efficiency of workflow processes (Wasi et al., 2025).

In addition to conventional types of artificial intelligence, generative AI technologies such as ChatGPT, DALL•E, and Claude are becoming a more prominent element in the supply chain. They enable the automatic generation of decision-making reports, the simulation of complex scenarios, and even assistance with strategic planning. These tools are already being tested in some companies to predict supply disruptions or create dynamic dashboards. However, their use raises new ethical issues, such as the opacity of generation processes, the risk of implicit bias in suggestions, and the difficulty of auditing automatically generated decisions (Dwivedi et al., 2023). In the warehousing context, AI regulates the complicated robots that are being used for order fulfillment operations, like autonomous transport vehicles and robotic manipulators. AI Automation of these processes lessens fulfillment times and reduces human error (Wamba et al., 2015). In the transport and logistics field, AI is crucial for repository and vehicle routing, as well as last-mile logistics. By using data like traffic jams/blockage, bad weather, and delivery restrictions, optimization algorithms tell the routes that should be used to minimize time and energy consumption, thereby lowering overall fuel expense, helping to achieve on-time performance, and reducing carbon impact (Bag et al., 2021). When it comes to relationship management, AI can be useful in (1) evaluating supplier and their performance, (2) identifying noticeable failure risks, and (3) improving procurement tactics. In addition, Dubey et al. (2021) state that the use of AI tools is being used by firms to search for hidden indications in the financials or logistics data, to anticipate any disruptions, and ensure they have security in their supply chain.

AI is being utilized by various industrial sectors to increase the effectiveness of their supply chains. For example, Walmart previously used predictive models to adjust their inventory levels to local purchasing patterns. And in the pharmaceutical industry, Pfizer used algorithms to predict supply chain disruptions during a health crisis. Amazon is employing computer vision, robots, and machine learning in its warehousing operations to allow for fully automated order fulfillment. These are representative examples of how AI is transforming logistics processes. While conventional uses of AI are one aspect, emerging stresses the need for using explainable AI (XAI) because organizations need an ability to understand and audit algorithmic decisions, and green AI is growing to minimize the carbon footprint of complex models (Arrieta, A. B. et al., 2020). Furthermore, AI is a major enabler for dealing with risks and resilience. It can help ensure operational continuity either by simulating numerous crisis scenarios or defining breaking points, or reorganizing flows disrupted by incidents. It became a valuable predictive capability when the COVID-19 crisis started, as many companies were able to use AI to pivot procurement plans almost overnight (Ivanov et al. 2019). In addition to assisting with organizational efficiency, AI also helps organizations to enhance sustainability in their supply chains. It acts to mitigate waste, maximize resources, and monitor environmental exposure, amongst other things. This means that organizations can improve and advance the ethical performance of their partners through a holistic analysis of ESG (environmental, social, and governance) data (Dubey et al., 2019).

While AI brings many advantages to supply chains, the use of AI raises ethical issues, including the risk of automating decision-making in selection from various suppliers and even algorithmic supplier selection that is inherently biased or discriminatory. While much of the literature has documented these sorts of problems, even Floridi and Mittelstadt (2020) have noted that biases derived from algorithms could reproduce discrimination that already existed while obfuscating where responsibility lies for faulty, algorithmically-driven decisions. Algorithmic examples are the use of surveillance or monitoring tools to evaluate employees in warehouses that have been automated, creating some social tensions. The ethical issues also include the climate change impacts of AI use, especially the energy usage that comes with data creation and training for complex algorithms, which seems significant within sustainable development strategies (Strubell et al., 2020; Mittelstadt, 2019). Despite the advancements, some obstacles with AI in supply chains persist as the costs of implementation are high, a lack of digital skills exist, there are conflicting information systems, and the algorithms for AI lack transparency. To meet these challenges, organizations need to invest in and maintain reliable digital infrastructures, strengthen their data practices, and enable trustworthy co-dependent working among the logistics, IT, and administrative functions (Wamba et al., 2015). AI is enhancing accuracy in the supply chain management. As shifts in technology grow, AI will integrate deeper than 'his/ her brain do the heavy lifting' and will start 'AI augmentation' decision-making in intricate strategic scenarios. Whatever the case, the adoption of AI in supply chains will pose an equal number of challenges in governance and accountability. The time is right to pause and think about the boundaries of ethics governing the design and use of AI of the hour. The crucial starting point is the ethical governance and the fundamental ethics which anchor the technology, accountability, and transparency on that AI is built on.

PRINCIPLES OF ETHICAL AI

The ascent of AI in recent years as a critical component in a supply chain creates ethical dilemmas due to the boundless advancement of the technology; Instead of musing solely on the actionable tasks of AI, we have to ponder on the very essence of the technology's purpose with respect to human rights, societal structures, and how the inner workings of the AI systems involve mental processes (Jobin et al., 2019). Ethical AI can best be described as the principles that guide the design, development, and use of artificial intelligence to ensure that they are designed, developed, and used responsibly and transparently with respect to human values. In 2019, the OECD suggested five principles of responsible AI: technical robustness, transparency, accountability, inclusion and human well-being, responsible governance. These principles are now commonly referred to in academic literature as core principles of AI governance (Ribeiro et al., 2025). The five principles have provided member states with an ethical basis. It should be noted that principles have similar foundations based on the principles of transparency, fairness, accountability, and privacy protection (Afroogh et al., 2024). The five responsible AI principles especially resonate with.... transparency and explainable AI, especially in supply chains as automated decisions impact workers, consumers, the environment, and economic interests. Transparency is one of the most critical foundations of ethical AI. In the context of supply chains, automated decisions must be understandable to the relevant societal stakeholders.

Transparency and Explainability

Algorithmic transparency is one of the cornerstones of ethical AI. It is a mandate for AI systems to provide users an understanding of the process it went through to come to a decision. Therefore, explainability is critical, especially in supply chains, specifically to decisions around supplier selection, logistics, or procurement (Samek et al., 2017). This kind of transparency instills trust in stakeholders and provides a way to audit the systems. Thus, it should come as no surprise that critical systems would rely on interpretable machine learning models (e.g., decision trees or SHAP methods) (Arrieta et al., 2020). In addition to transparency, fairness and non-discrimination represent another fundamental pillar to ensure fair and inclusive AI systems in supply chains.

Justice and Non-Discrimination

Fairness and the lack of discriminatory bias are also essential components of AI ethics. Algorithms trained on historical data can reproduce, or even amplify, systemic discrimination (Barocas et al., 2019). For example, in a supply chain, a biased AI system could disadvantage certain suppliers based on geographic, linguistic, or financial criteria. Researchers emphasize the importance of proactively managing bias, starting from the data collection phase, by ensuring sufficient representativeness and testing models on a variety of cases (Mehrabi et al., 2021). But ensuring fairness is not enough: it is equally essential to assign responsibility for decisions made by or with the support of artificial intelligence.

Responsibility and Accountability

An ethical AI system must be responsible and accountable. This implies clearly defining human responsibilities in the event of an AI error or malfunction. In supply chain management, this translates into mandatory human oversight, particularly for high-impact decisions (Floridi et al., 2018). Additionally, businesses must designate data ethics officers, create recourse procedures for impacted parties, and document the algorithm creation process Another primary principle of ethical AI concerns the protection of privacy. This is especially concerning for supply chains where there is a massive collection and sharing of data.

Respect for Privacy

Another cornerstone of ethical AI is protecting personal data. The nature of supply chains creates significant sensitive data related to suppliers, customers and employees. Legitimate use of that data, and more broadly, data that involves AI methods, requires full adherence to applicable regulatory frameworks, such as the General Data Protection Regulation (GDPR) of Europe. Furthermore, ethical governance models should embody privacy, transparency and accountability, as well as environmental and social considerations in the design and development of AI systems. (Samarawickrama, 2022). Data protection by design (or privacy by design) must be inclusive of all. This includes limiting the data collected to only the data necessary, along with ensuring the database was not compromised, and to a large extent, using anonymity. In addition to the technical and legal perspectives, the ethics of AI also requires the consideration of environmental and social consequences of embedding AI into logistics systems.

Sustainability and Social Responsibility

Finally, ethical ITAI also has to conceive environmental and social effects associated with technology uptake, but it is much tougher to measure the trajectory of the environmental impacts (Vinuesa et al., 2020). Training large AI models has a significant energy impact, and the many factors for the introduction of these tools into supply chains can cause undesirable externalities when regulations actually exist (Vinuesa et al., 2020). The aspiration is to produce AI that reflects their aspirations to engage and develop towards the UN's Sustainable Development Goals (SDGs), such as authentic green logistics, waste minimization, and resource usage improvement. These aspirations can also be examined through the perspective of the Triple Bottom Line (Elkington & Rowlands, 1999). The Triple Bottom Line seeks to examine organizational performance in light of reconciling three interlinked dimensions: economic (profit), social (people), and environmental (planet). Applied to AI, this implies that not only are systems to improve logistics efficiencies, they must show respect for human rights in their application, and maximise their actions to minimize their carbon footprint. Similarly, stakeholder theory (Freeman, 2010) calls for integrating ethics into the governance of AI systems, taking into account the expectations, impacts, and rights of all stakeholders: suppliers, employees, customers, NGOs, regulatory authorities, etc. This requires mechanisms for consultation, accountability, and transparency, which are essential to ensure the equitable adoption of artificial intelligence in supply chains. However, despite these normative frameworks, ethical approaches often remain difficult to operationalize. The lack of measurable indicators, commercial conflicts of interest, and information asymmetries between parties make the practical application of ethical AI complex, particularly in globalized supply chains (Morley et al. 2020).

SUSTAINABILITY IN SUPPLY CHAINS

Sustainability has become a strategic priority for companies operating in increasingly globalized supply chains and subject to growing environmental, social, and regulatory pressures. Firms that are trying to align their businesses with the United Nations Sustainable Development Goals (SDGs) consider sustainable development to be a greater business focus. Within the supply chain context, this incorporates environmentally conscious, socially responsible, and financially sustainable practices across the full life cycle of products and services.

Definition and Dimensions of Sustainability

The triple bottom line, which comprises ecological, social, and economic factors, is the foundation for SCM as sustainability entails managing supply chains along all three aspects with a drive to lessen environmental impact, promote social justice, and provide economic sustainability with respect to logistics operations (Panigrahi et al., 2019). The principles of ethical AI discussed above find direct application here, as digital technologies—if properly managed—can support sustainability objectives in supply chains.

Sustainability Practices in Supply Chains

The challenges associated with improving sustainability in supply chains ultimately drive organizations to pursue suppliers, often through the use of all the latest tools, processes, and initiatives, especially toward lowering their footprints. Some approaches to improving sustainability in all supply chain activities include ecodesign, which entails integrating environmental parameters from the product design stage to lessen the environmental footprint; ethical sourcing, which entails choosing contractors/suppliers with sustainable, socially responsible practices; sustainable logistics, which refers to improving logistics operations to lessen energy use and greenhouse gas emissions; and the circular economy, which refers to the application of the reuse and recycling of materials to extend product life (Farooque et al., 2019).

Sustainability Issues and Challenges

Despite its many benefits, integrating sustainability into the supply chain presents several challenges. First, the complexity of supply chains—characterized by numerous stakeholders and geographically dispersed operations—makes the coordination of sustainability initiatives difficult. Secondly, it would cost lots of money for green technologies to be utilized upfront, as many organizations do not do it because of the high upfront costs. Moreover, a lack of transparency around suppliers can restrict organizations from accurately analyzing their sustainability initiatives (Nguyen & Zuidwijk, 2025). The rise of AI technology and its integration on the supply chain has created ethical questions and challenges. It is of utmost importance to address the ethical implications of the technology on human rights and social structures instead of merely focusing on use. AI systems are different at the core. The thought processes involved in constructing this technology are profound and complex (Jobin et al., 2019).

Role of Technology in Sustainability

Technological approaches to the development of sustainable and environmentally friendly supply chains

- Artificial Intelligence (AI): Algorithms are used to enhance transportation routes and predict demand to minimize waste. With the increase of sustainability integration into supply chains, Artificial Intelligence (AI) technology will enhance current flow planning and resource efficiency processes and solve basic environmental problems (Teixeira et al., 2025).
- Blockchain: Product traceability and holding suppliers to account for their sustainable practices. Enhanced technology in traceability and Transparency Sustainability Accountability (TSA) blockchain or AI technology will more often assist in making efficient allocations with accountability (Weller, 2024).
- Internet of Things (IoT): Facilitating effective traceability of Product movements and minimizing waste which enhances sustainability of the supply chain through real-time monitoring of transportation and storage conditions (Taj et al., 2023).

It's important to note that each of these technologies also has limitations to its sustainability impact. AI processes, for example, may require excessive energy resources when training algorithms, and connected devices produce a significant digital footprint as a result. Hence, a critical perspective must be adopted that considers the environmental life cycle of digital tools (Strubell et al.., 2020).

Governance and Regulation

The resilience of supply chains is also influenced by the combination of government programs and regulatory frameworks. These frameworks accomplish important tasks such as providing standards, tracking compliance and assuring corporate accountability at each stage across the supply chain. The European Duty of Vigilance Directive, for example, requires larger commercial entities operating in the EU to take action to identify, prevent, and mitigate adverse human rights and environmental impacts resulting not only from their own business operations, but also from their subsidiaries and associated business partners, including suppliers and subcontractors

(Kutscher, S., 2025). The Directive, thus, pushes and requires companies to complete due diligence procedures, including risk assessment, consultations with stakeholders, and remedial plans. The Directive currently holds companies accountable for any violations occurring anywhere across the supply chain, thus moving issues from voluntary and ad hoc to a prescribed level of action. In an effort to reduce reputational damage and financial liability, this text will argue the profits gained from unethical business practices and the spenditure on business regulation fines will motivate companies to report fake data, practice labor exploitation and sourcing, and sustain a carelessly harmful business ethic. The touchpoints of governance, data privacy, and data security still non-pecuniary, and are under the direct influence of 'social credits,' and remain key in the growing environment of America toward the businesses and serious AI-driven social responsibility obligations.

CHALLENGES IN IMPLEMENTING ETHICAL AI

Utilization of ethical AI signifies a multitude of prospects for efficiencies, resilience, and sustainability for supply chains. Yet, there may also be ethical dilemmas that can arise as a result of the use of ethical AI. The objective of this section is to highlight the ethical challenges involved with the implementation of ethical AI in supply chains, where such challenges are informed by some contemporary scholarship. The ethical challenges that are articulated will influence not only technical feasibility, but also governance and ethical liability issues, and how organizations can navigate and mediate competing pressures of innovation, compliance, and social legitimacy.

Supply Chain Complexity and Shared Responsibility

Often supply chains today have a multi-actor structure and a multi-dimensional stakeholder scene in existence. Due to the complexity of changing and dynamic conditions, it becomes increasingly difficult to ascertain who is ultimately responsible for the consequences of decisions made by AI systems. Cobbe et al. (2023) found that algorithmic accountability becomes more challenging with multiple stakeholders and multiple processes contributing to the development and use of AI systems in the entirety of the supply chain. In this scenario, ethical governance becomes obscured. This complexity is accentuated in global chains, where the multiplicity of jurisdictions, contractual norms, and cultural expectations makes the attribution of ethical responsibilities even more blurred (Morley et al.., 2020).

Data Quality and Algorithmic Bias

The ethical performance of an AI system is fundamentally dependent on the training data: what kind of data is used, and how good is that data? Training an AI system on biased data or incomplete data could result in discriminatory or unjust decisions. Goswami et al. (2024) argue that data quality challenges, including those caused by information silos or inconsistencies, are significant challenges to embedding ethical AI in the supply chain context. Also, datasets that lack diversity exacerbate existing bias problems.

Lack of Transparency and Explainability

Artificial intelligence algorithms (AI), particularly those based on deep learning, are considered "black boxes," making it difficult to understand or trust the choices AI makes. Arrieta et al. (2020) addressed the level of transparency of artificial intelligence systems as one of the issues that can undermine stakeholder trust and make it difficult to identify ethically troubling behaviors.

Organizational Resistance and Skills Lack

The ethical application of artificial intelligence requires considerable organizational processes, which involve skills and culture within the organization. Rharoubi et al. (2023) pointed out that some Moroccan companies faced internal Barriers to AI, which chiefly stemmed from fear of change and a lack of technological awareness. Likewise, responsible implementation is inhibited by a lack of AI ethics skills at the level of technical teams. Internal resistance is not only technical; it is also cultural. The absence of ethical leadership, in-house training on responsible AI matters, and social dialogue channels further expose the disconnection between technological advancement and organisational acceptance (Morley et al., 2020).

Regulatory Frameworks and Compliance

The challenges an ethical AI faces in addressing regulatory frameworks comes from the fact that not only are there no common and clear regulations, but the fact that the stakeholders have no accountability. Khan et al. (2022) also mentioned that the absence of specific designated independent supervisory authorities in the absence of designated independent AI specific laws exacerbates the stakeholder accountability. When businesses operate in multiple countries it further complicates issues with complexity as countries continue to add unique and differing rules and laws to the legal reality of a business. This regulatory fragmentation and lack of commonality in rules has led to adaptive governance alternatives that respond to both ethical and practical challenges wite generative AI resulting in a framework for collaboration among stakeholders in accordant policies (Reuel & Undheim, 2024).

Cybersecurity and Data Privacy Risks

Integrating artificial intelligence into supply chains raises immediate issues related to the handling of sensitive data. Ismaeil (2024) indicates that AI could create new potential cyber risks that may compromise the privacy of individuals, compromise the access to and reliance on AI technologies. Therefore, we must first determine if personal data is safe, and that the supply chain is free from cyberattacks before any ethical use of AI can occur. In summary, ethics and AI in the supply chain could mean much more than a technical change. It may involve a more fundamental change of the organization itself, the law, and overall culture. This begs the question of how much we can trust these systems when they make critical decisions without human intervention. The next section will discuss the relationship between decision-making algorithms, algorithmic explainability, and the possibility of error or unethical misuse of an AI system.

DATA PRIVACY AND SECURITY CONSIDERATIONS

The rise of AI in supply chains brings larger conversations about data privacy and data protection. Ethics of AI is not solely made up of legal requirements or legal permissions; the legality is often part of the ethical AI "shield" mentioned above, but we often reconnect the conversation back to the legality of things placed outside of the ethics itself. Data abuse or data processing - legal or not - will ultimately end the technology due to consumer bias, violations of fundamental rights or even a breakdown of trust not related to legality. Due to these innovations, sensitive data is now being circulated rapidly through increasingly interconnected networks in unprecedented volumes - creating riskier environments, increasing vulnerabilities, bad actors exploiting data, and an increased amount of attacks. There is a strong need to properly manage compliance risk both as required by applicable rules and regulatory obligations, and to uphold the wider ethical obligations associated with building trust among parties. This section will review issues of vulnerabilities in systems, data governance, privacy by design, and how blockchain technology can provide security to data and be a good steward of data in connecting AI with the supply chain.

Vulnerabilities of AI Systems to Adversarial Attacks

While machine learning can increase robustness and contribute to the development of Artificial Intelligence (AI) systems, such systems can also be attacked using adversarial examples to change outputs or expose sensitive information. Papernot et al. (2017) demonstrate that adversarial examples can manipulate the outputs of machine learning models through small visible perturbations; to a human being, the perturbations may not be visible. The harm in this case would be information privacy and model accountability, and there are serious implications into the latter - for instance, consider someone using a machine learning model to assist in the legal adjudication of a case.

Risks Associated with the Collection and Processing of Sensitive Data

AI in supply chains involves the collection and processing of large amounts of data that may include sensitive data relating to customers, suppliers, and internal processes. Voigt and Von dem Bussche (2017), stated the same point- 'without rigorous governance structures to manage these data, we risk compromising privacy and the potential for misuse.' Therefore, organisations must build data safeguards; these protections require organizations to develop policies around data collection, storage, use, and dissemination.

Integrating Privacy by Design into AI Systems

The process of private-by-design starts with building effective data safeguards into the design of AI systems from the outset. Radanliev and Santos (2023) indicate that there are several processes that need to be included, such as differential privacy, homomorphic encryption, and federated learning, aimed at securing data while not interfering with the usability of the artificial intelligence models derived from the data. These best practices will also assist organizations in securing the access and usage of the data safely, so that the analytical quality of the data is not diminished. These best practices prompt ongoing tension between data protection and the usefulness of the data for AI applications. Approaches like homomorphic encryption or making assumptions to model a federated population, pose an impact to the "accuracy" or "size" of the model that can be constructed. This tension represents some of the ethical and dilemma-generating complexities that organizations using responsible AI would face. As Veale and Edwards (2018) comment, anticipated strategies to uphold privacy will result in "computational trade-offs that might otherwise constrain the system from obtaining its optimal solution."

Regulatory Compliance and Data Governance Challenges

Organizations functioning in a globally connected economy face the challenges of not only a climate with heightened regulatory compliance specific to data, but also a complicated and dynamic regulatory compliance environment. The challenges and complexities are heightened with the introduction of a number of newly enacted data legislation such as the General Data Protection Regulation (GDPR) in Europe and similar publicly enacted legislation in numerous countries around the world. In their work, Voigt and Von dem Bussche (2017) highlighted that organizations pursuing compliance with Data Protection legal requirements have compliance stipulations to establish structured data governance, have and designate formal data governance processes veterans for data protection and compliance, and undertake risk-based assessments. A further complicating factor for organizations will be the EU initiative, the 2024 Artificial Intelligence Act, which introduces additional data compliance obligations outside of GDPR. For example, organizations undertaking the use of high-risk AI systems, including applications of risk management within supply chains, will be obligated to provide disclosure obligations, ongoing risk assessments, metrics and process objectives, and documentation, and ongoing human oversight obligations. Organizations also have technical frameworks and standards to responsibly and safely govern information and knowledge use globally with ISO/IEC 27001 (Information Security) and ISO/IEC 42001 (AI Governance). ISO/IEC 42001:2023 allows organizations to use AI applications in an ethical and risk-free manner, while maximizing efficiencies and making decisions (Biroğul & Şahin, 2023). Blockchain provides compelling potential solutions to improve data security and tracking in a supply chain. Saberi et al. (2019) stated that blockchain has potential to improve data traceability and security in a supply chain and when coupled with AI will reduce the potential on manipulated data that prevent accountability and to offer ethical transparency in decisions.

AI ALGORITHMS AND THEIR IMPACT ON DECISION MAKING

Although artificial intelligence (AI) might change how decisions in supply chains are made with algorithms that can analyze data at scale compared to any person, and provide recommendations at near real time, it is also necessary to reflect on the ethical issues of transparency, accountability, and bias. As societies continue to delegate more of their decisions to algorithmic systems, some basic questions arise: who is accountable for the decisions? Who is liable for what happens? These questions become even more serious in the context of supply chains, where AI decision-making can impact suppliers or customers or employees on a large scale.

As Floridi and Cowls (2019) note in their unified framework of ethical AI, it is essential to develop the conduct and implications of automated decisions using clear principles: accountability, transparency, justice, beneficence, and explainability, which is necessary for ethical engagements.

Role of Algorithms in Supply Chain Optimization

Most AI solutions for supply chains rely on the application of machine learning and deep learning algorithms. These tools enable demand forecasting, inventory optimization, production process control, and real-time logistics tracking (Tirkolaee et al., 2021). These skills help reduce expenses and improve customer service (Ivanov et al., 2021). AI algorithms impact decision-making by providing simplified predictive analytics and by modeling complex patterns and correlations that are difficult to envision through traditional analytics methods (Shamsuddoha et al., 2025). However, they must be designed to be explainable to ensure user trust and ethical decision-making (Arrieta et al., 2020).

Automated Decision-Making and Transparency

However, the complexity of models, particularly deep neural networks, creates a "black box" problem that limits the understanding and traceability of decisions (Ribeiro et al., 2025). This opacity can lead to a lack of trust among stakeholders and complicates the detection of potential biases or errors in automated decisions.

Algorithmic Bias and Ethical Consequences

Algorithms will also, likely, have bias built into them from the composition of the training data, which can promote inequitable or discriminatory decisions. For instance, an algorithm giving preference to suppliers or geographical regions can heighten social inequities (Binns, 2018). To provide collective accountability and fairness in decision making, socially responsible organizations must consider proactively managing bias (Mehrabi et al., 2019).

Impact of Algorithms on Human Decisions

AI provides extensive support for human decisions, however, it never completely removes the human element. Automation bias, defined as the tendency to over-rely on automated recommendations, can detract from decision making awareness and lead to error (Parasuraman and Manzey, 2010). Therefore, we must frame the importance of developing and evolving mechanisms that can combine human decision making with algorithmic decision making. To limit reliance upon AI, effective human monitoring mechanisms should be implemented to ensure values, especially for high-risk decisions (Weber & North, 2023). The European AI Act 2024 also requires that any automated decision in critical contexts be subject to meaningful, competent, and informed human intervention.

Explainable Algorithms for Responsible Decision-Making

The intention of creating explainable algorithms (explanatory AI, XAI) is to allow for better transparency and understandability in the decision-making process. They create a more auditable decision process and can foster user confidence (Patil, 2024). EAI can allow the supply chain space to leverage technology to help make strategic decisions and operational decisions while ensuring ethical oversight (Guidotti et al, 2018). Explainable AI (XAI) approaches allow for better auditability of decisions, but remain limited in complex models (deep learning). It is therefore recommended to combine these techniques with human-algorithm co-construction approaches, which allow for enriching the decision without passively submitting to it. Doshi-Velez, F., & Kim, B. (2017). Ultimately, AI algorithms should not be considered as substitutes for human decision-making, but as support systems. The legitimacy of their use rests on their transparency, their explainability, and the assurance that the final decision remains under human control. The goal is less to automate the decision than to augment it.

THE FUTURE OF ETHICAL AI IN SUPPLY CHAINS

Artificial Intelligence is inherently disruptive. Within the scope of supply chain, we understand the function AI will play in potential reimagining towards capabilities that are efficient, resilient and sustainable to the system. As we consider the future use and uptake, therefore we must also ensure that ethical practices are embedded within the design and development of an AI algorithms and systems, that will enable, support and build responsible, equitable and transparent forms of behaviours. This is serious and engaged form because beyond the technical notion, it shapes a mindful - ethical consideration towards a properly inclusive ethical system that places these considerations of awareness and ethics in terms of accountability, transparency and social risk. Floridi and Cowls (2019) honorably frame our thinking of a trustworthy AI in terms of these intersecting principles; beneficence, non-maleficence, autonomy, justice and explicability.

Towards Responsible and Sustainable AI

Recent developments reflect a growing interest in purposeful ethical Artificial Intelligence and respect for human rights, humane, transparent and ethico-sustainable operations and considerations for a range of associated environmental issues (Floridi et al., 2018). As part of a commercial operation for supply chains, algorithms are being developed that include flows of goods and services, but also sustainability variables that including allowable CO2 emissions and similars for resource consumption being CO2 emissions (Dubey et al., 2023).

The Rise of Regulatory Frameworks and Ethical Standards

The expanding usage and acceptance of AI technologies goes hand in hand with the renewed regulatory and normative call to reign in AI technologies. One prominent example is the European Union's AI Act which intends to provide a regulatory framework for safety and accountability to promote transparency and safeguarding fundamental rights (Kutscher, 2025). The settlement movement will add pressure to the use of ethical AI in global supply chains (Jobin et al., 2019). During the next several years, we will be needed several different governance and behavioural regulation approaches to allow for the development of ethical AI. ICAP 42001 is an example. ISO/IEC 42001 provides a comprehensive framework for AI governance to help organizations knowingly go beyond compliance to emergent practices while planning an appropriate approach to responsible governance of AI (Biroğul & Şahin, 2023). In addition, new work exists and requires new and better adaptive governance frameworks and new collaboration will be needed to achieve coherence about establishing norms for ethical AI from locality to locality around the world (Reuel & Undheim, 2024).

Explainable Artificial Intelligence and Ethical Audits

Additionally, when considering the opportunity for AI in supply chains, we must think about improved transparency, including Explainable AI (XAI). The utilization of these tools will enable companies to make automated decisions and to understand, justify, and audit those decisions, providing greater assurance for stakeholders and support for compliance to ethical standards (Adadi & Berrada, 2018).

Multidisciplinary Collaboration and Stakeholder Involvement

The effective creation and use of ethical AI require collaborative systems between developers, regulators, decision-makers, and other AI users. Raji et al., (2020) suggest that collaborative systems are essential to identify and mitigate biases, provide accountability, and promote equity in terms of adoption through the different orientations of the relevant stakeholders in the development and use of AI. However, the level of collaboration that is possible to use to work collaboratively is extremely limited. Civil society and stakeholders from the global south/developing countries are inadequately represented and inadequately involved in the design and/or other processes. Raji et al., (2020) argues that the unequal allocation of governance of AI systems could lead to systemic bias and a loss of ethical legitimacy.

Integrating AI into Sustainability and Circular Economy Initiatives

A more overarching perspective for ethical AI is within the broader frames of corporate social responsibility (CSR) and the circular economy. Ethical AI can offer possibilities for collaboration along the supply chain with respect to sustainability, and those possibilities are to be understood in the same regard of ethics of use (Ghobakhloo, 2022). Again, the question is not whether AI is capable of effectively impacting the supply chain. Rather, the question is whether, and if so, how socially-responsible, ecologically sustainable and ethically governable, can AI be used. This returns us to the belief that appropriate use, that includes the type of audits, clarity of algorithms, stakeholder engagement - should be - important in order to develop a meaningful trust that is needed for intelligent systems to be commonly used in society in the future.

STAKEHOLDER PERSPECTIVES ON ETHICAL AI

In order to more clearly understand the complexities of the ethics of AI in supply chains, we must think through the contexts of the various actors involved: regulators, businesses, and consumers. Each actor presents different implications for an individual's acceptance of, understanding of, and support for the uptake of ethical principles. While the principles of ethical AI are increasingly debated at the conceptual level, their interpretation and implementation largely depend on the stakeholders involved. Indeed, businesses, consumers, and regulators do not have the same objectives or the same constraints, which sometimes creates tensions or misunderstandings. Understanding their respective representations is essential to fostering truly responsible adoption of AI.

Perspectives from Businesses

According to Choi, Wallace, and Wang (2018), businesses believe that artificial intelligence and advanced analytics are important mechanisms to optimize the performance, flexibility, and sustainability of their supply chains. However, businesses also recognize the ethical quandaries of AI, including algorithmic discrimination, violation of privacy, and lack of transparency (Mikalef et al., 2020). For this reason, many businesses are instituting internal policies that will regulate their use of AI through several guiding principles including codes of ethics and upfront periodic ethics checks (Benjamins, 2020). Furthermore, businesses regard ethics as a strategic asset, trusting it will improve consumer trust and brand value (Shin et al., 2023). However, some critics point to a growing trend towards instrumental ethics, where charters are used more as marketing tools than as binding operational frameworks. According to Mittelstadt (2019), corporate ethical commitments often remain non-binding, without real accountability mechanisms.

Perspectives from Consumers

From a consumer point of view, ethical awareness around AI is on the rise, especially about the clarity of decisions, data governance, and social consequences of technologies (Martin et al., 2021). According to Gursoy et al. (2020), there are expectations of businesses to use AI ethically, respecting their rights and privacy. The trust deficit may cause the avoidance of AI-led products and services, which makes it even more important for businesses to maximize transparency (Lamberton and Stephen, 2016). Moreover, consumers are clearly becoming more concerned about ethical and sustainable supply chains, often relying on these factors in their purchasing decisions (Kumar et al., 2022).

Perspectives from Regulators

Regulators are ramping up their work to create a range of legal and normative frameworks for AI, emphasizing the protection of rights, liability, and safety (Binns, 2018). Governments have ambitions to advance from fairness standards to limit harm from discrimination or misuse of data (Veale and Binns, 2017) to vet regulation, which may provide some assurance on aspects of traceability, transparency and sustainability of supply chains (Cobbe et al., 2023). Nevertheless, the regulatory landscape at the international level is highly fragmented and may create uncertainty for companies that operate across jurisdictions. This heterogeneity reinforces the need for collaboration between the institutions, businesses, and civil society. It has been noted that "a substantive ethical governance strategy must involve the co-construction of standards with all relevant stakeholders to ensure longevity and legitimacy" (Floridi et al., 2018). If stakeholders, regulators, businesses, and consumers are to fully exploit the rapid developments in technology, they need to collaborate in developing, and shaping standards that are suitable and proportionate (Cath et al., 2018).

METRICS FOR EVALUATING ETHICAL AI PERFORMANCE

To assess the efficacy of ethical artificial intelligence in supply chains, it is important to create, and apply metrics that assess the compliance of AI systems to basic ethical principles of transparency, fairness, accountability, and sustainability. This will show if, and how much, AI solutions are improving operational efficiency, and meeting access and ethical obligations. However, methods for assessing AI ethics remain highly heterogeneous across contexts. There is still no universally accepted framework for assessing ethical performance in a standardized manner, which limits the comparability of practices across companies or sectors (Mökander et al., 2021).

Measuring Transparency and Explainability

Transparency allows stakeholders and users in supply chains to understand (and thus, control) how the algorithms make decisions (Doshi-Velez and Kim, 2017) and is a fundamental component of ethical AI. In this instance, and to confirm model clarity, explainability metrics have been developed, with particular emphasis on measuring the simplicity of the rules and understanding the ability to present comprehensible explanations (Ribeiro et al., 2025), which contributes to user trust and the ability to audit the automated decisions in supply chains (Arrieta et al., 2020).

Assessing Fairness and Absence of Bias

Algorithmic bias can result in injustice and discrimination, which threatens the ethical imperative of artificial intelligence systems (Mehrabi et al., 2019). Several fairness measures can be utilized, including equal opportunity, demographic parity, or statistical refereeing, to evaluate the level of fairness of artificial intelligence procedures towards various groups or individuals (Verma and Rubin, 2018). Integrating print measures and indicators into supply chains is necessary as biased decisions can cause unfair outcomes, particularly concerning people management or supplier relationships. Ultimately, integrating fairness measures and indicators into supply chains is a necessary decision to avoid unfair outcomes in the first place, especially when it comes to people management or supplier relationships. It should also be noted that some indicators of justice are sometimes mutually incompatible. For example, aiming for both equality of opportunity and demographic parity can lead to complex ethical trade-offs (Binns, 2018).

Measuring Responsibility and Traceability

Accountability is defined as the ability to detect the final responsible party for mistakes and/or breaches of ethics and trace decision histories (Wieringa, 2020). If AI systems are to integrate accountability as a method of traceability, these systems will be required to have more robust logging functionality and regular audits of operations (Selbst and Barocas, 2018). Such mechanisms will support ethical compliance and to mitigate issues.

Data Protection Indicators

In Europe, the regulations (e.g., GDPR) that safeguard personal data (Voigt & Von dem Bussche, 2017) serve as helpful proxies for tracking compliance with data minimization, consent, and security, as well as when measuring the overall effectiveness of encryption or anonymization algorithms built into AI systems (Shokri et al., 2017).

Sustainability and Environmental Impact Assessment

With CO2 emissions measuring metrics, determining the algorithmic consumption of resources and measuring the energy costs of algorithms and models, an organizations can account for the environmental impact of AI systems (Strubell et al., 2020). These metrics can help supply chains deliver on their sustainability commitments, while respecting ethical and sustainable principles surrounding transportation logistics and disposal. In summary, to further enhance ethical signals, a matrix of quantitative (e.g. traceability, fairness, energy use, etc.) metrics combined with audits of qualitative human review, charters and ethics committees are required to determine if the use of AI in supply chains is responsible and accountable.

ETHICAL AI AND RISK MANAGEMENT

Ethical AI aims to anticipate, recognize, and mitigate these risks, evaluating systems to increase the likelihood of decreased risk to individual rights, harm, and trust. There are many ethical risks that arise in the supply chain which arise from the adoption of artificial intelligence (AI) which require a thoughtful and cautious attitude to management. The general aim of ethical AI is to provide some reasonable assurances that are programmed into automated systems do not introduce risks in the supply chain, respect individual rights, do not cause harm to anyone, and provide trust to all stakeholders. As Morley et al. (2020) indicate, this latent method of ethical risk management is typically a reactive and ad hoc method. Presently organizations have almost no standard tools to evaluate the second effects of their AI systems, especially in a context of modern supply chains that are characterized by complexity and globalization.

Identifying Risks Associated with AI in the Supply Chain

Floridi et al. (2018) identify the most anticipated ethical concerns that arise from Artificial Intelligence in the supply chain context as: issues of transparency, algorithmic discrimination, privacy concerns, and irresponsible decision making. By way of example, Raji et al. (2020) point out that if an algorithm is poorly designed, it can place employees, customers, or suppliers in a position of unfairness (e.g. design unjust choices that impact them). In addition, the information obtained from an AI system that has the potential to collect large amounts of private information could represent another area of concern regarding privacy and data security risks. Studies purport the unclear data governance model and lack of transparency regarding data collection and use factors into raising ethical concerns, especially with AI-enabled supply chains (Abhulimen & Ejike, 2024).

Integrating Risk Management into Technical AI

Morley et al. (2020) proposed that organizations take best practices of an operational mechanism, which embeds ethical principles, as a way of potentially managing any obviously identifiable risk(s) from the implementation of an AI. They state that risk management is an imperative feature with respect to ethically sound AI applications for supply chains and risks with data privacy and regulations. In order to set ethical risks, organizations ought to act or operate in governance constructs (such as transparency, accountability, and fairness; Jobin et al., 2019). Given this governance construct, organizations may explore the ethical impacts of their algorithmic use, if not possibly using an environmental impact assessment to establish algorithmic failures (Mittelstadt, 2019). Further, establishing multi-disciplinary teams may add different perspectives to manage ethical risks. In this vein, organizations are searching for formalized mechanisms, as "Ethical Impact Assessment" (EIA), or "Algorithmic Impact Assessment" (AIA), to biomedical or environmental ethical approaches (Reisman et al., 2018). These approaches enable organizations to use ethical principles as aspects of design, even before an ethical risk is considered.

Risk Mitigation Tools and Methods

To mitigate and manage ethical risks in regards to AI there are a number of tools such as improved security mechanisms with the intention of safeguarding personal private data and information (Sun,2020); ongoing monitoring systems to identify non-compliance of ethics codes or unusual behaviours (Tirkolaee et al., 2021) ; explainability methods/techniques to promote accountability and share information transparently (Lipton, 2018); and of course, algorithm audits to diagnose and mitigate biases (Raji et al., 2020). Nevertheless, technical tools alone are not sufficient. These tools need to be placed in a broader governance framework with assigned responsibilities to steward and monitor, as well as a process for remedies for affected stakeholders (Jobin et al., 2019).

Importance of Regulatory Compliance

Data protection regulations are governing standards with binding requirements that relate to transparency, consent and data protection (Voigt and Von dem Bussche, 2017). Data protection standards that have binding effects, such as the General Data Protection Regulation (GDPR) in Europe, are legally binding and can be considered an ethical risk management principle.

Impact on Resilience and Trust

Reactive ethical risk management not only facilitates compliance and fosters supply chain resilience, but contributes to trust-building with customers, regulators, and suppliers which is a precursor to business sustainability (Ivanov et al. 2019). It also assists, in particular, to mitigate disruptive moments stemming from scandals or litigations. To summarize, ethical management of AI risk in the supply chain is more than compliance. It is a proactive, integrated, multi-disciplinary process, referencing technical tools, regulatory standards, and specific ethical governance. Without these elements, the resilience and trust they sought would be ephemeral.

COLLABORATIVE APPROACHES TO ETHICAL AI ADOPTION

Cooperative approaches that involve diverse stakeholders are instrumental in responsibly and effectively adopting artificial intelligence (AI). Cooperation among regulators, businesses, suppliers, consumers, and researchers, enables sharing of best practice, co-construction of solutions that incorporate ethical principles, and harmonization of standards.

Co-Design and Stakeholder Integration

It is important to involve all relevant stakeholders from the design stage, to clarify and establish that AI systems work towards ethical duties and operational needs (Sanders and Stappers, 2008). Through this co-design, context-specific ethical risks can be identified, algorithms can be adapted to meet end-user expectations, and transparency and fairness criteria can be defined (Dignum, 2019). This ongoing contact ensures good technology adoption and less resistance to change. However, implementing co-design encounters organizational obstacles. On the one hand, stakeholders do not always have the same level of skills or interest in ethical issues. On the other hand, the time and resources required for dialogue sometimes slow down technological deployment. These tensions must be anticipated from the early stages of the project (Schiff et al., 2020).

Corporate Networks and Strategic Alliances

The spread of ethical principles and the diffusion of assessment tools among partners are facilitated by corporate networks and strategic alliances (Christopher, 2016). In the supply chain, collaboration between suppliers, distributors, and customers helps establish common standards for ethical AI, share compliance costs, and ensure a more responsible value chain (Wieland & Handfield, 2013). These coalitions also contribute to joint innovation around sustainable and ethical technologies Global investigations have shown how multi-actor initiatives and partnerships develop solid AI ethic guidelines like non-discrimination, explainability and social impact assessment (Jobin et al., 2019; Fjeld et al., 2020).

Public-Private Partnerships and Regulatory Framework

Partnerships between public and private actors play a key role in structuring the adoption of ethical AI (Calo, 2017). Regulators, in collaboration with businesses and universities, can develop flexible regulatory frameworks that foster innovation while ensuring the protection of fundamental rights (Cath et al., 2018). These partnerships are also essential for training, certification, and raising awareness of ethical issues.

Communities of Practice and Knowledge Sharing

Supply chain experts and professionals share their ethical AI challenges and solutions within an informal forum of communities of practice (Wenger, 1998). These networks encourage sectoral self-regulation to complement official regulations (Brown and Duguid, 2001). They also contribute to the rapid diffusion of ethical innovations, collective learning, and skills updating. However, the value of these communities lies in mutual trust and transparent exchanges. Without a common ethical vocabulary or institutional support, their impact remains limited to organizations that are already aware (Winfield et al., 2019).

Joint Training and Capacity Building

Custom designed training courses which, at the same time, generating decision-makers and technical teams are needed to facilitate the advancement and development of ethical competencies around AI (Morley et al., 2020). Jobin et al. (2019) advised that the progress in collective understanding of issues raised and the collective uptake of good ethical practices can be stimulated by the collaborative experiences brought together from a diversity of profiles and sectors. Progressing and strengthening competencies is key in providing responsible AI processes in the procurement process. In the end, the ethical use of AI will require sensible collaborative dynamics and stakeholder engagement across the private, public and science sectors, and the accountability, as we consider sustainable and inclusive solutions that are genuinely reflective of industry's realities.

TECHNOLOGICAL INNOVATIONS SUPPORTING ETHICAL AI

To promote ethical AI in supplychains requires a greater understanding of technology. Unique innovations that embed ethical principles of fairness, accountability, safety and transparency, in the nature of artificial intelligence technologies represent unique progress. And subsequently those innovations will assist in developing equitable, automated decision processes provide improved traceability and increase compliance of supply chains.

Explainable AI (EAI) Frameworks

Explainable Artificial Intelligence (EAI) is a focus area of ethical AI. EAI enables end-users to easily understand how and why algorithms make decisions, ultimately improving transparency and trust from stakeholders, and minimizing bias (Adadi and Berrada, 2018). In a supply chain applications, these discussions are meaningful when discussing improvements in optimization, deciding on suppliers and making pricing decisions. EAI will also bring value in an automated supply chain context, whereby difficult AI decision-making processes can be made more transparent (Adadi and Berrada, 2018). Unfortunately, explainability techniques such as SHAP or LIME, for instance, currently often available demonstrate significant complexity or limited success on deep models, at which point it may hinder non-specialists' understanding and therefore limit their usefulness as effective transparency (Rudin, 2019).

Blockchain Technologies for Ethical Traceability

Blockchains are considered as an emergent technology to evidence flows authenticity and transparency integrity in the supply chain. The immutable, decentralised and contextual knowledge affordances of blockchain allows it to repeatably affirm compliance to environmental and social standards, product provenance, and the accountability of the parties involved (Saberi et al., 2019). The recognition of blockchain's capability as an evidentiary instrument leads to the proliferation of integrating artificial intelligence (AI), as an attestation to supply chain trust and ethical trust protocols. Morkunas, Paschen, and Boon (2019) concluded that making AIs compatible with blockchains, notably using smart contracts, could enable underlying ethical governance.

Green and Eco-Efficient AI

Green artificial intelligence is described as another emergent area, that produces energy-efficient AI systems and is part of what describes environmental ethics in that some (AI) models consume significant amounts of energy (Strubell et al., 2020). Examples of green AI innovation sees algorithm optimization, decentralized computing, to minimize energy costs, and smaller models. Schwartz et al. (2020) asserted that reducing the environmental footprint of AI in supply chains fits within algorithmic ethics. Logistics companies such as DHL, UPS, and others are already testing optimization models based on low-carbon AI in logistics to contribute to minimizing emissions associated with route planning. These efforts showcase the operationalization of green AI in a practical industrial context (Schwartz et al., 2020).

Automated Audit and Validation Tools

Automated audit systems are increasingly monitoring the ethical principle compliance of artificial AI systems. Automated audit systems use standardized metrics (e.g. IEEE or ISO), sophisticated dashboards, and bias detectors (Raji et al. 2020); a supply chain can integrate and therefore offer predictive models. As auditing systems could be utilized right now to change the mode of delivery in order to meet regulations and societal expectations. Raji et al. (2020) argue that auditing systems provide a significant advancement in the ethical governance of smart supply chains. On the contrary, Raji et al. (2020) discuss that automated auditing systems could be limited; some bias or technical-ethical malfunctions may escape the algorithmic approach. Auditing systems must be paired with human workflow in order to verifiably address the ethical dimension.

Federated Artificial Intelligence

Organizations employ federated AI technology to maintain high levels of algorithm performance while safeguarding customer and supplier data by deploying artificial intelligence models in a federated manner (Kairouz et al., 2021). The technology is new and has been implemented in logistics economies and around the world for optimizing flows, leveraging constraints of information security to their advantage. Breakthroughs in federated AI are being developed to enhance the security and ethics of systems optimizing collaborative logistics. In summary, while innovations in technology are valuable levers for ethical AI, they should not be viewed as 'magic bullets'. The success will depend on their application alongside versatile organizational processes, along with an ethical culture among public and private stakeholders.

CULTURAL SHIFTS TOWARDS ETHICAL PRACTICES

The implementation of ethical artificial intelligence (AI) within supply chains is not solely a technological change; it will necessitate a cultural transformation in how attitudes, stakeholder expectations, management standards, and organizational principles are regulated. Companies should be looking past compliance, even compliance with standards, moving toward a proactive ethical culture that focuses on transparency, accountability, and sustainability. However, recent work shows that the transition from declarative ethics to an integrated ethical culture remains difficult, particularly in large organizations where performance logic still dominates accountability objectives (Martin et al., 2020).

A New Culture of Algorithmic Responsibility

Companies using AI as part of their logistics strategies are recognizing that not only is it critical to establish a culture of algorithmic accountability, ethical governance cannot be a superfluous consideration. Ethical governance runs much deeper than developing ethical governance mechanisms, finding potential biases, or incorporating ethical principles into the development of algorithms; it also includes ethical principles being integrated into the development of algorithms, identifying potential biases, and developing ethical governance mechanisms. Ethical governance should also be considered from the design phase of the AI systems being used and this requires an organizational culture that is aligned with shared accountability.

Employee Training and Awareness

Setting up deliberate and sustained employee development is crucial to achieving the norm change associated with algorithmic ethics. The idea of tackling bias, transparency, and algorithmic fairness issues may serve as a medium for ethical behavior (Morley et al. 2020). Today, leading companies are developing AI ethics modules for their executives, data scientists, and logistics managers. To promote the responsible use of technology, it should be an essential component of a company's culture to educate about AI ethics. Prior research indicates that structured ethics programs are crucial to promoting ethical AI practice in organizations (Long & Magerko, 2020; Fjeld et al., 2020). These actions demonstrate a structural commitment to an ethically responsible organizational culture.

Ethics as a Lever for Competitive Advantage

A cultural shift is also taking place among executives, who now perceive ethics as a competitive advantage. Consumers, increasingly concerned about corporate responsibility, value brands that are transparent about their use of AI (Huang and Rust, 2021). This social pressure is pushing companies to adopt a culture of ethics to meet growing societal expectations. “AI ethics is becoming a strategic differentiator in markets driven by trust and transparency” (Huang and Rust, 2021).

Inclusion and Diversity in Model Development

Culturally advanced organizations tend to integrate multidisciplinary and diverse teams into algorithm design. This helps minimize biases associated with cognitive or social uniformity. Diversity is seen not only as an integration issue, but also as a way to strengthen the ethical soundness of models (Cowls et al., 2021). Encouraging diversity in algorithm development is critical to developing ethical artificial intelligence. However, inclusive approaches can face internal barriers, such as institutional biases or a lack of diversity skills. It is therefore essential to combine inclusive recruitment policies with ongoing training in applied ethics (Fosch-Villaronga & Poulsen, 2022)

Ethics Integrated into Governance

As a last point, the most sophisticated companies are operationalizing governance, which revolves around ethics committees, Chief Ethics Officers (CEOs), and internal policies. For ethics to be sustainable in technology value chains, it must be incorporated into governance structures. This is a culture that is institutionalized ethics, as opposed to ethics based on the will of the individual (Floridi et al., 2018). Therefore, implementing ethical AI in supply chains cannot be limited to technical or regulatory adjustments. Creating a deep-level, interoperable, and institutionalized cultural change requires the agenda of weaving together a value system of accountability, inclusion, and transparency into every aspect of the organization.

RECOMMENDATIONS FOR IMPLEMENTING ETHICAL AI IN SUPPLY CHAINS

In line with the discussions in this chapter, there are several initiatives that may help inform the way forward to responsibly and sustainably utilize artificial intelligence in a responsible and sustainable align with the supply chain. The steps in responsibly and sustainably deploying artificial intelligence hs with several considerations, drawing from the discussions in this chapter. First, formal ethical governance structures such as internal ethics committees and AI ethics officers, should be established so that there are governance structures that can proactively protect transparency, fairness, and accountability in processes that incorporate artificial intelligence. When ethical governance strucutres are established, they would assist institutionalized ethical governance structures in organizations and make ethical governance part of the continual systematic proactive decision making process. Second, is to enhance interdisciplinary collaboration with diverse stakeholders. A corporation must ensure it captures and embeds throughout the entire organization a diversity of commentary consisting of ethicalists, data scientists, logistics specialists, lawyers, and social scientists. In completing these initiatives, organizations and our knowledgable staff will somo for a substantive review of the humane implications of the technology with ethical constraints in mind to apply the technology. This approach takes a comprehensive view and is more beneficial than having different fields only look at the edges of the issue. Plus, engaging in conversations across various disciplines can help lower the chances of misunderstandings. These indicators provide a solid framework for organizations to compare themselves with one another and evaluate their practices to strengthen them.

One key recommendation that's emerging is to embrace human-centered AI training as a vital part of organizational growth. This means that every employee should receive training focused on the ethical aspects of AI, whether for technical understanding or to align their actions with the organization's shared commitment to responsible innovation. By doing so, they can actively participate in evaluating ethical considerations. For small and medium-sized enterprises (SMEs), this approach will also help clarify existing metrics and risk assessments in a way that's easy to understand and apply. It's equally crucial to harmonize regulatory frameworks and motivate organizations to reach a global consensus on international standards for ethical AI practices in logistics. This is essential for fostering trust and ensuring interoperability while safeguarding public health in global supply chains. Moreover, to truly establish ethical AI, organizations need to undergo a cultural shift towards moral and ethical considerations. Ethics shouldn't just be a set of guidelines; they need to be woven into the fabric of daily operations and business practices. If organizations want their ethical AI to be built on trust, they must cultivate a culture of transparency, inclusivity, and continuous learning—where ethics aren't left to chance or randomness. Instead, ethics should drive an innovative culture. Ultimately, the goal is to be open, truthful, and honest in all communications. Stakeholders, suppliers, consumers, and regulators need to understand how AI systems function and how decisions are made. Creating clear channels for communication and feedback fosters accountability between organizations and builds trust over time. In the end, these recommendations aim to help organizations adopt AI more effectively while genuinely committing to a culture of sustainable and shared ethical responsibility.

CONCLUSION

The integration of ethical artificial intelligence in the supply chain is essential for creating responsible and sustainable logistics systems. In this chapter, we've explored different aspects of this transition by familiarizing ourselves with the ethical standards that govern artificial intelligence, as well as the associated risks, technology, governance, and cultural considerations.

A Necessary Convergence between Ethics, Performance, and Sustainability

While AI has undeniably demonstrated its potential to improve flows, enhance information sharing, streamline decision-making, and boost resilience (Riad, Naimi, & Okar, 2024), it also brings along several risks, such as the need for algorithmic transparency, the potential for bias in decision-making, and concerns over privacy violations (Mittelstadt, 2019).

Towards Human-Centered AI

Human-centricity, responsibility, transparency, and fairness are the necessary foundations on which ethical systems of artificial intelligence depend (Morley et al., 2020). This implies establishing appropriate governance and reorganizing management techniques, taking into account the perspectives of various stakeholders, including regulators, employees, and customers. AI ethics requires and imposes quantifiable and concrete processes, and is no longer limited to a set of theoretical principles (Morley et al., 2020).

A Cultural and Collaborative Transformation

Organizational culture is extremely important. Ethical instruments will remain formal and superficial if internal principles, attitudes, and methods are not completely transformed. The goal of organizational culture is to encourage interdisciplinary collaboration, establish a culture based on algorithmic responsibility, and foster lifelong learning. Furthermore, collaboration between institutions, companies, public authorities, and researchers is a necessary element for the evolution towards ethical AI. Individual approaches are no longer sufficient; only concerted actions will harmonize, monitor, and assess ethics in AI systems dedicated to logistics. Collaboration between actors and stakeholders across organizational boundaries is the key to the success of algorithmic ethics (Jobin et al., 2019).

Future Challenges and Research Avenues

Establishing international standards for ethical AI will likely require critical work in the coming years. Integrating emotional intelligence into algorithms, developing powerful ethical performance indicators, and understanding the geopolitical and cultural influence of AI models internationally all require further work. Furthermore, it is crucial to strengthen regulatory accountability arrangements by establishing evolving compliance frameworks that take into account the dynamic nature of technologies (Jobin, Ienca, & Vayena, 2019). En fin de compte, la création d'une IA éthique dans les chaînes d'approvisionnement nécessite plus que la conformité: elle exige un effort systémique, inclusif et continu pour intégrer les valeurs humaines à chaque couche de la transformation technologique.

ACKNOWLEDGMENT

"This work was carried out with the support of the CNRST under the 'PhD-Associate Scholarship-PASS' program, for which we express our sincere gratitude."

REFERENCES

Abhulimen, A. O., & Ejike, O. G.Angela Omozele AbhulimenOnyinye Gift Ejike. (2024). Ethical considerations in AI use for SMEs and supply chains: Current challenges and future directions. *International Journal of Applied Research in Social Sciences*, *6*(08), 1653–1679. DOI: 10.51594/ijarss.v6i8.1391

Afroogh, S., Mostafavi, A., Akbari, A., Pouresmaeil, Y., Goudarzi, S., Hajhosseini, F., & Rasoulkhani, K. (2024). Embedded ethics for responsible artificial intelligence systems (EE-RAIS) in disaster management: A conceptual model and its deployment. *AI and Ethics*, *4*(4), 1117–1141. DOI: 10.1007/s43681-023-00309-1

Arrieta, A. B., Díaz-Rodríguez, N., Del Ser, J., Bennetot, A., Tabik, S., Barbado, A., & Herrera, F. (2020). Explainable Artificial Intelligence (XAI): Concepts, taxonomies, opportunities and challenges toward responsible AI. *Information Fusion*, *58*, 82–115. DOI: 10.1016/j.inffus.2019.12.012

Benjamins, R. (2020). Towards organizational guidelines for the responsible use of AI. *arXiv preprint arXiv:2001.09758*. DOI: 10.3233/FAIA200433

Binns, R. (2018, January). Fairness in machine learning: Lessons from political philosophy. In *Conference on fairness, accountability and transparency* (pp. 149-159). PMLR.

Biroğul, S., & Şahin, Ö. Exploring the Impact of ISO/IEC 42001. (2023). AI Management Standard on Organizational Practices. *Advances in Artificial Intelligence Research*, *5*(1), 14–22.

Choi, T. M., Wallace, S. W., & Wang, Y. (2018). Big data analytics in operations management. *Production and Operations Management*, *27*(10), 1868–1883. DOI: 10.1111/poms.12838

Cobbe, J., Veale, M., & Singh, J. (2023, June). Understanding accountability in algorithmic supply chains. In *Proceedings of the 2023 ACM Conference on Fairness, Accountability, and Transparency* (pp. 1186-1197). DOI: 10.1145/3593013.3594073

Culot, G., Podrecca, M., & Nassimbeni, G. (2024). Artificial intelligence in supply chain management: A systematic literature review of empirical studies and research directions. *Computers in Industry, 162*, 104132. DOI: 10.1016/j.compind.2024.104132

Doshi-Velez, F., & Kim, B. (2017). Towards a rigorous science of interpretable machine learning. *arXiv*.

Dwivedi, Y. K., Kshetri, N., Hughes, L., Slade, E. L., Jeyaraj, A., Kar, A. K., Baabdullah, A. M., Koohang, A., Raghavan, V., Ahuja, M., Albanna, H., Albashrawi, M. A., Al-Busaidi, A. S., Balakrishnan, J., Barlette, Y., Basu, S., Bose, I., Brooks, L., Buhalis, D., & Wright, R. (2023). Opinion Paper:"So what if ChatGPT wrote it?" Multidisciplinary perspectives on opportunities, challenges and implications of generative conversational AI for research, practice and policy. *International Journal of Information Management*, *71*, 102642. DOI: 10.1016/j.ijinfomgt.2023.102642

Elkington, J., & Rowlands, I. H. (1999). Cannibals with forks: The triple bottom line of 21st century business. *Alternatives Journal*, *25*(4), 42.

Farooque, M., Zhang, A., Thürer, M., Qu, T., & Huisingh, D. (2019). Circular supply chain management: A definition and structured literature review. *Journal of Cleaner Production*, *228*, 882–900. DOI: 10.1016/j.jclepro.2019.04.303

Feizabadi, J. (2022). Machine learning demand forecasting and supply chain performance. *International Journal of Logistics*, *25*(2), 119–142. DOI: 10.1080/13675567.2020.1803246

Fjeld, J., Achten, N., Hilligoss, H., Nagy, A., & Srikumar, M. (2020). Principled artificial intelligence: Mapping consensus in ethical and rights-based approaches to principles for AI. *Berkman Klein Center Research Publication*, (2020-1).

Floridi, L., & Cowls, J. (2019, July). A unified framework of five principles for AI in society. *Harvard Data Science Review*, *1*(1).

Floridi, L., Cowls, J., Beltrametti, M., Chatila, R., Chazerand, P., Dignum, V., Luetge, C., Madelin, R., Pagallo, U., Rossi, F., Schafer, B., Valcke, P., & Vayena, E. (2018). AI4People—An ethical framework for a good AI society: Opportunities, risks, principles, and recommendations. *Minds and Machines*, *28*(4), 689–707. DOI: 10.1007/s11023-018-9482-5 PMID: 30930541

Fosch-Villaronga, E., & Poulsen, A. (2022). Diversity and inclusion in artificial intelligence. *Law and artificial intelligence: Regulating AI and applying AI in legal practice*, 109-134.

Freeman, R. E. (2010). *Strategic management: A stakeholder approach.* Cambridge University Press. DOI: 10.1017/CBO9781139192675

Guidotti, R., Monreale, A., Ruggieri, S., Turini, F., Giannotti, F., & Pedreschi, D. (2018). A survey of methods for explaining black box models. *ACM Computing Surveys*, *51*(5), 1–42. DOI: 10.1145/3236009

Ivanov, D., Dolgui, A., & Sokolov, B. (2019). The impact of digital technology and Industry 4.0 on the ripple effect and supply chain risk analytics. *International Journal of Production Research*, *57*(3), 829–846. DOI: 10.1080/00207543.2018.1488086

Ivanov, D., Dolgui, A., Sokolov, B., Werner, F., & Ivanova, M. (2021). A digital supply chain twin for managing the disruption risks and resilience in the era of Industry 4.0. *Production Planning and Control*, *32*(9), 761–777. DOI: 10.1080/09537287.2020.1768450

Jobin, A., Ienca, M., & Vayena, E. (2019). The global landscape of AI ethics guidelines. *Nature Machine Intelligence*, *1*(9), 389–399. DOI: 10.1038/s42256-019-0088-2

Kutscher, S. (2025). The EU AI Act: Law of Unintended Consequences? *Technology and Regulation*, *2025*, 316–334. DOI: 10.71265/krne7205

Lamarre, E., Smaje, K., & Zemmel, R. (2023). *Rewired: the McKinsey Guide to Outcompeting in the Age of Digital and AI*. John Wiley & Sons.

Lipton, Z. C. (2018). The mythos of model interpretability: In machine learning, the concept of interpretability is both important and slippery. *Communications of the ACM*, *61*(10), 36–43. DOI: 10.1145/3233231

Long, D., & Magerko, B. (2020, April). What is AI literacy? Competencies and design considerations. In *Proceedings of the 2020 CHI conference on human factors in computing systems* (pp. 1-16). DOI: 10.1145/3313831.3376727

Mittelstadt, B. D. (2019). Principles alone cannot guarantee ethical AI. *Nature Machine Intelligence*, *1*(11), 501–507. DOI: 10.1038/s42256-019-0114-4

Mökander, J., Morley, J., Taddeo, M., & Floridi, L. (2021). Ethics-based auditing of automated decision-making systems: Nature, scope, and limitations. *Science and Engineering Ethics*, *27*(4), 44. DOI: 10.1007/s11948-021-00319-4 PMID: 34231029

Morley, J., Floridi, L., Kinsey, L., & Elhalal, A. (2020). From what to how: An initial review of publicly available AI ethics tools, methods and research to translate principles into practices. *Science and Engineering Ethics*, *26*(4), 2141–2168. DOI: 10.1007/s11948-019-00165-5 PMID: 31828533

Nguyen, L. T., & Zuidwijk, R. (2025). Sustainable supply chain governance: A literature review. *Business Ethics, the Environment & Responsibility*, *34*(2), 541–564. DOI: 10.1111/beer.12668

Papernot, N., McDaniel, P., Goodfellow, I., Jha, S., Celik, Z. B., & Swami, A. (2017). *Practical black-box attacks against deep learning systems using adversarial examples.* In *Proceedings of the 2017 ACM on Asia Conference on Computer and Communications Security* (pp. 506–519).

Parasuraman, R., & Manzey, D. H. (2010). Complacency and bias in human use of automation: An attentional integration. *Human Factors*, *52*(3), 381–410. DOI: 10.1177/0018720810376055 PMID: 21077562

Patil, D. (2024). Explainable Artificial Intelligence (XAI) for industry applications: Enhancing transparency, trust, and informed decision-making in business operation. *Trust, And Informed Decision-Making In Business Operation (December 03, 2024).*

Raji, I. D., Smart, A., White, R. N., Mitchell, M., Gebru, T., Hutchinson, B., & Barnes, P. (2020). Closing the AI accountability gap: Defining an end-to-end framework for internal algorithmic auditing. *In Proceedings of the 2020 Conference on Fairness, Accountability, and Transparency (FAT '20)**, 33–44. *ACM*. DOI: 10.1145/3351095.3372873

Reisman, D., Schultz, J., Crawford, K., & Whittaker, M. (2018). *Algorithmic impact assessments: A practical framework for public agency.* AI Now Institute.

Reuel, A., & Undheim, T. A. (2024). Generative AI needs adaptive governance. *arXiv preprint arXiv:2406.04554.*

Riad, M., Naimi, M., & Okar, C. (2024). Enhancing supply chain resilience through artificial intelligence: Developing a comprehensive conceptual framework for AI implementation and supply chain optimization. *Logistics (Basel)*, *8*(4), 111. DOI: 10.3390/logistics8040111

Ribeiro, D., Rocha, T., Pinto, G., Cartaxo, B., Amaral, M., Davila, N., & Camargo, A. (2025). Toward Effective AI Governance: A Review of Principles. *arXiv preprint arXiv:2505.23417.*

Rudin, C. (2019). Stop explaining black box machine learning models for high stakes decisions and use interpretable models instead. *Nature Machine Intelligence*, *1*(5), 206–215. DOI: 10.1038/s42256-019-0048-x PMID: 35603010

Saberi, S., Kouhizadeh, M., Sarkis, J., & Shen, L. (2019). Blockchain technology and its relationships to sustainable supply chain management. *International Journal of Production Research*, *57*(7), 2117–2135. DOI: 10.1080/00207543.2018.1533261

Samarawickrama, M. (2022). AI governance and ethics framework for sustainable AI and sustainability. *arXiv preprint arXiv:2210.08984.*

Schiff, D., Biddle, J., Borenstein, J., & Laas, K. (2020, February). What's next for AI ethics, policy, and governance? A global overview. In *Proceedings of the AAAI/ACM Conference on AI, Ethics, and Society* (pp. 153–158). DOI: 10.1145/3375627.3375804

Schwartz, R., Dodge, J., Smith, N. A., & Etzioni, O. (2020). Green AI. *Communications of the ACM, 63*(12), 54–63. DOI: 10.1145/3381831

Shamsuddoha, M., Kashem, M. A., Nasir, T., Hossain, A. I., & Ahmed, M. F. (2025). Quantum Computing Applications in Supply Chain Information and Optimization: Future Scenarios and Opportunities. *Information (Basel), 16*(8), 693. DOI: 10.3390/info16080693

Strubell, E., Ganesh, A., & McCallum, A. (2020, April). Energy and policy considerations for modern deep learning research. *Proceedings of the AAAI Conference on Artificial Intelligence, 34*(09), 13693–13696. DOI: 10.1609/aaai.v34i09.7123

Sun, P. (2020). Security and privacy protection in cloud computing: Discussions and challenges. *Journal of Network and Computer Applications, 160*, 102642. DOI: 10.1016/j.jnca.2020.102642

Taj, S., Imran, A. S., Kastrati, Z., Daudpota, S. M., Memon, R. A., & Ahmed, J. (2023). IoT-based supply chain management: A systematic literature review. *Internet of Things : Engineering Cyber Physical Human Systems, 24*, 100982. DOI: 10.1016/j.iot.2023.100982

Teixeira, A. R., Ferreira, J. V., & Ramos, A. L. (2025). Intelligent supply chain management: A systematic literature review on artificial intelligence contributions. *Information (Basel), 16*(5), 399. DOI: 10.3390/info16050399

Tirkolaee, E. B., Sadeghi, S., Mooseloo, F. M., Vandchali, H. R., & Aeini, S. (2021). Application of machine learning in supply chain management: A comprehensive overview of the main areas. *Mathematical Problems in Engineering, 2021*(1), 1476043. DOI: 10.1155/2021/1476043

Veale, M., & Binns, R. (2017). Fairer machine learning in the real world: Mitigating discrimination without collecting sensitive data. *Big Data & Society, 4*(2). DOI: 10.1177/2053951717743530

Veale, M., & Edwards, L. (2018). Clarity, surprises, and further questions in the Article 29 Working Party draft guidance on automated decision-making and profiling. *Computer Law Review International, 19*(4), 97–104. DOI: 10.9785/cri-2021-220402

Voigt, P., & Von dem Bussche, A. (2017). *The EU General Data Protection Regulation (GDPR): A Practical Guide*. Springer. DOI: 10.1007/978-3-319-57959-7

Wamba, S. F., Akter, S., Edwards, A., Chopin, G., & Gnanzou, D. (2015). How 'big data'can make big impact: Findings from a systematic review and a longitudinal case study. *International Journal of Production Economics, 165*, 234–246. DOI: 10.1016/j.ijpe.2014.12.031

Wasi, A. T., Anik, M. A., Rahman, A., Hoque, M. I., Islam, M. D., & Ahsan, M. M. (2025). A Theoretical Framework for Graph-based Digital Twins for Supply Chain Management and Optimization. *arXiv preprint arXiv:2504.03692*. DOI: 10.2139/ssrn.5293091

Wellbrock, W., Malinovska, M., & Ludin, D. (2025). Ethical implications and potential opportunities and risks of artificial intelligence in supply chain management. *Discover Sustainability*, *6*(1), 886. DOI: 10.1007/s43621-025-01808-3

Weller, F. (2024). Blockchain Technology for Secure and Transparent Supply Chain Management. *International Journal of Computing and Engineering*, *6*(3), 15–28. DOI: 10.47941/ijce.2138

Winfield, A. F., Michael, K., Pitt, J., & Evers, V. (2019). Machine ethics: The design and governance of ethical AI and autonomous systems [Scanning the issue]. *Proceedings of the IEEE, 107*(3), 509–517. DOI: 10.1109/JPROC.2019.2900622

KEY TERMS AND DEFINITIONS

AI Governance: Frameworks, policies, and processes established to guide the ethical, legal, and responsible use of artificial intelligence technologies within organizations.

Algorithmic Bias: Systematic and repeatable errors in AI decision-making that result in unfair outcomes, often caused by biased training data, flawed algorithms, or inappropriate model assumptions.

Artificial Intelligence (AI): Computational systems capable of performing tasks that typically require human intelligence, such as learning, reasoning, problem-solving, and decision-making, using algorithms and data models.

Data Privacy: The protection of personal and sensitive information from unauthorized access, misuse, or disclosure, ensuring compliance with relevant regulations and ethical standards.

Ethical AI: The design, development, and deployment of artificial intelligence systems in ways that respect human rights, ensure fairness, avoid bias, promote transparency, and uphold societal values.

Responsible Innovation: An approach to innovation that anticipates potential societal impacts, engages stakeholders, and ensures that technological advancements align with ethical and sustainable objectives.

Supply Chain Optimization: The process of improving the efficiency, cost-effectiveness, and performance of the supply chain through strategic decision-making, process redesign, and technology adoption.

Supply Chain: A network of organizations, resources, activities, and technologies involved in the creation and delivery of a product or service from suppliers to end customers.

Sustainable Supply Chain: A supply chain that integrates environmental, social, and economic considerations into its operations, aiming to reduce negative impacts and create long-term value.

Chapter 8
Ethical Intelligence in Motion:
Leveraging AI for Responsible Sourcing in Supply Chain and Logistics

Md Mehedi Hasan Emon
https://orcid.org/0000-0002-6224-9552
American International University-Bangladesh, Bangladesh

ABSTRACT

This chapter explores how artificial intelligence (AI) is reshaping responsible sourcing within contemporary supply chain and logistics networks. Moving beyond efficiency-driven narratives, it argues for the integration of ethical intelligence where AI systems are harnessed to uphold labor rights, environmental standards, and supply chain transparency. Drawing on global frameworks, empirical case studies, and recent regulatory developments, the chapter examines both the promises and perils of AI in procurement. It critically engages with issues of algorithmic bias, data ethics, and governance, emphasizing that technological innovation must be matched by normative reflection. By situating AI within a broader ethical architecture, the chapter offers a roadmap for organizations seeking to align digital transformation with social responsibility and sustainable sourcing practices.

INTRODUCTION

The chapter Ethical Intelligence in Motion: Leveraging AI for Responsible Sourcing in Supply Chain and Logistics the chapter explores a critical and timely

DOI: 10.4018/979-8-3373-3104-1.ch008

intersection between artificial intelligence (AI) and ethical sourcing and the current arena of supply chain management. The popularization of AI in the procurement sphere, supplier selection, and logistics has led to an increase in doubts regarding the ethical aspects of AI-based tools. Beyond the continuous conversation that seems to have focused on efficiency and optimization, the current chapter indicates a vital need to include AI in the sourcing system in a responsible and visibly transparent manner. The debate places ethical sourcing at the center stage of international business, which has been enhanced by the rising legal requirements, stakeholder pressure, and increasing environmental and work-related commitments. To consider ambivalence and transformational nature of AI, the chapter illustrates both increased visibility and risk reduction benefits to monitor the entire complex supply chain and sets new challenges of addressing algorithmic bias, data ownership, and socio-environmental responsibility. The discussion would be based on theoretical knowledge, conceptualized cases by the apparel industries/electronics industries and agribusinesses to benefit both the opportunities and dangers of AI in the enforcement of ethical sourcing, especially in the supplier auditing processes, labor rights enforcement, environmental conformance, and tracking. In a complementary way, the chapter questions the governance arrangement and normative systems such as the ESG metrics, and new due diligence law, which shapes the ethical use of AI in global supply chains. The chapter promotes the balanced and thoughtful approach, in which the technology is not to be framed as a panacea, but rather another factor of ethical intelligence that is highly strategic and organizational competence. However, is that the value of AI in responsible sourcing is not only laden on automation or cost savings, but whether it can aid in making equitable, transparent, and otherwise value-added to society decisions. The chapter ends with providing recommendations as well as best practices to integrate ethical AI into supply chain processes, such as mitigation of biases, stakeholder-based design of AI, privacy of data, and proper compliance methods. It adds a unique dimension to the emerging narrative on ethical AI, not as a luddite, but as one of the cornerstones of digitalization of supply chains.

SITUATING ETHICAL INTELLIGENCE IN THE DIGITAL SUPPLY CHAIN

The supply chain of the twenty-first century is no longer transactional, one-way transmission of goods between point of origin and end customer. Rather, it has become an extensively informed, multi-layered, global distributed network growing more data centric and automated decision making. Artificial Intelligence (AI) is at the heart of this change- the technologies that allow the system to learn using data, choose, and evolve with time. The use of AI in the field of procurement, logistics, warehousing, and supplier management is very significant and is gaining pace. But, simultaneously with its business advantages, AI also poses several ethical questions, especially in the subject of responsible sourcing. The broad definition of responsible sourcing as the procurement of goods and services taking into consideration human rights, environmental sustainability and social impact has been a key issue in the governance of global supply chains. In a climate of increasing consumer awareness, tough regulatory systems, and demands of corporate responsibility, companies are now required to go behind the first-tier suppliers to the whole value system (Tao et al., 2023). It is in this regard that the idea of emotional intelligence, which is incorporating ethical thinking into algorithmic systems, becomes an essential corporate skill.

In contrast to the generations of technologies that came before it in the supply chain world, AI is not here just to aid human decision-making; instead, it is actively replacing it in several critical components of the supply chain, namely supplier selection, forecasting, and contract negotiation. Patterns, anomalies as well as sourcing suggestions may qualify in real-time by training algorithms based on historical data. However, automation in these areas has the risk of blurring the social and environmental observations to which responsible sourcing is aimed. Indicatively, AI systems which are configured to focus on the price or speed without giving due regards to the working conditions or carbon emissions, end up supporting unethical practices unknowingly (Pawar et al., 2025). Moreover, it is also not bias free, that is algorithmic decision-making. History purchases may constitute types of overriding inconsistencies, i.e., absence of minority created built partners or the institutional negligence of adjacent subject to corrupt ecological regulation. When these kinds of datasets train a system of AI, it can potentially join and compound existing biases without being specifically trained to spot and eradicate them (Hanna et al., 2025). These increase the burning question: Who is creating the algorithms? What are their values coded with as regards decision rules? What interests are catered for? The ethical risks do not only end with data bias. The topic of explainability and accountability is a problem with AI systems, as well. In this contest, their blacklisting through algorithm can render it challenging among stakeholders to recognize why an issue known as the black box problem confounds (Calacci & Pentland, 2022). This transparency raises suspicion, particularly when it comes to life and death circumstances such as those of Human Rights due diligence. Further, fragmented and outsourced supply chains obscure the chain of accountability as no entity can monitor the impacts of the adoption of AI tools in terms of sourcing decisions across the tires. And it is against this backdrop that ethical intelligence requires a shift toward a compliance-driven approach to a value-based governance. This entails such ethical parameters in both the results of the AI-tools but also in the design, application, and supervising of the AI Tools. Firms should adopt ethical systems of AI as they focus on fairness, inclusiveness, and caring towards the environment. The importance of businesses to be proactive in order to mitigate risks against human rights abuse and environmental damages may be measured by how the European Union (EU) and Germany are putting some regulatory demands of consultations in this regard under two laws: the Corporate Sustainability Due Diligence Directive (CSDDD) and the German Supply Chain Act, respectively, which are laws that are increasingly touching on digital technology (dos Santos Sequeira, 2024).

Figure 1. Ethical Intelligence Framework for AI in Responsible Sourcing

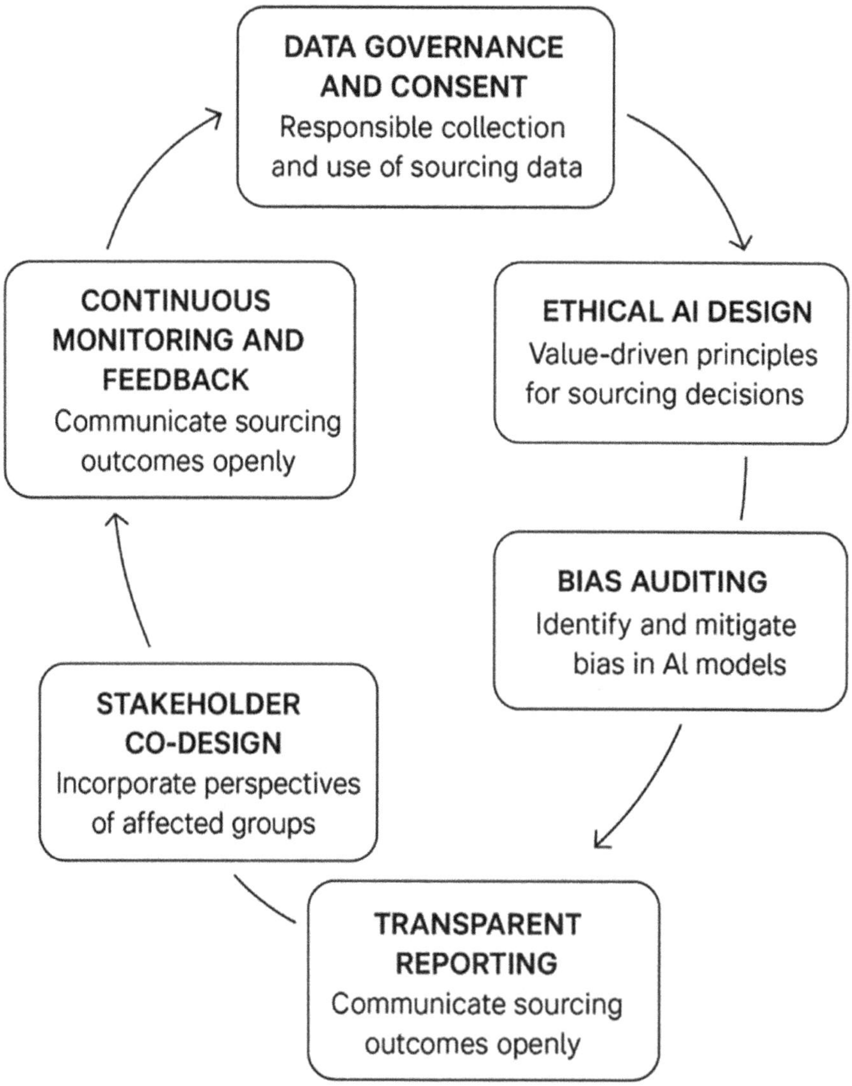

The depicted framework determines a circular approach to integrating ethical intelligence into the responsible sourcing environments across the AI life cycle. The six components that are part of the model include Data Governance and Consent, ethical AI design shaped by value-driven principles, bias auditing, co-design involving the stakeholders, including the perspectives of the groups affected by the process, continuous monitoring and feedback system, and open reporting, which highlights that ethical intelligence is not an event but a dynamic and iterative process..

AI can help in enjoying responsible sourcing. Machine learning and in-depth analytics can give real-time transparency throughout the supply chain, reveal the potential abuse of human rights, and foresee environmental risks in geography. These early warning indicators regarding misconduct on the side of suppliers can be extracted using Natural language processing (NLP) software on audit reports, news articles, and NGO database. By supporting AI with blockchain, the aspect of traceability can be ensured between raw materials extraction to a final assembly, giving the immense transparency (Emon & Chowdhury, 2025; Patel et al., 2024). In case it is integrated with ethical criteria, these technologies can also introduce risk management as another preventive and anticipation-based job. However, this can only be achieved through the deliberate organization efforts. The AI governance structures should incorporate ethical sourcing, in which cross-functional teams composed of procurement experts, data analysts, ethical teams, and compliance officers, should be involved. The suppliers are also eligible to engage in such discussions but not as mere transport carriers of data, but as partners in ethical innovation. It has a massive importance, the co-design principle: regarding the ethical AI systems, they are to be designed in a collaborative fashion and with an input of different stakeholders, which will include the most-affected ones, in terms of the decision on the supply chain (Rezaei et al., 2024).

ETHICAL SOURCING IN CONTEMPORARY SUPPLY CHAIN DISCOURSE

There is a paradigm shift in the discourse of the present concerning supply chains: a shift of the periphery to the center of corporate strategy, stakeholder relations, and process. Ethical sourcing, previously the target of voluntary corporate social responsibility, initiative, or marketing distinction, has grown into an essential component of the supply chain governance. In the economy of the increased connectivity and watchdog effects, the call of the ethical products is now being met by the legalities, pressure of the stakeholders and civil society pressure. This metamorphosis is particularly visible in the wake of scandals on the very top of the corporate world exposing exploitative conditions of work, destruction of environment, and violation of human rights in corporate supply chains. The Rana Plaza tragedy in Bangladesh, the cobalt mining controversy in the Democratic Republic of Congo has chimed in the price in lives and eco-systems about the hazy and price-determined approaches to sourcing. The brands use in the likes of apparel, electronics, agribusiness even pharmaceutical industries have not only been facing the backlash because of the immediate action(s) taken but also due to lack of oversight of upstream and downstream activities in the supply chains (Khan et al., 2024; Thomas et al., 2023). In this connection, ethical sourcing has ceased damage control. It is a dynamic strategic approach to ensuring that goods and services are acquired in line with global expectations, human decency, and ecological continuity. In essence, ethical sourcing implies the integration of social, environmental and governance issues into the purchasing processes. This involves among others provision of fair labor conditions as well as avoiding the exploitation of child and forced labor, minimization of carbon emission, land rights respecting, and supplier diversity fostering. It is also concerned with open supplier relationships, the traceability of raw materials, and the significant contribution of stakeholders in the procurement life cycle (Centobelli et al., 2022). Conversely, the traditional concept of procurement, which is dominated by cost-effectiveness and delivery time positions the value-based procurement in the first plane that would serve the long-term corporate values and expectations of society.

The involvement of AI in ethics is both exciting and a scary perspective. On the one hand, AI enhances the capability of companies to monitor, audit, and analyze unbelievably large chain network of supplies in real-time. The satellite imagery could be analyzed using machine learning algorithms and reveal the instances of illegal deforestation. The use of natural language processing tools will enable them to review supplier communications to warn against noncompliance, the predictive analytics will be able to predict which suppliers will violate the labor laws based on habits in previous years (Garg et al., 2021). The AI tools have another additional power of being end-to-end traceable powered by blockchain, meaning that firms will be able to track the presence of the origins of raw materials, and in doing so wash away the allegations of the fraudulent purport of ethical compliance. However, ethical sourcing does not have dilemmas about the use of AI that do not exist. Left wisely developed and applied, AI tools can reinforce the already present bias in the collection of fostering data, ignoring equally complex socio-political eases of the areas of acquisition and the encouragement of an untrue illusion of disinterested decisions utilizing subjective criteria. As another example, black box algorithms may produce supplier risk scores that would discriminate against small-sized and/or minority-owned businesses without the digital infrastructure and the ability to see their past data (Emon & Khan, 2025c; Richey et al., 2023). Moreover, ethical monitoring can be accomplished by automating ethical audits through AI, which eliminates the need of human auditors, who are better placed to understand cultural and contextual peculiarities and thus turn ethical monitoring into a purely technical task. The risks highlight the importance of what scholars and practitioners are eagerly starting to embrace as the ethical intelligence the ability to add ethical reasoning to the implementation, use and management of the AI systems installed to enable sourcing and supply chain management (Khoury et al., 2023). Ethical intelligence is a cross-functional venture to which data scientists, procurement officials, ethicists and community stakeholders can work.

Figure 2. Evolution of Ethical Sourcing from CSR to Legal Mandate

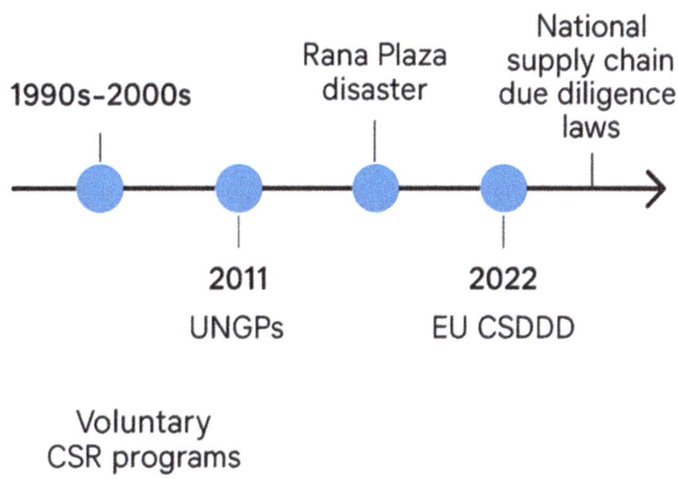

A chronological timeline is presented in the figure, which reflects the historical development of ethical sourcing, starting with the early voluntary Corporate Social Responsibility (CSR) promises in the 1990s and 2000s, followed by the high-profile scandals in the supply chain, including, in 2013, a tragedy in the Rana Plaza factory, and leading to the formulation of binding regulatory frameworks, including the United Nations Guiding Principles on Business and Human Rights (UNGPs, 2011), the European Union Corporate This chronology shows that ethical sourcing gradually shifted in orientation both formally (as a CSR program) and in practice (as a statutory and strategic necessity) due to the heavy weight of outside forces and changing standards of stakeholder expectations that have made ethical sourcing a functional necessity in the public domain of supply-chain regulation.

THE ROLE OF AI IN ENHANCING SUPPLY CHAIN VISIBILITY AND ACCOUNTABILITY

Supply chain in the twenty first century is a complex net of interconnected processes that spans geopolitical borders and is arranged and managed via a widespread network of suppliers whose legislative frames vary significantly. In this setup, the need regarding visibility has transformed into an essential one as without firm transparency, corporations cannot engage in ethical sourcing. Traditional tools such as supplier audit, document verification, and compliance reporting lack the ability to question the specifics of the present-day global value chain. AI has therefore come up as an agent of change that can provide better visibility. Its strength lies in the ability to work with large amounts of data at unprecedented speeds with a high degree of fineness concerning contextual awareness. The inputs, such as shipping manifests, the history of invoices, IoT sensor data, procurement-platform data, things declared in customs, or even the weather patterns can be structured and unstructured and fed into machine-learning algorithms to see the anomaly that may indicate ethical misconduct or regulatory non-compliance (Singh et al., 2024). They can be expressed as unusual routes, mismatched pricing history, or super-affluent production between manufacturers, which can give an insight into widespread labor exploitation, counterfeit, or environmental contamination.

Natural-language processing (NLP) can boost the analytical capabilities of AI because it helps to analyze qualitative texts: third-party reviews, media research, whistleblower testimonies, and social-media materials. NLP provides real-time, flexible layers of risk management through sentiment analysis, entity extraction, and the classification of the risk level. One example is, a supplier is mentioned in an article on forced labor or environmental mistreatment, the reminder can be issued by NLP, the protocols to investigate the situation can be set forth in a due-diligence framework. The combination of blockchain technology is a more sophisticated manifestation of AI-based supply-chain responsibility. Decentralized ledgers produce irreversible chains of transaction histories, certificated and provenance information and in combination with AI analytics deliver auditable, tamper resistant chains of custody. The IBM Food Trust platform demonstrates such synergy: with the capability to track agriculture produce along the farm-consumption chain, this technology secures food security and, at the same time, guarantees ethical sourcing procedures (Xu et al., 2020). In the same vein, the diamond-traceability system offered by Everledger uses blockchain and AI to track the origins of precious stones in a manner that addresses longstanding industry qualms on whether these precious stones are sourced out of conflict free zones.

Despite these developments, the use of AI-powered visibility tools comes amidst many ethical implications. One phenomenon that continues to be a stumbling block in data quality. Suppliers Supplying companies in many parts of the Global South have to work with limited internet connectivity, decentralized documentation, limited digital infrastructure. This data conditions makes their outputs incomplete or erroneous, and this possibly skews the resulting models trained on those dataset on AI (Behera & Sharma, 2024; Khan & Emon, 2025a). Furthermore, though AI systems will be able to identify the differences in patterns of output among supplier facilities and cluster weather-induced delays, they will not be able to gather the cultural background required to understand the labor experience on the ground, the consent of communities or the unofficial practices of negotiation. Scholars such as MacFarlane, (2020) assert that the need to complement techno-centric governance with critical ethnographic research and participatory design in case the ethical complexities governing supply-chain governance are not to be turned into computational rules.

Figure 3. Integration of Blockchain and AI for Ethical Traceability

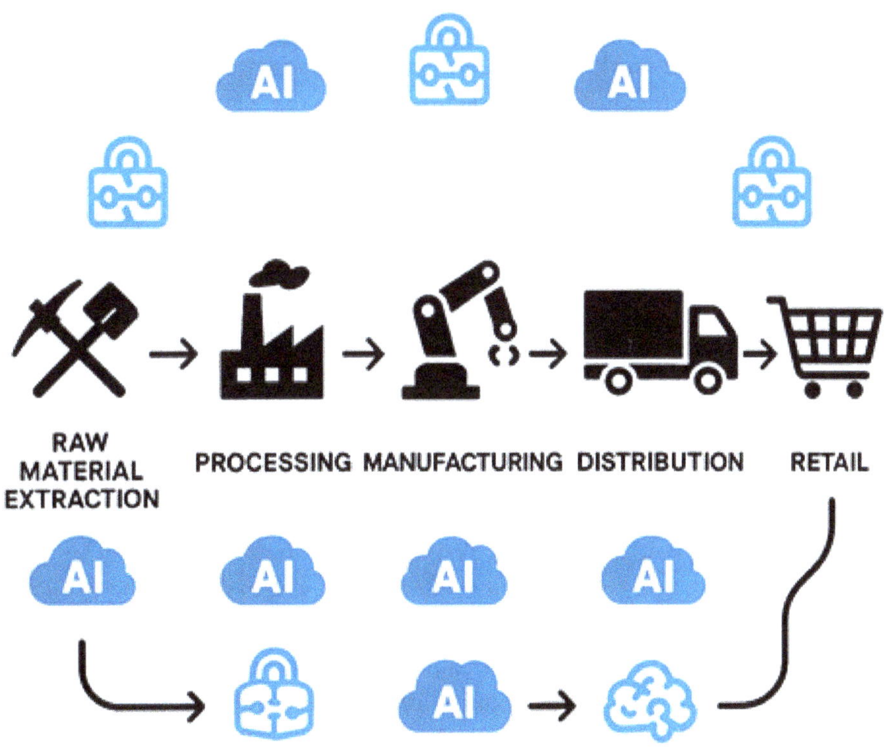

The current figure shows a layering process flow graph that shows how blockchain and artificial intelligence technologies can be implemented along a global supply chain. The diagram describes the main steps between the raw materials extraction and processing and the manufacturing, distribution, to the retail delivery while the blockchain icons are put on each step to show that each immutable record which will be captured, whether as certification or as transactions will result in a transparency and accountability level. At the same time, AI icons overlie these stages showing the use of analytics, such as anomaly detection, sentiment analysis, and so on that allow the real-time control of the process and ethical-compliance verification. The diagram shows how tamper-proof data of blockchain and AI when applied together enhance traceability as may be seen with the Everledger system of monitoring diamonds and such efforts of IBM Food Trust in the agricultural supply chains.

ALGORITHMIC BIAS AND ETHICAL PROCUREMENT DECISIONS

As the use of data-driven approaches to defining the efficiency and agility of procurement has increasingly gained prominence, the corresponding issue of algorithmic bias in the work of AI-based systems brings deep concerns about organizational decision-making ethics. Due to the growing use of AI by enterprises in making supplier choices, risk assessment, and contracting, the assumption of systematic neutrality overclouds the baked-in discrimination within the built-in datasets and design logics. This development, in turn, presents not only ethical procurement-relying on the principles of fairness, inclusiveness, and sustainability-with an inflection point but it also demands that technological innovation be improved along with an intensive examination of the biases to which AI can contribute or exacerbate. Algorithmic bias in the sphere of procurement has three interconnecting roots: shoddy historical data, unclear model design, and various institutional priorities. In practice, several suppliers can base their algorithms on past performance indicators, logistics financials, conformance history, and shipping records to predict future output or give a supplier score (Asif et al., 2022; Emon, 2025c). In cases where this historical evidence reflects systemic exclusion or discriminatory tastes such as chronic neglecting of the small-scale providers, female-owned businesses, or Global South businesses, then the algorithm will probably replicate these trends. As Bibri & Allam, (2022) has claimed, algorithms are not inserted into a vacuum, or, as phrased by the authors, algorithms simply reflect and reproduce the values and assumptions of the societies develop them.

An example of the same is the use of recommendation engines in procurement and vendor platforms. These systems are characterized by a tendency to focus on either cost-efficiency, delivery speed, and scale which is not in their favor, especially small or medium sized enterprises (SMEs) with a tendency towards social or environmental responsibility. This creates a feedback loop in which suppliers that do not conform to the cost-based norms are left out not because they were underperforming but because of an implicit value system that was traditionally in place (Alioto, 2022; Khan & Emon, 2025b). Another issue is the danger of systematic exclusion of the enterprises that are headed by women, racial minority, and indigenous peoples who were historically under-represented in the procurement networks hence their omissions also contribute to training data distortion. The result is a digital form of marginalization that supposedly prospers efficiency but recreates already present inequalities. There are also geographic assumptions which make things even more difficult. Blocks on potential suppliers in Asia, Africa or Latin America can be based on the risk or compliance criteria determined by algorithms fed on primarily North American or European data where delay in responsiveness is less likely to be an indictment of supplier incompetency but rather a logistical constraint or due to political unrest. When it comes to inter contextual inertness and local calibration, these systems may cause worldwide chain-of- supply disparities under the guise of immense robotic decision-making (Mosallanezhad et al., 2023).

The issue of explainability is not enough as an ethical solution when the implicit assumptions go unstudied. Instead, the companies need to ensure an active diversification of the suppliers who sign the training datasets: cooperatives, women enterprises, social enterprises, etc. to reduce systematic bias in model prediction (Popescu et al., 2022). A methodological route towards this is Fairness-aware Machine Learning (FAML) approaches whereby engineered constraints help guarantee fair performance of the model across groups. In addition, the requirements of governance of responsible AI in procurement should be institutionalized. Ethical oversight organizations ought to observe not only compliance procedures but also social justice considerations of algorithmic choices, store ethical procurement KPIs and retain human control in key AI-assisted operations. On their part, procurement managers must have a combination of technical skills and ethics-based reasoning that may allow them to challenge algorithmic decisions, as opposed to unquestionably accepting them. The interdisciplinary partnership between data scientists and those in the procurement field can integrate ethical limits into the quantitative models, thus the democratization of the governance and revealing any potentially unaddressed biases. Finally, there is the procedural advancement in regulation. The Artificial Intelligence Act and Corporate Sustainability Due Diligence Directive (CSDDD), proposed by the European Commission, promote risk-based governance, transparency and human rights due diligence as requirements prior to applying AI to business practices. These tools offer the policy framework required to establish ethical protections in algorithmic systems of purchasing (Pria et al., 2024).

Figure 4. Sources of Algorithmic Bias in Procurement Systems

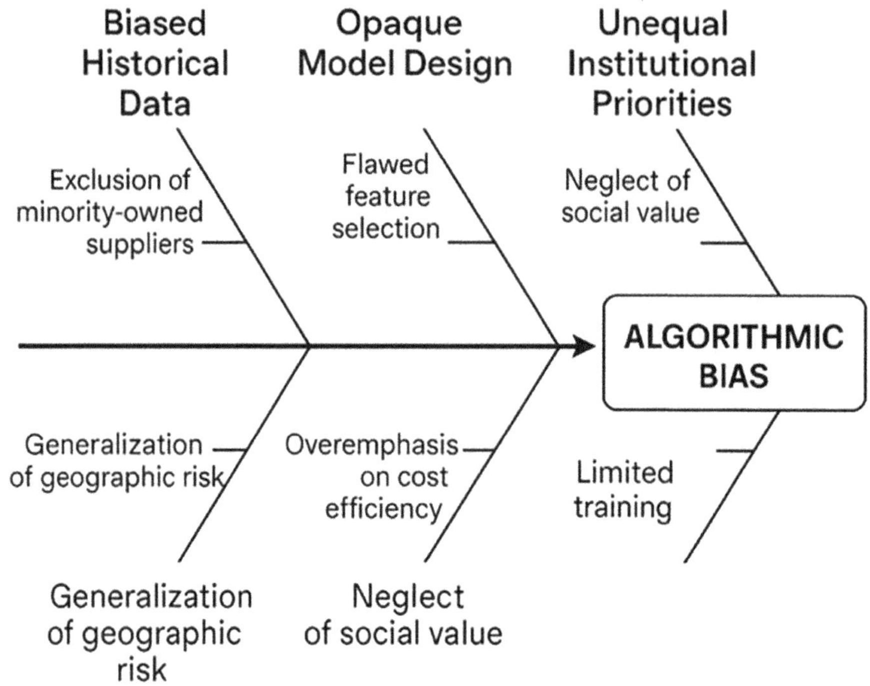

The figure shows a cause and effect fishbone diagram that illustrates how three major causes of algorithmic discrimination inside procurement systems include the use of biased historical data, the lack of transparency in model design and unbalanced institutional priorities. Specific examples of its bias are also divided into empirical categories of each main branch, that is, the exclusion of minority owned suppliers, perverse feature selection, excessive focus on cost-effectiveness, and lack of social value. This scheme helps understand how these inter-related variables contribute to the propagation of algorithm bias, as well as prepares the readers to grasp a complex side to the AI-based procurement processes.

AI-POWERED RISK ASSESSMENT FOR LABOR AND ENVIRONMENTAL COMPLIANCE

The use of artificial intelligence (AI) as a form of supply-chain governance represents a critical break with the traditional systems of compliance that are usually based on periodic audits, self-reporting, or independent certification. In comparison, AI enables a new, uninterrupted, and proactive paradigm: predictive analytics, machine-learning algorithms, and geospatial technologies allow observing supplier behaviour and environmental footprint in real-time. These kinds of capabilities will help organizations to retain labor standards and environmental sustainability within the supply systems globally. The ability to use large and disparate datasets, such as satellite imagery, sensor data, anonymized employee comments, and external news stories, and generate indicators of non-compliance at early stages is among the first contributions of AI. As an example, the geospatial and environmental data-driven machine-learning models can discover unauthorized changes in the land use, including deforestation, illegal mines encroachment on the preservation territories. This has led to the emergence of projects, like that of a global forest watch which uses artificial intelligence enabled analysis of satellite images, that have become staple in the management of multinational corporations, NGOs and other interested stakeholders in evaluating their supply chains risks and raw material sources (Delabre et al., 2020). On the grounds of labor compliance, using data processing with the help of AI, it is possible to study wages in factories, shifts, and records of working-time to detect anomalies that signal to overtime, underpayment, or forced labor. Moreover, it is possible to check anonymously filled employee questionnaires, discuss social media and grievance systems, and detect the mood of the workers and possible infringements of human rights, as these arise in advance to imminent loss of reputation or regulatory disaster (Horne, 2023). Accordingly, AI-based tools do not only provide a real-time picture of existing compliance but also provide predictive insights when there will be other areas of non-compliance. One of the examples of such capabilities is the Microsoft AI for Humanitarian action programme, which partners with civil society organizations and local non-governmental organizations to identify areas of the increased risk of labor exploitation. The system combines socioeconomic metrics, satellite, and historical labor-violation statistics to produce heat maps and consequently illustrate procurement departments in assigning sourcing priorities. These collaborations are good examples of the way technological innovation, domain knowledge, and local knowledge can support each other, especially when the latter are limited in algorithmic assessments (Kougkoulos et al., 2021). However, the introduction of risk-assessment tools based on AI requires minute attention on the ethical side. Another key issue is technocratic imperialism- a tendency to base a decision solely on the results of the algorithm to rely on it in place of human reasoning or a democratic process when there is a need to make a decisive decision. Even advanced AI systems are black boxes and as such can be trusted only as far as the data and assumptions that helped create them. The availability of real-time data in many locations in the Global South is low and cultural/contextual complexity (e.g. informality of labor, language heterogeneity, complex local regulations), can easily overcome the capacity of sophisticated analytic engines. (Kougkoulos et al., 2021).

DATA ETHICS IN SUPPLY CHAIN AI: CONSENT, TRANSPARENCY, AND GOVERNANCE

Data ethics form one of the main governance issues in the context of the growing prominence of AI along supply chain operations. Specifically, there are normative concerns regarding consent, transparency and governance, which are pressing in relation to the amalgamation and analyzing, as well as the implementation of data to make informed decisions involving procurement, labor compliance, and environmental monitoring. These are especially relevant in cases where the data involves vulnerable groups of stakeholders, factory workers, small-scale suppliers, and communities that are exposed to environmental degradation, whose rights and agency can be easily violated by transparent data systems. The principle of consent is perceived as a cornerstone of data ethics but is usually muddled in international supply chain arrangements. In a world where corporate data policies can attest to the importance of consent as a fundamental principle, the mechanisms used to secure procedural understanding of informed consent among those who comprise supplier firms are often identifying or perfunctory. This syndrome is especially troublesome where digital literacy is low, where existing power disparities are more favorable to either the recipient of the supply or where poorly structured data insurance laws are in place. Employees, as another example, are becoming more exposed to biometric tracking, surveillance of the internal communications and monitoring in their workplaces in the name of maximizing productivity, but without clear understanding of the intentions of these data operations and usually without an opportunity to withdraw themselves out of the relevant databases (Chisita et al., 2025).

The transparency issue qualifies as a powerful challenge to ethical application of AI to the supply chain setting. Opacity is an issue in a variety of machine learning models applied to risk assessment, supplier scoring, and logistics optimization; this has been widely referred to as a subset of the black box problem. Ensemble and complex neural network structures can be configured to make authoritative suggestions with no explanation of the variables or analysis logic behind them. Such opaqueness destroys interorganizational trust, weakens accountability and in certain cases leads to discriminatory or prejudiced results. Demand of algorithmic accountability, therefore, is not limited to technical explanations, but also offers an ethical consideration (von Eschenbach, 2021). Strong governance data regimes should thus form the pinnacle of any sourcing initiative that involves AI. A fidelity to data provenance: the capacity to watch the roots of datasets, test their validity and prove their ethical gathering occupies the heart of such structures. The origin should be clear, verifiable, and in line with the laws such as the GDPR in the EU or, the laws in development, like the AI Act in the EU (Emon & Khan, 2025d; Labadie &

Legner, 2023). Without such protection mechanisms, there is a risk that AI regimes will contribute to the existing inequities instead of reducing it.

It is also important to ensure that there are decision audit trails, and this should flow the input, decision variables, and output in a way that can be reviewed by internal compliance functions and external auditors. Decision registries are institutionalized, assist with redress in the event of an error or damage and enhance the ability to demonstrate due diligence to regulators and civil society. The low chance of obtaining procedural justice can be minimized by using explanatory techniques like SHAP (Shapley Additive Explanations) or LIME (Local Interpretable Model-Agnostic Explanations) to explain why a certain supplier is considered high-risk or ineligible to participate in the procurement processes (Sreenivasan & Suresh, 2022) Moreover, privacy-enhancing technologies (PETs) can offer a framework of resolving the needs of data and ethical compulsions. As an example, with differential privacy, statistical noise is added to data so that aggregate conclusions may be made without sacrificing the privacy of individuals. The federated learning enables collective model training in neighborhood copyrighted data, thereby enabling AI models to be refined without data transmission between sites and therefore reducing commercial or political exposure (Uvet et al., 2025). These initiatives highlight an ethical concern of the informational rights of all parties regardless of their respective power or regulation.

REAL-WORLD APPLICATIONS: CASE STUDIES IN ETHICAL AI SOURCING

The recent debate on ethics in sourcing as articulated by artificial intelligence (AI) has taken a new shape in that it has moved out of science fiction and into a new direction regarding empirical studies. Organizations are also facing more complex challenges in supply-chart, increased compliance to various bodies, and more pressure of accountability among the various stakeholders. The reaction is that a specially chosen group of enterprises is beginning to use AI-powered platforms not only to optimize the logistics or compress expenses, but to imbue sustainability and ethics into the structuring of procurement. This discussion reviews three case-studies as an illustrative example to explore the ways in which such AI implementation can be used to implement the principles of ethical sourcing: Patagonia (apparel), Fairphone (electronics), and Unilever (agribusiness). This review reaffirms the fact that the eventual effectiveness of AI will rest on sound corporate governance, stakeholder engagement, and conformity to a top-layer ethical agenda irrespective of the transparency and traceability provided.

Apparel Industry: Patagonia's Hybrid Model of Automated and Human Oversight

In the past, Patagonia has branded itself as an industry leader in the creation of sustainable and ethically thought-out clothes. Over the past years, the organization has incorporated AI-powered analytics to create dynamic supplier scorecards that can evaluate their alignment with labor and social standards. They incorporate digital audit records, satellite photographs, and the answers of employees. The machine-learning models will be trained to detect anomalies related to a bade working hours underreport, high numbers of overtime hours, and an incorrect number of wages noted in a record as they are often associated with exploitative working practices (Alogogianni & Virvou, 2023). The Patagonia AI architecture therefore has access to structured and unstructured data connecting it to its international supply chain and offers supervision over the compliance audits of the factories, chemical-usage registers as well as external information sources. The company has uniquely decided to combine algorithmic review and continuous human attention. The result of this hybrid structure is that the considerations of ethics are now possible because of contextual, cultural, and relational dimensions, considerations that are typically missing when attempting to determine ethics solely through AI systems. An example is that in cases where a supplier is given a poor compliance rating after predictive analysis, Patagonia sends regional compliance officers to make physical follow-ups before acting with curative or capital sanctions (Golchha & Nagariya, 2025). In addition, the company has recognized the weaknesses of its AI tools, not relying on technological determinism. However, Patagonia uses AI mainly to prioritize risk without providing determinative judgment and therefore shows how ethical governance can be promoted through technology without being indifferent to their sociopolitical consequences.

Electronics Sector: Fairphone and AI-Driven Traceability for Conflict-Free Minerals

Digital electronics industry is one of the most ethically complicated sectors in which the acquisition of crucial raw materials, mainly minerals, is concerned not only because of the opaque nature of the industry but also because of its deeply fragmented supplier base. Technological leader Fairphone is a Dutch social enterprise that has ventured out into innovative territory by putting ethical sourcing front and center of its business model by way of a general focus on transparency and a direct referral to stakeholder cooperation. The most logical way to put these principles to work within a business framework is based on the use of AI-powered traceability systems that would allow the firm to keep constant track of the cobalt, tungsten, and tin minerals, often mined under unsustainable practices in the sub-Saharan area (Mvile & Bishoge, 2024). The main AI functionalities, such as natural language processing and geospatial analysis, are used in collaboration with the electronic mine certifications, satellite imagery, and the ground reporting of non-governmental organizations, which determine whether each material is indeed conflict-free. Further, a data layer that is enabled by blockchain is incorporated to produce an unchangeable history of the provenance of raw materials across all the nodes of a supply chain (Vijaykumar et al., 2024). It is revealed in the study that the described sourcing strategy is not kept in a vacuum, rather supported by a comprehensive supplier empowerment and community development agenda. The cooperation of Fairphone with local mining cooperatives and associations is also tied to providing quality data enrichment, consent by the side of the information owner to release data into the public domain and develop local indigenous technical expertise. Therefore, the AI infrastructure is not intended to be a surveillance tool but rather a participatory model that aims at redistributing the informational power among the peripherals of supply-chains. Overall, Fairphone is evidence of the two-fold promise of AI: it can not only identify unethical behavior but also motivate financially disenfranchised suppliers to be responsible.

Agribusiness: Unilever's AI-Driven Environmental Monitoring and Risk Management

Unilever, a multinational company operating in consumer goods and acquiring businesses, has enthusiastically invested heavily in artificial intelligence to foster sustainability performance. The key point of this effort is to implement AI-empowered platforms used to conduct systematic environmental risk identification in a wide supplier network. The organization combines satellite images and sophisticated machine-learning algorithms to keep track of palm-oil-related deforestation, evaluate soil erosion, and track the seasonal water use dynamics on various agricultural facilities (Mangla et al., 2022). The AI infrastructure used by Unilever was conceptualized in partnership with technology companies and environmental NGOs to predict potential (or hotspots), generated by means of the aggregation of land-use dynamics, biodiversity loss, and climate vulnerability risk factors, contributing to procurement strategy, supplier outreach, and policy design. Secondly, the company has also committed to release real-time alerts about deforestation on publicly available sources, including Global Forest Watch, expanding transparency and leading to external responsibility (Emon & Khan, 2025b; McGrath et al., 2021). In more recent times, Unilever has started to explore federated learning paradigms to reduce data privacy issues arising when sensitive supplier data is gathered. It is a non-centralized architecture that stores the data on site in national territorial jurisdictions but contributes to global risk analytics, especially to the field of agribusiness, where data privacy is a contentious topic that meets land tenure, indigenous sovereignty and overall geopolitics.

Synthesis and Implications

The above three cases exhibit the fact that AI can significantly contribute to ethical sourcing but only if so. First, there is need for organizational will in terms of giving emphasis to ethics and efficiency as well. In the absence of this commitment, AI tools peril to serve as instruments for the strengthening of cost-saving logics at the expense of the ethical ends. Second, these two things, that is; data integrity and consent, are non-negotiable. Validity of AI insight lies in the reliability of collection of data and reverence of stakeholder autonomy either through supplier scorecards, traceability tools or through environmental monitoring systems. Third, hybrid models combining AI with human oversight, participatory governance, and contextual knowledge have a more satisfactory performance than purely automated ones. Moreover, these case studies highlight that ethical AI sourcing is not a prescription method but a tailor-made process. Ethical risks and technological affordances differ greatly in industries. In apparel, labor rights dominate; in electronics, origin of minerals and conflict areas are critical, in agribusiness, environmental degradation and issues of land governance are top. Therefore, in terms of the deployment of AI, it needs to be adjusted not only to the architecture of the supply chain, but to the individual moral economies and regulatory scenery under which organizations are functioning. Finally, AI promises tools to strengthen the visibility of the supply chain and ethical performance, its success calls for a wider ecosystem of accountability grounded in corporate values, voice of stakeholders and unyielding desire for justice. These are early zeitgeists of the cases that give necessary lessons to the companies to make their digital innovation synonymous with their role in ethical responsibility in this global sourcing.

REGULATORY LANDSCAPES AND ETHICAL COMPLIANCE STANDARDS

The ethical compliance environment in global sourcing has undergone a paradigm shift in the few years, resulting from coming together of regulatory pressure and stakeholder activism along with swift technological evolution. With artificial intelligence (AI) systems becoming ingrained in procurement and supply chain management, firms are growing under the crosshairs of public pressure for them to uphold their technological adoption in line with what is legal and ethical. This alignment is no longer an issue of corporate social responsibility but becomes a legal necessity in many jurisdictions. At the center of this emerging regime, are revolutionary legislative tools like the European Union's Corporate Sustainability Due Diligence Directive (CSDDD), Germany' s Supply Chain Due Diligence Act, and the United Kingdom's Modern Slavery Act While the Supply Chain Due Diligence Act (LkSG) adopted by Germany is one of the strictest legal infrastructures in terms of ethical procurement to date, it went into effect in 2023. It requires companies with more than 3,000 employees to conduct risk assessments and introduce preventative measures and to publish their due diligence work done alongside their entire supply chain, i.e., including indirect suppliers. Remarkably, the law stipulates that the firms must ensure monitoring risks of forced labor, child labor, occupational safety, and degradation of the environment (Schilling-Vacaflor & Lenschow, 2022).

The EU's Corporate Sustainability Due Diligence Directive (CSDDD), provisional agreement of 2024 and to be in effect by 2026, dramatically enlarges the regulatory expectations for ethical sourcing. It requires the firms who operate in the EU with 500 employees or €150 million in net turnover to ensure that they address deleterious outcomes of human rights and the environment all over their supply chains (Emon, 2025b; Schacherer, 2024). What is especially pertinent in the context of AI is prominence given to transparency and traceability – two abilities, which AI systems allegedly can enhance. Yet it is not enough to use algorithmic tools solely: companies cannot do that. The directive emphasizes the need for stakeholder engagement, grievance mechanisms, and board-level accountability for due diligence outcomes emphasizing the need for AI to be enacted in a larger governance arrangement. The UK's Modern Slavery Act, even though it is older and narrower, is an integral part of the landscape of ethical sourcing. The Act obliges companies, working in the UK with a turnover of at least £36 million to submit annual statements describing efforts made to prevent slavery and human trafficking to take place in their supply chain. Although it does not have as strong enforcement mechanisms as with the LkSG or CSDDD, recent amendments and proposed updates do indicate a shift towards higher levels of expected compliance, including more precise direction on risk assessment methodologies and how they may lead to penalties for non-compliance (Georgiadis & Poels, 2022). When applied to evaluate high-risk areas or supplier profiles, AI tools employed in evaluating conditions of labor needs thus not only to be efficient but also require ethical trustworthiness.

Apart from governing legislation, voluntary frameworks still are a dominant force to be reckoned with in guiding corporate ethics in AI-enabled sourcing. The revised in 2023 OECD Guidelines for Multinational Enterprises advanced explicit mention of the challenges of the emerging technologies e.g. AI. They incentivize firms to take on responsible business conduct by socializing transparency, fairness, and stakeholder inclusion into any technological deployment (Riedmann-Streitz et al.,

2025). Similarly, UN Global Compact also advocates for the ten principles of human rights, labor, environment, and anti- corruption in which the normative direction for firms on the path to harmonizing their AI practices with global sustainability goals is provided. The significance of these voluntary frameworks is not limited to their moral clout but also is expanding in terms of shaping the expectations of investors and measurement of reputational indicators.

BEST PRACTICES FOR EMBEDDING ETHICAL AI IN SUPPLY CHAINS

The role played by such artificial intelligence (AI) systems in determining procurement strategies, management of the supplier, and compliance monitoring has significantly swelled up as AI systems found their place in the global supply chain. Nevertheless, the proper utilization of these tools is a still undecided and poorly explored field. AI is not ethically neutral, it represents data that it is trained in, assumptions of its designers, and socio-economic conditions in which it is implemented. Therefore, integration of moral principles into the life cycle of AI systems applied in sourcing and procurement is not only a normative obligation but a strategic requirement in the emerging transparent and accountability-based world economy. Based on insights from cross-sectors, this discussion describes the five best interrelated practices that have become essential for operationalizing ethical AI in supply chains. These practices are not technocratic super settings but/and key elements of the responsible governance of algorithmic systems influencing thousands of suppliers, workers, and communities around the world.

Establishing Cross-functional Ethical Review Boards

The establishment of cross-functional ethical review boards is one of the most essential moves to take in embedding ethical AI. These boards pool together varied talents from AI developers, procurement gurus, legal advisors, ethics and civil society representatives to study the ethical impact of AI tools throughout their utility lifetime. It is to make sure that procurement algorithms will not bring efficiency at the expense of fairness, inclusivity, and human rights. Unlike the usual oversight bodies that tend to act reactively, these review boards should weigh in at the stage of design of models of AI, both in terms of technical performance and social impact. For example, Microsoft and Unilever have embedded ethical oversight frameworks into their AI governance frameworks to check their sourcing algorithm regularly on unintended consequences (Padmapriya & Jayamala, 2025). Importantly, such boards need to be authoritative and autonomous in a way that should allow stopping deployment or requiring a complete design overhaul at the first signs of potential ethical risks.

Conducting Regular Bias Audits and Applying Fairness Metrics

Procurement algorithms that measure performance of suppliers, calculate risk, or suggest sourcing plans can be affected by biases in the form of the past. For instance, if the previous procurement datasets discriminate large multinational suppliers or discriminate in a way that does not include women- or minority-owned businesses, AI models trained over such data are likely to reproduce and institutionalize the patterns (Emon & Khan, 2025a). To fight this, organizations should ensure to carry out bias audits regularly as well as incorporate fairness metrics into their algorithmic evaluation systems. These audits ought to have statistical bias and disparity in outcomes between demographics of suppliers considered. Tools like SHAP (Shapley Additive Explanations) or LIME (Local Interpretable Model-Agnostic Explanations) enable procurement teams to find out about the variables that are causing the algorithm to make decisions and whether the drivers are ethically justifiable or not.

Collaborative Engagement with Suppliers and Local Communities

AI systems applied in the context of supply chains tend to have an impact on stakeholders outside the design and decision-making process, especially in the context of the Global South. These involve suppliers, factory workers, and local communities who could be alienated from procurement by limitations of data, problems of social-political bias, or infrastructural divide. This can be overcome through Ethical AI deployment that involves participatory design process of involving such stakeholders. Participatory design of AI (a long-emphasized feature in HCI and development scholarship), means that the objective, assumptions and constraints of AI are co-developed with those that AI will affect. For instance, Fairphone's traceability systems for conflict-free minerals in sub-Saharan Africa were developed in partnership with local cooperatives and NGOs whereby the model accommodates the contextual realities and creates trust. But involving local actors also enhances data quality and relevance of the model which leads to enhancement of outcomes in terms of ethics and operational.

Ensuring Transparency in AI-Driven Sourcing Decisions

Transparency is one of the fundamental ethics of AI, but it needs to extend beyond the technical explanations to institutional and procedural clarity. In the context of procurement, it means ensuring that AI systems are documentation of how the systems impact sourcing decisions, the criteria they utilize, and avenue for stakeholders to challenge or appeal algorithmic outcomes being clearly documented. Global regulatory frameworks are already starting to demand such disclosures. For example, the proposed Artificial Intelligence Act by the European Union requires organizations to publish the impact assessments and to keep audit trails for high-risk AI applications, including those in the supply chains (Brintrup et al., 2024). Companies that have already gone ahead of this curve like IBM and SAP are already publishing AI transparency reports and building internal dashboards, so that procurement professionals can interrogate the basis for automated recommendations. In addition, transparency mechanisms must provide grievance redressal systems to enable suppliers and workers to challenge decisions of negative impact on them. This builds a feedback mechanism that strengthens accountability and enables a constant improvement of AI systems.

CHALLENGES AND FUTURE TRAJECTORIES

The application of ethical AI into global SCM has alluring potential with caution, but also in face of a set of constant and changing challenges. Although many corporations and regulatory regimes realize the use of AI systems to ensure transparency, traceability, and accountability in the procurement and sourcing, the implementation of ethical AI is ridden with structural, technical, epistemic, and philosophic challenges. These issues make it more challenging to translate the ethical principles into practice and bring up wider ones regarding the possibility of embedding the ethics into the systems, which are originally designed to be efficient and optimizing.

Persistent Challenges in Operationalizing Ethical AI

In the front line of these challenges are data asymmetries, which simply refers to the uneven distribution of access to quality data among stakeholders up and down the supply chains. Large multinational corporations tend to have huge data infrastructures, while small suppliers especially in the Global South may not have digital capacity to provide or manage real time data (Emon, 2025a). This imbalance is the way in which AI systems assess and rank suppliers, instilling structural inequalities in the name of algorithmic impartiality. Such asymmetries cannot be left to dumb chance, and may lead to the marginalization of suppliers, who are technically compliant, but digitally invisible. A second and associated problem can be found in the technological monopolization of the development of AI. Some tech firms, mostly into the Global North, control the creation of supply chain AI platforms leaving people worried about epistemic colonialism and lack of sensitivity of context in AI solutions. These platforms could be designed to assume certain cultural assumptions for/from which the dominant groups might not be aware of or prioritize. The socio-economic realities of various sourcing regions may not aptly exist in such culture. In addition, by proprietary nature of these systems, transparency and auditability are regularly inhibited, preventing the stakeholder scrutiny and democratic governance.

Future Directions: Toward a Reflexive and Inclusive AI Ethic

To overcome these challenges, the future paths for AI ethics should be aimed at the three interlinked priorities: participatory design, context-sensitive auditing, and inclusive innovation. These priorities speak of a turning point in moving from the technocratic rule towards the more democratic and reflexive models of AI development and deployment. Participatory design ensures that all the stakeholders that have been affected by the AI systems are actively involved in the development and deployment of the AI systems. It includes suppliers, labor unions, representatives from the community, and civil societies. This model understands that the majorly affected by algorithm decisions should have a voice in the structure of these decisions. These methods are particularly important in sourcing settings where decisions about procurement may influence livelihoods, labor, and well-being of the community (Esmaeilzadeh, 2024). Co-design processes are also valuable in increasing the relevance, usability and legitimacy for AI, developing trust and accountability within the system from the beginning. Context-sensitive auditing implies the creation of AI evaluation frameworks considering the norms, rules of law, and socio-economic conditions of the location. Instead of the universal measures of fairness or compliance, auditing should vary to accommodate the diversity of global supply chains. Finally, inclusive innovation means the wider reconfiguration of AI ecosystems to enable greater participation by its users in the design and governance of AI tools. This entails nurturing AI hubs in the Southern world, as well as funding local data infrastructures along with joint training in technology, ethics, and purchasing. Ethical AI cannot be maintained simply by regulation alone. It needs institutional contexts which incentivize ethical leadership, inter-sector collaboration, and public deliberation (Khan & Hasan Emon, 2024). Research institutions, development bodies, as well as corporate consortiums all have work to do in incubating such ecosystems.

RECOMMENDATIONS

This chapter combines the above discussion of ethical intelligence in AI-enabled supply chains and suggests a catalog of essential suggestions that may further be used to develop responsible sourcing practices. The implications of the findings concerning the initial characterization of ethical intelligence as a socio-technical issue that exists within the context of digital supply networks and the subsequently used demonstration of the strengths of AI in terms of transparency, risk assessment, and bias mitigation highlight the importance of understanding that ethical AI is a complex activity instead of a single technological solution. Therefore, a multi-dimensional approach that incorporates ethics as a pillar of AI development and deployment is needed. First, there should be formation of cross-disciplinary governance frameworks to gather technical experts, contracting leaders, ethicists, legal attorneys and persons' representing impacted communities. It is only by keeping heterogeneous views under persistent watch that one can guarantee that AI tools will maximize, where possible, not only efficiency but social justice, environmental sustainability, and human dignity (Khan et al., 2024). These governing institutions must have clear and decisional powers to make interventions at important stages throughout the AI lifecycle, such as its design, its deployment, and its monitoring and continual improvement. Considering that there is always the threat of algorithmic bias, regular bias testing, and fairness evaluation should be a part of the regular AI-based procurement tool administration. These explainable AI approaches to bias audits support discovering the dark sides of object-based decision-making criteria and safeguard against biased approaches toward suppliers, particularly those marginalized by not being available to the commercial procurement process (Murikah et al., 2024). The need to regularly update the data stewardship to reflect the socio-economic diversity and changing ethical standards through updates on training datasets is therefore essential which takes it beyond the shortfalls of the once off audits with no sustainability.

Aware of the potential of AI to cut through layers of political and bureaucratic complexity, companies ought to institutionalize new participatory design regimes through effective involvement of suppliers, laborers, and neighborhood locals. This type of inclusivity complements its legitimacy and contextual significance and can result in the development of trust and common ownership of the ethical sourcing targets (Díaz-Rodríguez et al., 2023). Reflective cycles and collaborative spaces provide the platform to discuss, redress grievances, and improve themselves, and thus reduce the risk in technocratic tendencies that could alienate even the most complicated relations to the algorithmic realm. Transparency is an essential component of ethical AI deployment, and it should not concern explainable algorithms only but also full transparency of AI impact on the procurement decisions and risk assessment. Accountability to the regulators and civil society is evidenced within governance processes of reporting to the public, ethical impact assessment, and supplier feedback channels. Proactive transparency places companies in a situation where they can follow and adhere to the emerging regulatory system of the EU AI

Act and the Corporate Sustainability Due Diligence Directive, which focuses on creating strict documentation and auditability of the social and environmental effects of AI (Dempere et al., 2024). Ongoing learning architecture that is integrated into AI system is also critical to achieving matching between model delivery and evolving ethical values, environmental environment, and labour rights environments. The idea of ethical AI must be understood as the living ability which develops in parallel with the larger socio-technical environment.

Policy wise, governments and multilateral should focus on inclusive policies and training opportunities to SMEs and suppliers within the less-digitized territories though AI standards. The reduction of digital divides and meaningful contribution to the sphere of AI governance prevents an advent of new marginalization and contributes to the equitable spread of the benefits of ethical AI. Moreover, by encouraging ethical sourcing-specific open-source AI tools, such financing will help to make it more democratic and develop a culture where innovation responds to different cultural and natural conditions. Finally, research community plays an extremely important role in the further development of context-sensitive auditing and participatory design approaches, thus finding the compromise between technical precision and socio-cultural complexity. Hybrid frameworks of assessment that incorporate quantitative measures of fairness alongside qualitative measures of impacts on stakeholders promise to go beyond the reductive thinking about ethical AI. The combination of computer scientists, ethicists, supply-chain experts, and social scientists in a joint effort is essential to produce practical knowledge and convert research on action.

CONCLUSION

In the era characterized by the coalescence of globalized trade, fragility of the environment, and increasing calls for social accountability, this chapter examined critically how artificial intelligence cannot be only a tool of optimization, but rather transformation of ethics in supply chains. The use of AI in procurement and logistics if sensitively designed vis a vis the inclusive principles can change the notional sourcing practices long ignoring the labor welfare, ecological stewardship and stakeholder justice. Through the different thematic conversations that ranged from AI informed risk assessments, data ethics, regulatory environments, and real-world case studies, it comes out that there is no such thing as a checklist for ethical sourcing that can be further automated by tech wizardry. AI, in the strongest of its incarnations, works as a catalyst: it enhances visibility, accountability, and responsiveness, but it does not substitute the requirement for human judgment, local context, as well as normative deliberation. Technology can uncover risk patterns, locate blind spots, and suggest interventions, but the validity of those interventions depends on the values that support them, and the voices of those making such interventions. This chapter has also brought out the fact that the ethical use of AI also depends on the development of institutional and cultural capabilities. Transparent algorithms, bias audits, participatory systems design and strong governance frameworks are not bells and whistles; they are the ways to deliver responsible AI. They are the framework on which ethical intelligence is mounted. Ethical AI in supply chains is not actually a destination in the first place, but an ongoing process of negotiation, reflection and recalibration, almost a "motion" of ethics that ambles on parallel with advancement in technology and changing global expectations. Yet, significant challenges remain. Data asymmetries, epistemic injustice, low algorithmic literacy of the procurement actors, and the economic barriers faced by the smaller firms in the Global South make the ideal of equitable AI complicated. And not only, the philosophical debate on optimization vs. ethics allows considering the question of whether our current AI paradigms are truly in sync with the relational, contextual, and frequently non-quantifiable nature of ethical sourcing.

REFERENCES

Al Khoury, G., Thrassou, A., Papasolomou, I., & Vrontis, D. (2023). The requisite role of emotional intelligence in customer service in the retail banking sector. *The International Journal of Organizational Analysis*, *31*(7), 3315–3347. DOI: 10.1108/IJOA-04-2022-3229

Alioto, M. (2022). From Less Batteries to Battery-Less: Enabling A Greener World through Ultra-Wide Power-Performance Adaptation down to pWs. ESSCIRC 2022-IEEE 48th European Solid State Circuits Conference (ESSCIRC), 33–40. DOI: 10.1109/ESSCIRC55480.2022.9911465

Alogogianni, E., & Virvou, M. (2023). Interpretable Machine Learning for Undeclared Work Prediction. 2023 14th International Conference on Information, Intelligence, Systems & Applications (IISA), 1–8. DOI: 10.1109/IISA59645.2023.10345943

Asif, M., Searcy, C., & Castka, P. (2022). Exploring the role of industry 4.0 in enhancing supplier audit authenticity, efficacy, and cost effectiveness. *Journal of Cleaner Production*, *331*, 129939. DOI: 10.1016/j.jclepro.2021.129939

Behera, J., & Sharma, R. (2024). Understanding superstar firms' phenomenon. *Competitiveness Review*. Advance online publication. DOI: 10.1108/CR-09-2023-0231

Bibri, S. E., & Allam, Z. (2022). The Metaverse as a virtual form of data-driven smart cities: The ethics of the hyper-connectivity, datafication, algorithmization, and platformization of urban society. *Computational Urban Science*, *2*(1), 22. DOI: 10.1007/s43762-022-00050-1 PMID: 35915731

Brintrup, A., Kosasih, E., Schaffer, P., Zheng, G., Demirel, G., & MacCarthy, B. L. (2024). Digital supply chain surveillance using artificial intelligence: Definitions, opportunities and risks. *International Journal of Production Research*, *62*(13), 4674–4695. DOI: 10.1080/00207543.2023.2270719

Calacci, D., & Pentland, A. (2022). Bargaining with the Black-Box. Proceedings of the ACM on Human-Computer Interaction, 6(CSCW2), 1–24. DOI: 10.1145/3570601

Centobelli, P., Cerchione, R., Del Vecchio, P., Oropallo, E., & Secundo, G. (2022). Blockchain technology for bridging trust, traceability and transparency in circular supply chain. Information \& Management, 59(7), 103508.

Chisita, C. T., Durodolu, O. O., & Rusero, A. M. (2025). Data capitalism in the milieu of the surveillance economy: What can libraries do? *IFLA Journal*, *51*(2), 339–349. Advance online publication. DOI: 10.1177/03400352241286170

Delabre, I., Alexander, A., & Rodrigues, C. (2020). Strategies for tropical forest protection and sustainable supply chains: Challenges and opportunities for alignment with the UN sustainable development goals. *Sustainability Science*, *15*(6), 1637–1651. DOI: 10.1007/s11625-019-00747-z

Dempere, J., Udjo, E., & Mattos, P. (2024). The Entrepreneurial Impact of the European Directive on Corporate Sustainability Due Diligence. *Administrative Sciences*, *14*(10), 266. DOI: 10.3390/admsci14100266

Díaz-Rodríguez, N., Del Ser, J., Coeckelbergh, M., López de Prado, M., Herrera-Viedma, E., & Herrera, F. (2023). Connecting the dots in trustworthy Artificial Intelligence: From AI principles, ethics, and key requirements to responsible AI systems and regulation. *Information Fusion*, *99*, 101896. DOI: 10.1016/j.inffus.2023.101896

dos Santos Sequeira, A. C. (2024). A Step Towards Corporate Sustainability—The New Directive on Due Diligence (pp. 17–33). DOI: 10.1007/978-3-031-77531-4_2

Emon, M. M. H. (2025a). The Mediating Role of Supply Chain Responsiveness in the Relationship Between Key Supply Chain Drivers and Performance: Evidence from the FMCG Industry. *Brazilian Journal of Operations & Production Management*, *22*(1), 2580. DOI: 10.14488/BJOPM.2580.2025

Emon, M. M. H. (2025b). Leveraging AI on Social Media & Digital Platforms to Enhance English Language Teaching and Learning. In AI-Powered English Teaching (pp. 211–238). IGI Global Scientific Publishing. DOI: 10.4018/979-8-3373-1952-0.ch007

Emon, M. M. H. (2025c). Overcoming Hurdles: Navigating Challenges in the Adoption of Smart Logistics. In Emerging Trends in Smart Logistics Technologies (A. Hlali, pp. 197–228). IGI Global Scientific Publishing. DOI: 10.4018/979-8-3373-2434-0.ch007

Emon, M. M. H., & Chowdhury, M. S. A. (2025). AI and IoT-Powered Smart Logistics: Transforming Supply Chains for Efficiency and Sustainability. In Emerging Trends in Smart Logistics Technologies (A. Hlali, pp. 45–72). IGI Global Scientific Publishing. DOI: 10.4018/979-8-3373-2434-0.ch002

Emon, M. M. H., & Khan, T. (2025a). The mediating role of attitude towards the technology in shaping artificial intelligence usage among professionals. *Telematics and Informatics Reports*, *17*, 100188. DOI: 10.1016/j.teler.2025.100188

Emon, M. M. H., & Khan, T. (2025b). The transformative role of Industry 4.0 in supply chains: Exploring digital integration and innovation in the manufacturing enterprises. *Journal of Open Innovation*, *11*(2), 100516. DOI: 10.1016/j.joitmc.2025.100516

Emon, M. M. H., & Khan, T. (2025c). The Rise of Circular Supply Chains: Recycling and Closed-Loop Systems Could Mitigate Inflationary Pressures. In Supply Chain Disruptions and Impact on Global Inflation (pp. 195–226). IGI Global. DOI: 10.4018/979-8-3373-1982-7.ch008

Emon, M. M. H., & Khan, T. (2025d). Corporate Social Responsibility for Sustainable Development: A Systematic Review of Business Contributions to Address Global Challenges. *Review of Business and Economics Studies*, *13*(2), 6–39. DOI: 10.26794/2308-944X-2025-13-2-6-39

Esmaeilzadeh, P. (2024). Challenges and strategies for wide-scale artificial intelligence (AI) deployment in healthcare practices: A perspective for healthcare organizations. *Artificial Intelligence in Medicine*, *151*, 102861. DOI: 10.1016/j.artmed.2024.102861 PMID: 38555850

Garg, R., Kiwelekar, A. W., Netak, L. D., & Bhate, S. S. (2021). *Potential Use-Cases of Natural Language Processing for a Logistics Organization.*, DOI: 10.1007/978-3-030-68291-0_13

Georgiadis, G., & Poels, G. (2022). Towards a privacy impact assessment methodology to support the requirements of the general data protection regulation in a big data analytics context: A systematic literature review. *Computer Law & Security Review*, *44*, 105640. DOI: 10.1016/j.clsr.2021.105640

Golchha, C., & Nagariya, R. J. (2025). Exploring Consumer Choices and Shopping Patterns. In Cultural Sensitivity, Cross-Border Logistics, and E-Commerce in Global Marketing (pp. 181–214). IGI Global. DOI: 10.4018/979-8-3373-0528-8.ch009

Hanna, M. G., Pantanowitz, L., Jackson, B., Palmer, O., Visweswaran, S., Pantanowitz, J., Deebajah, M., & Rashidi, H. H. (2025). Ethical and Bias Considerations in Artificial Intelligence/Machine Learning. *Modern Pathology*, *38*(3), 100686. DOI: 10.1016/j.modpat.2024.100686 PMID: 39694331

Horne, J. (2023). Towards a Solution: The Responses (pp. 323–395). DOI: 10.1007/978-3-031-25444-4_12

Khan, T., & Emon, M. M. H. (2025a). Preparing for the Next Disruption: Lessons Learned From Past Crises and Building a Playbook for Future Supply Chain Shocks. In Supply Chain Disruptions and Impact on Global Inflation (pp. 253–286). IGI Global Scientific Publishing. DOI: 10.4018/979-8-3373-1982-7.ch010

Khan, T., & Emon, M. M. H. (2025b). The role of digital supply chain practices in enhancing firm performance: Insights from the manufacturing sector of Bangladesh. *Brazilian Journal of Operations & Production Management, 22*(2), 2493. DOI: 10.14488/BJOPM.2493.2025

Khan, T., Emon, M. M. H., & Nath, A. (2024). Quantifying the Effects of AI-Driven Inventory Management on Operational Efficiency in Online Retail. 2024 27th International Conference on Computer and Information Technology (ICCIT), 2092–2097. DOI: 10.1109/ICCIT64611.2024.11021996

Khan, T., Emon, M. M. H., & Rahman, S. (2024). Marketing Strategy Innovation via AI Adoption: A Study on Bangladeshi SMEs in the Context of Industry 5.0. 2024 6th International Conference on Sustainable Technologies for Industry 5.0 (STI), 1–6. DOI: 10.1109/STI64222.2024.10951050

Khan, T., & Hasan Emon, M. M. (2024). Determinants of AI Image Generator Adoption Among Marketing Agencies: The Mediating Effects of Perceived Usefulness. 2024 IEEE 3rd International Conference on Robotics, Automation, Artificial-Intelligence and Internet-of-Things (RAAICON), 177–182. DOI: 10.1109/RAAICON64172.2024.10928548

Kougkoulos, I., Cakir, M. S., Kunz, N., Boyd, D. S., Trautrims, A., Hatzinikolaou, K., & Gold, S. (2021). A Multi-Method Approach to Prioritize Locations of Labor Exploitation for Ground-Based Interventions. *Production and Operations Management, 30*(12), 4396–4411. DOI: 10.1111/poms.13496

Labadie, C., & Legner, C. (2023). Building data management capabilities to address data protection regulations: Learnings from EU-GDPR. *Journal of Information Technology, 38*(1), 16–44. DOI: 10.1177/02683962221141456

MacFarlane, J. M. (2020). The Techno-Centred Imagination. In *Transhumanism as a New Social Movement* (pp. 205–233). Springer International Publishing., DOI: 10.1007/978-3-030-40090-3_8

Mangla, S. K., Kazançoğlu, Y., Yıldızbaşı, A., Öztürk, C., & Çalık, A. (2022). A conceptual framework for blockchain-based sustainable supply chain and evaluating implementation barriers: A case of the tea supply chain. *Business Strategy and the Environment, 31*(8), 3693–3716. DOI: 10.1002/bse.3027

McGrath, P., McCarthy, L., Marshall, D., & Rehme, J. (2021). Tools and Technologies of Transparency in Sustainable Global Supply Chains. *California Management Review, 64*(1), 67–89. DOI: 10.1177/00081256211045993

Mosallanezhad, B., Hajiaghaei-Keshteli, M., Smith Cornejo, N. R., & Rodríguez Calvo, E. Z. (2023). An IoMT platform for an integrated sustainable energy-efficient disaster relief supply chain to prevent severity-driven disruptions during pandemics. *Journal of Industrial Information Integration*, *35*, 100502. DOI: 10.1016/j.jii.2023.100502

Murikah, W., Nthenge, J. K., & Musyoka, F. M. (2024). Bias and ethics of AI systems applied in auditing - A systematic review. *Scientific African*, *25*, e02281. DOI: 10.1016/j.sciaf.2024.e02281

Mvile, B. N., & Bishoge, O. K. (2024). Mining industry's potential for community development, challenges, and way forward in the East Africa community. A review. *Local Development & Society*, •••, 1–24. DOI: 10.1080/26883597.2024.2423948

Padmapriya, S. S., & Jayamala, C. (2025). *Strategic Policy Recommendations and Implications for Effective Management in Organizational and Regulatory Frameworks.*, DOI: 10.1007/978-3-031-84909-1_12

Patel, D., Sahu, C. K., & Rai, R. (2024). Security in modern manufacturing systems: Integrating blockchain in artificial intelligence-assisted manufacturing. *International Journal of Production Research*, *62*(3), 1041–1071. DOI: 10.1080/00207543.2023.2262050

Pawar, V., Chavan, P., Vhatkar, A., Khang, A., & Gawankar, S. (2025). Green Transportation and Moral Licensing: Navigating Ethical Challenges with Artificial Intelligence (AI) and Automation (pp. 527–562). DOI: 10.1007/978-3-031-72617-0_28

Popescu, A., Polat-Erdeniz, S., Felfernig, A., Uta, M., Atas, M., Le, V.-M., Pilsl, K., Enzelsberger, M., & Tran, T. N. T. (2022). An overview of machine learning techniques in constraint solving. *Journal of Intelligent Information Systems*, *58*(1), 91–118. DOI: 10.1007/s10844-021-00666-5

Pria, S., Al Rubaie, I., & Prasad, V. (2024). *Enhancing Business Intelligence Through AI-Driven Integration of Sustainability Metrics via ESG Factors.*, DOI: 10.4018/979-8-3693-2185-0.ch004

Rezaei, M., Pironti, M., & Quaglia, R. (2024). AI in knowledge sharing, which ethical challenges are raised in decision-making processes for organisations? *Management Decision*. Advance online publication. DOI: 10.1108/MD-10-2023-2023

Richey, R. G.Jr, Chowdhury, S., Davis-Sramek, B., Giannakis, M., & Dwivedi, Y. K. (2023). Artificial intelligence in logistics and supply chain management: A primer and roadmap for research. *Journal of Business Logistics*, *44*(4), 532–549. DOI: 10.1111/jbl.12364

Riedmann-Streitz, C., Streitz, N., Antona, M., Marcus, A., Margetis, G., Ntoa, S., Rau, P.-L. P., & Rosenzweig, E. (2025). How to Create and Foster Sustainable Smart Cities? Insights on Ethics, Trust, Privacy, Transparency, Incentives, and Success. *International Journal of Human-Computer Interaction*, *41*(4), 2491–2522. DOI: 10.1080/10447318.2024.2325175

Schacherer, S. (2024). Building responsible and sustainable supply chain frameworks: Limits of international investment law and the CSR initiatives taken by the EU and China. *Asia Pacific Law Review*, *32*(2), 347–369. DOI: 10.1080/10192557.2024.2349825

Singh, R. K., Agarwal, A., Dwivedi, D. K., & Yadav, S. (2024). Ethical Claims and Enforcement in Legal Metrology in India. In Handbook of Quality System, Accreditation and Conformity Assessment (pp. 1–23). Springer Nature Singapore. DOI: 10.1007/978-981-99-4637-2_51-1

Sreenivasan, A., & Suresh, M. (2022). Modeling the enablers of sourcing risks faced by startups in COVID-19 era. *Journal of Global Operations and Strategic Sourcing*, *15*(2), 151–171. DOI: 10.1108/JGOSS-12-2020-0070

Tao, R., Wu, J., & Zhao, H. (2023). Do Corporate Customers Prefer Socially Responsible Suppliers? An Instrumental Stakeholder Theory Perspective. *Journal of Business Ethics*, *185*(3), 689–712. DOI: 10.1007/s10551-022-05171-5

Thomas, J., Patil, R. S., Patil, M., & John, J. (2023). Addressing the Sustainability Conundrums and Challenges within the Polymer Value Chain. *Sustainability (Basel)*, *15*(22), 15758. DOI: 10.3390/su152215758

Uvet, H., Park, A., Dickens, J., Oh, J., & Hazen, B. (2025). How blockchain technology utilization influences corporate social responsibility through supply chain transparency and the role of supplier risk. *International Journal of Logistics Management*, *36*(4), 1195–1221. Advance online publication. DOI: 10.1108/IJLM-04-2024-0245

Vijaykumar, V., Mercy, P., Lucia Agnes Beena, T., Leena, H. M., & Savarimuthu, C. (2024). Convergence of IoT, Artificial Intelligence and Blockchain Approaches for Supply Chain Management. In Blockchain, IoT, and AI Technologies for Supply Chain Management (pp. 45–89). Apress. DOI: 10.1007/979-8-8688-0315-4_2

von Eschenbach, W. J. (2021). Transparency and the Black Box Problem: Why We Do Not Trust AI. *Philosophy & Technology*, *34*(4), 1607–1622. DOI: 10.1007/s13347-021-00477-0

Xu, J., Guo, S., Xie, D., & Yan, Y. (2020). Blockchain: A new safeguard for agri-foods. *Artificial Intelligence in Agriculture*, *4*, 153–161. DOI: 10.1016/j.aiia.2020.08.002

KEY TERMS AND DEFINITIONS

Ethical Sourcing: Purchasing of goods and services in a way that recognizes and upholds human rights, proper working standards, and sustainable use of the environment.

Artificial Intelligence (AI): Various tools and techniques that allow machines to show similar skills to humans like learning, analyzing, and making decisions.

Supply Chain Visibility: The capability to keep tabs on products, components, and suppliers all along the supply chain to maintain openness and follow the rules.

Algorithmic Bias: When the design or the information behind AI algorithms is unfair, discrimination may happen during procurement and in the selection of suppliers.

Responsible Procurement: When making purchase decisions, including ethical, social, and environmental factors are considered.

Data Ethics: The guidelines for handling data in a responsible manner, mainly focusing on getting consent, protecting privacy, and providing transparency.

Sustainability: Using resources and processes in a way that addresses present needs and still allows future generations to meet their own.

Governance Frameworks: The framework and policies responsible for guiding and making sure AI and ethical rules are included in the supply chain.

Section 4

AI Ethics and the Financial Sector

Chapter 9
Optimizing Tax Expenditure and Mobilizing Tax Revenue Through Ethical Artificial Intelligence

Lahcen Ettahiri
https://orcid.org/0009-0000-1002-168X
Ibn Tofail University, Morocco

Lotfi Benazzou
Ibn Tofail University, Morocco

ABSTRACT

Tax policy is an essential tool for promoting economic growth and social development, with tax expenditures playing a central role in stimulating investment and supporting development projects. However, empirical studies show that these expenditures often fail to achieve their objectives and result in significant revenue losses for the state. In this context, artificial intelligence (AI) offers innovative solutions to optimize tax expenditures and improve tax revenue mobilization, but raises ethical issues regarding the protection of taxpayers' privacy and confidentiality. This paper examines how the integration of ethical principles moderates the relationship between tax expenditure optimization and tax revenue mobilization, using the autoregressive distributed lag (ARDL) model. Our results confirm that tax expenditure optimization has a limited impact on tax revenue mobilization and that the integration of AI ethics has a moderating role in this relationship in the positive direction to optimize tax expenditures and maximize tax revenue.

DOI: 10.4018/979-8-3373-3104-1.ch009

INTRODUCTION

In light of the rapid changes the world is experiencing across economic, social, and technological dimensions, digital transformation is emerging as a central driver in reshaping public policies and guiding development strategies. This is particularly true for developing countries, which face a dual challenge: on one hand, the need to mobilize financial resources effectively and efficiently to finance development and social policies; and on the other, the imperative to rationalize public spending to enhance its effectiveness, efficiency, and impact on citizens' well-being.

There is no doubt that tax policy is one of the most important instruments of fiscal policy. This is evident in the crucial role taxes play in financing public expenditure, redistributing income, and influencing market behavior—whether by encouraging or discouraging certain economic actions (OECD, 2023). The development of the tax system has been a cornerstone in achieving the objectives of the new development model, which emphasizes social justice, economic efficiency, and sustainability. One key instrument available to governments in this regard is tax expenditure.

Tax expenditure refers to exceptional fiscal measures adopted by states to promote social and economic well-being, even though such measures often deprive the state budget of substantial financial resources (IMF, 2019). Despite this cost, tax expenditures remain vital in shaping the economic and social landscape, aiming to enhance overall well-being. They are commonly used to stimulate investment and support social initiatives. However, studies have shown that these measures frequently result in foregone revenue without necessarily achieving their intended objectives (Masiya et al., 2024).

In this context, there is an urgent need for innovative tools that can improve the efficiency of tax expenditures and enhance the mobilization of financial resources. Recently, growing attention has been given to the relationship between Artificial Intelligence (AI), tax expenditure, and tax revenue mobilization. Scholars, researchers, and public authorities alike are increasingly exploring how AI can play an interventionist role in helping rationalize fiscal policies and achieve goals comparable to those of direct public spending (Bratic, 2006).

Tax expenditure is a crucial tool used by governments to stimulate specific economic sectors or achieve social objectives. However, in many cases, it proves to be inefficient or poorly targeted. This raises the question of the true financial impact of tax expenditure on overall tax revenues (Caiumi, 2011).

In this context, the emergence of new technologies—particularly Artificial Intelligence (AI)—offers promising opportunities to optimize both tax expenditures and revenue mobilization. AI plays a transformative role in institutional and administrative reform, enabling the rapid analysis of vast datasets, the identification of complex patterns and behaviors, and the facilitation of data-driven decision-making.

Leading countries such as Estonia, India, and Brazil have begun integrating AI technologies at various stages of tax system management, including in the analysis of the economic effects of tax policies. These nations have achieved tangible results in expanding their tax base, improving the targeting of tax exemptions, and enhancing the evaluation of their effectiveness (CIAT, 2021).

AI empowers tax authorities to exploit large-scale data, identify inefficiencies and loopholes, detect unproductive tax expenditures, and mitigate the issue of ineffective fiscal spending. It opens new possibilities for restructuring tax systems and enhancing their capacity to achieve policy objectives more effectively. Countries that have adopted AI in tax administration have reported improved compliance, broader tax bases, and more efficient and equitable resource allocation.

However, integrating artificial intelligence (AI) into tax policy is not just a technical process; it requires profound institutional and structural transformation. This includes taking ethical considerations into account, particularly with regard to data protection and taxpayer privacy, establishing adequate digital infrastructure, training human resources, and modernizing tax administration practices. Indeed, the adoption of AI in the tax field raises ethical challenges related to issues of transparency, fairness, accountability, and the protection of taxpayers' rights.

This article aims to demonstrate how AI can improve the return on tax expenditures and tax revenue mobilization by applying the ARDL (Autoregressive Distributed Lag) model, based on official data and institutional reports (Saba & Monkam, 2025). It also seeks to examine the relationship between artificial intelligence, tax expenditure efficiency, and tax revenue mobilization in a broad international context, without focusing on any particular country and to show that the effectiveness of AI on tax revenue mobilization and tax expenditure optimization depends on the level of ethics applied.

This article analyzes how AI ethics can improve the performance of tax expenditures and the efficiency of tax revenue collection. By examining the interaction between AI and ethical issues, the study shows how responsible technological innovation can optimize tax policies and strengthen the overall sustainability of the tax system.

BACKGROUND

The main objective of this work is to present all existing publications concerning the importance of the ethical dimension in the effectiveness of AI in tax revenue mobilization and, subsequently, tax expenditure optimization. In-depth theoretical and empirical studies have made it possible to comprehensively assess the impact of artificial intelligence on the link between tax expenditures and tax revenue mobilization.

1. Conceptual Framework

The concept of tax revenue mobilization involves optimizing income and profit taxation to enhance tax potential and maximize the tax effort exerted by tax administrations (Mirrlees, 1971). It depends on factors such as the tax base, the level of taxpayer compliance, and the efficiency of tax administration.

The concept of tax expenditures has received widespread acceptance among economists, financial theorists, and international financial organizations. Perhaps the most influential definition comes from Professor Stanley Surrey, who first formally defined tax expenditures in 1967 as "government programs that aim to provide financial assistance through tax provisions rather than direct spending" (Shoup, 1975).

The Organization for Economic Co-operation and Development (OECD) defines tax expenditures as all tax benefits granted by legislators to encourage certain sectors or categories of taxpayers based on predefined criteria. Tax expenditures represent a significant share of public revenue losses, yet their real contribution to economic or social development is not always clear (OECD, 2010).

Moreover, mobilizing tax revenues remains a major challenge for many countries, often due to complex tax systems, tax fraud, evasion, and limited administrative capacity. However, there is no doubt that artificial intelligence presents a significant opportunity to reform tax expenditure policies and improve revenue efficiency. These findings offer decision-makers valuable recommendations for leveraging AI to optimize tax systems, public spending, budget transparency, and fiscal sustainability.

Discussions about artificial intelligence (AI) date back to the 1950s with John McCarthy, widely regarded as one of the founding fathers of AI (Andresen, 2002). AI is increasingly applied across many fields, with its technologies exerting profound effects on physical and social systems that governments and policymakers must consider.

AI focuses on creating machines that mimic human intelligence. Despite early computational limitations, the digital revolution of the 20th century accelerated AI development, leveraging advances in computer processing, big data, and machine learning algorithms. Today, AI encompasses a suite of technologies enabling machines to perform tasks that typically require human intelligence, including learning, perception, reasoning, and decision-making.

The adoption of artificial intelligence will help simplify administrative procedures within tax administrations and improve their efficiency by optimizing controls and the management of tax expenditure in an effective and rapid manner (Djellaba, and al.,2024).

The example of Uganda provides a concrete illustration of how African countries are mobilizing public digital infrastructure (PDI) to enhance the efficiency of their tax policy, while laying the groundwork for the future integration of artificial intelligence. Although AI itself is still in its infancy in this country, several digital tools deployed by the Uganda Revenue Authority (URA) foreshadow the intelligent automation of the tax system (Sanday & Nalweera, 2025).

The integration of artificial intelligence into tax policies is not limited to fraud detection or compliance. It offers a strategic opportunity to:

- Optimize the design of tax expenditures: by simulating the economic and social effects of different tax schemes, AI can help decision-makers better target measures, reduce waste, and maximize the impact of incentives.
- Improve revenue collection and mobilization: by combining predictive analysis, control automation, and service personalization, AI helps increase collection rates while reducing administrative costs.
- Strengthen tax governance: the ability to analyze tax data in real time enables greater transparency, early detection of risks, and more agile, evidence-based decision-making.

In countries such as Italy and Estonia, AI tools have already been deployed to identify undeclared income, optimize VAT collection and monitor tax credit abuse. These applications are part of the EU's wider digital tax agenda and illustrate the transformative potential of AI in rationalizing tax expenditure and improving tax revenue mobilization (Ilieva,2025).

Therefore, the integration of AI into tax policy is not simply a technological change, but a strategic reform that is essential to address long-standing challenges in tax revenue mobilization and tax incentive optimization. However, this integration requires consideration of a very important aspect, namely the ethical dimension of AI.

However, the literature has paid limited attention to the complex interaction between AI, tax expenditures, and revenue mobilization, particularly in the context of comprehensive structural reforms (Saba & Monkam, 2025). The relationship between AI and tax expenditures remains an underexplored area, especially within the framework of modern developmental models.

The concept of AI ethics refers to the principles and standards aimed at ensuring that the implementation and use of artificial intelligence technologies respect fundamental human values, such as transparency, justice, fairness, accountability, and privacy protection (Radanliev, 2025).

The figure below shows the ethical principles of AI. It synthesizes the fundamental principles necessary to develop, deploy and use AI systems ethically.

Figure 1. AI ethics principles

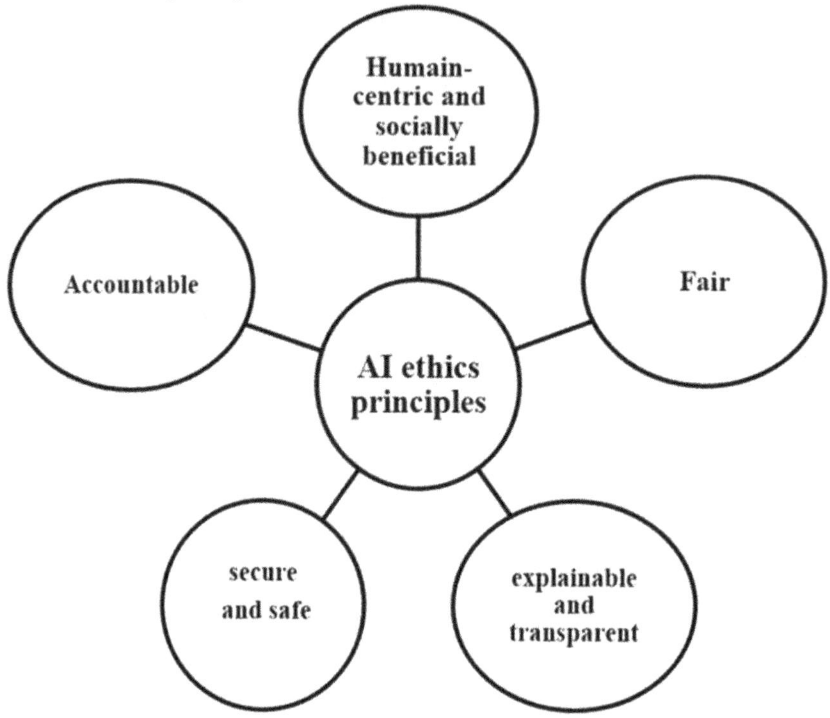

Unlike a purely technological approach, the ethics of artificial intelligence takes into account the social, economic, and political impacts that these technologies can have on individuals and society as a whole. With the development of AI in the field of taxation, its morality is being questioned. The use of algorithms in sovereign functions requires transparent governance, guarantees regarding the protection of personal data (Floridi & Cowls,2019).

These risks can undermine confidence in the tax system because citizens perceive AI as a tool for opacity, arbitrariness, or excessive surveillance. L'AI is a vehicle for transforming the relationship between taxpayers and the government, which will have a negative impact on tax revenues and tax expenditure optimization (Cantens & Tourpe,2025).

Indeed, despite these advances, several challenges remain in the implementation of AI solutions in taxation:

- Protection of personal data: the massive exploitation of sensitive data imposes strict safeguards to protect privacy and ensure taxpayers' trust.

- data privacy: AI is often used to analyze sensitive personal data of taxpayers, which raises privacy concerns.
- Technological and human complexity: the success of these systems also depends on the training of tax agents and the ability of institutions to integrate these tools into their processes.
- Digital inequalities: it is essential to ensure that digitization and AI do not reinforce inequalities in access to public tax services.

Thus, the integration of AI into tax policies is no longer just a technological choice, but a strategic necessity, the effectiveness of which depends on ethical oversight. This oversight ensures that AI improves the efficiency of public spending, promotes tax fairness, guarantees taxpayers' rights, and ensures the sustainability of tax revenues, in line with the requirements of the modern economy.

2. Theoretical Framework

Several theories provide the foundation for analyzing the relationship between AI, tax expenditure, and revenue mobilization.

a. Modern Public Finance Theory (Musgrave & Musgrave, 1989)

it views tax expenditure as a policy tool that requires careful justification regarding its objectives and expected outcomes by using AI. Technological advances like Big Data and Cloud Computing are significantly transforming tax administrations worldwide (OECD, 2016)

b. Tax Incidence Theory

It analyzes tax system performance by determining who ultimately bears the tax burden, beyond the entity legally responsible for paying the tax. For instance, a tax levied on corporations may ultimately affect consumers or workers through price or wage adjustments (Krauss & Johnson, 1972). AI ethics can model shifting effects across consumers, workers, and firms, revealing unintended outcomes.

c. Governance Theory (Hanisch et al., 2023)

emphasizes accountability and participation. AI ethics strengthens tax expenditures and tax revenue mobilization by improving transparency, risk detection, and tax engagement.

d. Social Contract Theory

The social contract theory considers that citizens agree to give up part of their freedom (for example, paying taxes) in exchange for protection, justice, and public services guaranteed by the government (Huemer,2013). The application of social contract principles via artificial intelligence (AI) in the tax system has several important implications for tax legitimacy. Furthermore, improving taxpayer confidence is another major implication (Chrug & Schiff,2025).

3. The Ethical Challenges of Artificial Intelligence

We note that AI improves the efficiency of tax revenue mobilization and tax expenditure optimization, but its integration raises important ethical questions. Issues of fairness, accountability, and confidentiality must be addressed to ensure that the adoption of advanced AI technology does not undermine tax justice or erode the trust of taxpayers, citizens, and users in tax authorities.

The application of artificial intelligence in the field of taxation, while offering significant advantages in terms of efficiency, speed, and transparency, also raises a number of ethical issues. this application poses several ethical challenges that must be addressed to ensure its responsible use (Allahrakha,2024). The issue of privacy and consent is paramount. The use of AI to analyze taxpayers' tax and behavioral data raises significant concerns regarding citizens' privacy (Mantelero,2018).

It is imperative that the collection of tax revenues and the management of tax expenditure policy, which require the collection, storage, and analysis of sensitive data, be carried out in full compliance with the right to privacy and with the informed consent of the individuals concerned. This helps to avoid excessive surveillance or exploitation of taxpayer data, thereby preserving their confidentiality and individual freedom. In addition, transparency and the application of AI algorithms represent a major challenge. AI systems can be extremely complex and operate in an opaque manner, creating an algorithmic “black box.” This opacity poses challenges for the transparency and explainability of decisions made by these systems (Doshi-Velez & Kim, 2017).

Artificial intelligence can perpetuate or even amplify existing biases in the data on which they are trained (Barocas, Hardt & Narayanan, 2019), which can lead to discriminatory decisions. It is essential to ensure that AI algorithms used in the tax field are fair and do not discriminate against certain groups of taxpayers. This requires constant monitoring and corrective mechanisms to identify and mitigate any potential bias in AI systems.

These issues require constant monitoring and operational mechanisms for AI systems to identify and mitigate any behavior that undermines the ethical dimension of AI.

MAIN FOCUS OF THE CHAPTER

Research highlights AI's potential to modernize tax administration by providing advanced technologies that optimize tax expenditure and boost tax revenue mobilization. AI fundamentally transforms modern financial systems, offering tools to maximize tax expenditures and encourage revenue mobilization (Farooq & Vatyan, 2025). This renewed focus is driven by the urgent need for transparent and effective fiscal policies, especially in low- and middle-income countries where mobilizing national resources is vital for achieving sustainable development goals (Varela et al., 2022).

Since independence, fiscal expenditures have been integral to public finance, as demonstrated by numerous laws incorporating exceptional fiscal measures (Hagemann, 2012). Recent OECD studies reveal that tax expenditures are costly and lead to significant revenue losses relative to benchmark tax systems (Hagemann, 2012).

This research aims to explore the impact of Artificial Intelligence on rationalizing tax expenditures and subsequently on the level of tax revenues collected from both direct and indirect taxes. It also seeks to propose avenues for leveraging AI's advantages in the tax expenditure and revenue mobilization nexus.

By integrating ethical considerations, the research explores in greater detail how AI ethics can enhance transparency, fairness, and trust in tax systems, thereby strengthening the relationship between tax administrations and taxpayers (Social Contract), improving tax revenues, and ensuring the socioeconomic impact of tax expenditures.

Thus, the ethical dimension of AI emerges as a moderating factor that ensures its role in rationalizing tax expenditures and improving revenue mobilization.

Issues, Controversies, Problems

Discussing the advantages of using artificial intelligence in the field of taxation, particularly to improve tax revenue collection and subsequently optimize tax expenditures, is a subject that provokes much debate, but when we add the ethical aspect of achieving these advantages while respecting taxpayers' rights, At this point, this issue is accompanied by a series of controversies and difficulties, mainly related to the ethical issues that arise from it. This explains the interest in their implementation and their impact on the ground, due to the demanding nature of efficiency, fairness, transparency, and governance.

Based on this summary, the initial debate on the role of AI in the tax field (tax expenditure policy and tax revenue mobilization mechanisms) highlights the need for a critical reassessment of the ethical dimension. Indeed, while artificial intelligence offers unprecedented opportunities to improve efficiency, combat tax fraud, and optimize tax expenditures, its effectiveness ultimately depends on the integration of ethical principles into its deployment.

Indeed, the role of AI in strengthening tax revenue mobilization and, consequently, in streamlining tax expenditures is conditioned by its ability to respect taxpayers' rights, ensure transparency, and guarantee the fairness of tax practices. Without an ethical framework, the use of AI risks exacerbating inequalities and eroding public trust in tax institutions. Therefore, the use of AI in the tax field requires not only technological innovations, but also a normative foundation that guarantees ethics in the deployment of data collected and processed by AI.

PROBLEM STATEMENT AND HYPOTHESES

It is therefore timely to problematize this subject in order to treat it scientifically and present recommendations. We formulated the following research question: "How does the ethics of artificial intelligence strengthen and moderate the relationship between tax expenditure optimization and tax revenue mobilization?"

To address this, we propose two hypotheses:

- **Hypothesis One (H0):** Tax expenditures have not an impact on improving tax revenue mobilization.
- **Hypothesis Two (H1):** AI ethics plays a positive moderating role in the relationship between tax expenditure optimization and tax revenue mobilization.

Our study highlights the crucial importance of integrating artificial intelligence into tax expenditure rationalization policy in order to improve tax revenue mobilization. This integration appears to be an essential response to current challenges, as it aims to reconcile two major objectives: increasing tax revenue while reducing the overall tax burden. This dual challenge highlights the need to adopt AI systems in a responsible approach. In other words, there is a requirement to include in-depth reflection on the ethics of artificial intelligence to ensure transparency, fairness, and efficiency in tax management.

RESEARCH METHODOLOGY

1. Epistemological Positioning

The choice of epistemological positioning legitimizes the foundations of the subject under study and is formed by selecting an epistemological paradigm and mode of reasoning. Moreover, deduction "is above all a means of demonstration that helps to extract a consequence from a general rule put to the test of facts. It is characterized by the fact that, if the hypotheses initially formulated (premises) are true, then the conclusion must necessarily be true" (Scandura & Williams, 2000). Deduction is the reasoning underlying the hypothetico-deductive approach.

Thietart insists on a reflective approach which is not limited to the collection of data, but which engages a coherent internal logic on the nature and production of knowledge to ensure the scientific rigor and credibility of the conclusions. (Turki & Triki, 2023).

In research aimed at generating knowledge, the researcher must consider what knowledge is, the principles governing their conception of knowledge, and the foundations that confer value on the knowledge produced.

Piaget's succinct definition clarifies that epistemology encompasses methodology—the study of the constitution of knowledge. For a researcher, clarifying the epistemological stance in a research project is equivalent to articulating the fundamental hypotheses that guide the process of developing and justifying knowledge within the project. Epistemological questioning thus includes methodological questioning but is not limited to it (Jiang & Li, 2020).

It is important not to confuse epistemology with methodology. While epistemology concerns the framework of knowledge and its justification, methodology refers to the study of methods used to produce knowledge, detailing how the researcher conducts investigations and the tools used for data collection and analysis. Methodology can be seen as a subset of epistemology, providing the basis for methodological inquiry.

Our research adopts a positivist epistemology combined with hypothetico-deductive reasoning (Haddar, 2010). This approach aims to study the causal relationship between an independent variable and a dependent variable, considering a moderating variable—Artificial Intelligence. Our research consists of formulating several hypotheses and testing them against empirical reality to verify their validity.

2. Presentation of the Econometric Model

A mediating or intermediary variable helps explain how or why the independent variable affects the dependent variable. It is particularly useful for researchers seeking to understand the mechanism linking two variables, where the independent variable influences the mediating variable, which in turn affects the dependent variable.

In our thesis, introducing mediating variables allows for a clearer analysis of the causal relationships between the independent and dependent variables. Baron and Kenny (1986) explain that mediation helps illuminate the mechanism underlying statistical relationships. For this reason, we have adopted a quantitative research approach using econometric analysis.

Specifically, to identify the impact of Artificial Intelligence ethics on the relationship between and tax revenues and tax expenditures, our research builds on the foundational work of Anderson et al. (1971) and Jurion (2005), who demonstrated that public authorities face budget constraints in their tax expenditure policies (Anderson, 1971; Jurion, 2004).

Our study suggests that AI technologies, through advanced data analysis, predictive capabilities, and algorithmic optimization, can enable public authorities to better assess the costs of tax incentives and optimize revenue losses. However, simply adopting AI does not guarantee the desired outcomes; its impact depends on the integration of AI ethics.

Ethical considerations such as transparency, accountability, and fairness act as a moderating factor, determining the extent to which AI-based systems translate tax data into effective decisions. The ARDL approach assumes that the two complementary time series behave similarly over time, establishing a long-term equilibrium relationship. Cointegration refers to this phenomenon where a linear combination of the series has a zero mean and constant variance.

In other words, while AI can improve the optimization of tax expenditures and mobilize the full fiscal potential, the presence of the ethical dimension ensures that these improvements are both fair and sustainable. From a theoretical perspective, economic theory provides insights into the long-term equilibrium relationships between variables. To empirically test these dynamics, the autoregressive distributed lag (ARDL) model is particularly suitable. Unlike simple long-term equilibrium models, ARDL allows for the estimation of short- and long-term effects while accounting for mixed integration orders. Its distinct analytical characteristics make it an appropriate framework for assessing how AI ethics moderates the relationship between tax mobilization and tax expenditure optimization

Due to limitations in earlier cointegration models, Pesaran and Shin (1999), Pesaran et al. (1999), and later Pesaran et al. (2001) introduced the ARDL model—a hybrid of Distributed Lag and Autoregressive models (Jurion, 2004).

This model accounts for the lagged influence of explanatory variables on the dependent variable, meaning that the effect of the explanatory variable at time t depends not only on its current value but also on its values in previous periods (t - i). It is applicable regardless of whether the regressors are integrated of order I

(0), I (1), or a mixture of both. However, it is not suitable if any of the series are integrated of order I (2).

Economic theory thus offers elements to understand long-term equilibrium relationships. This dual-level analysis is crucial for comprehending the structuring role of tax intelligence in optimizing tax expenditures and, consequently, improving the performance of tax revenue mobilization.

The ARDL model represented as follows:

$$y_t = \alpha + \sum_{\{i=1\}_i^{\{p\}\gamma} y_{\{t-i\}}} + \sum_{\{j=0\}_j^{\{r\}\beta} x_{\{t-j\}}} + \delta w_t + \epsilon_t \tag{1}$$

This model belongs to the class of dynamic models. Indeed, a dependent variable (yt) can be explained both by its own lagged values own lagged values (yt-i), by present values of the independent variables (xt) and their time-lagged values (xt-i).

Figure 2. Method selection for times series data

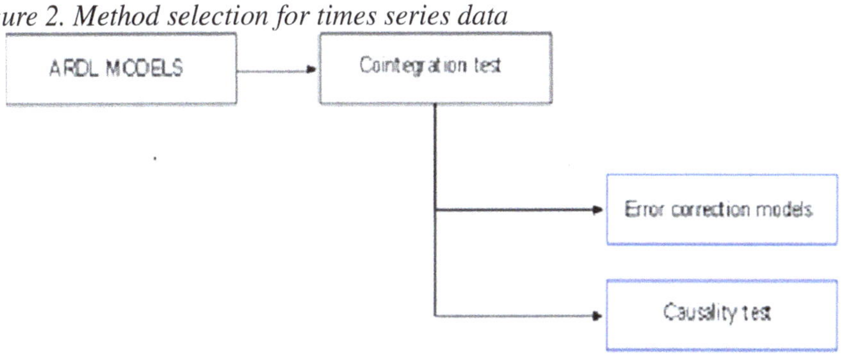

In this study, we employ the Autoregressive Distributed Lag (ARDL) model introduced by Johansen (1988). This method serves as an alternative cointegration test that examines both long-term relationships and dynamic interactions between variables (Pesaran and Shin, 1999). It offers a flexible and robust approach, particularly well-suited for small sample sizes.

The ARDL approach is especially appropriate for our research, which aims to analyze the long-term relationships and short-term dynamic interactions among artificial intelligence, tax expenditures, and tax revenue mobilization. This model enables us to assess both the immediate and long-term effects of adopting artificial intelligence on the optimization of tax incentive policies, which in turn impacts the overall performance of the tax system.

Given its widespread use in numerous studies and the nature of our data and hypotheses, we have chosen to apply this model in our analysis. Based on our literature review, we identified the following key variables for our study:

Figure 3. Definition of Variables

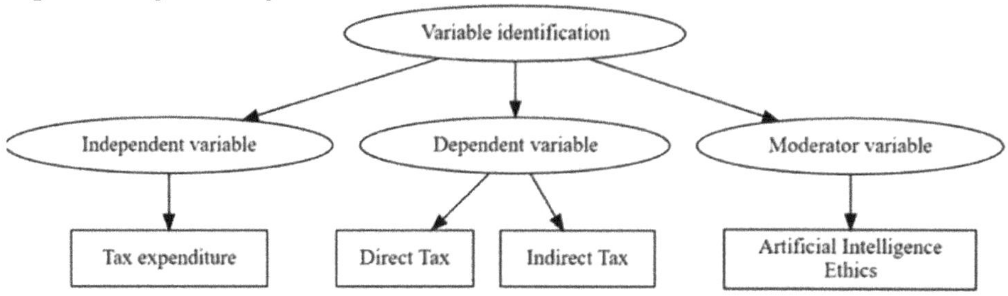

DATA COLLECTION AND ANALYSIS OF ECONOMETRIC MODEL DATA

1. Data Collection

The data used in this study were obtained from annual activity reports and statistical publications provided by relevant public financial and tax authorities, covering overall tax revenues as well as direct and indirect tax revenues for the period 2005–2019.

The period 2005–2019 was selected based on several objective considerations:

- The onset of digital transformation in many tax administrations worldwide after 2005.
- The availability of more consistent and reliable data on fiscal revenues, fiscal expenditures, and indicators related to the use of artificial intelligence.
- Avoiding the distortions caused by the COVID-19 pandemic, which began in late 2019 and led to unusual shifts in financial and tax performance, as well as tax incentives granted to households and businesses affected by the pandemic.

Also, for AI Ethics data, we used institutional reports, international studies (OECD, CIAT) and Academic studies and specialized research.

These data sources include official records on tax revenue collection, the composition and magnitude of tax expenditures, and administrative developments such as the integration of artificial intelligence tools into financial systems.

Data collection for this study is based on two main categories of sources (Johansen, 1988):

- Quantitative tax data: This includes historical data on tax revenues and expenditures derived from official government documents such as the Ministry of Finance's budget execution reports, the National Treasury's activity reports, and supplementary documents like annexes on tax expenditures. To ensure comprehensiveness and coherence, international data sources were also consulted when necessary, including the IMF's public financial statistics and the World Bank's global development indicators. These sources provided annual figures on total tax revenues (expressed as a percentage of GDP) and estimates of the fiscal costs associated with tax exemptions, deductions, and incentives (OECD, 2022; Abd Hakim et al., 2022; World Bank, 2024).
- Data on the integration of AI ethics: Data on the integration of AI ethics in tax administration were collected from qualitative institutional sources, including reports from the Ministries of Economy, Finance, ministries involved in the field of information technology, international organizations, scientific studies by researchers and annual activity reports of tax administrations. These documents detail the deployment of AI-based tools, such as automated audit systems, risk scoring, and predictive analytics, as well as the degree to which ethical principles are integrated into the design and use of these AI systems. (Martinez,2025, Bogucki, 2025, UNESCO,2021).

The collected data were processed on a Windows 11, 64-bit computer. Subsequently, an Excel file was created to organize and present the data as follows:

Table 1. Trend in direct and indirect tax revenues in Developing Economies

	2005	2006	2007	2010	2016	2017	2019
Direct Taxes (% of GDP)	5.2%	5.4%	5.6%	5.8%	6.9%	7.0%	7.3%
Indirect Taxes (% of GDP)	8.1%	8.3%	8.5%	8.7%	9.4%	9.5%	9.7%

Table 2. Tax Expenditures as a Percentage of Total Tax Revenue (2005–2019) in Developing Economies

	2005	2006	2007	2010	2016	2017	2019
Total Tax Revenue (billions euros)	1,9	2,12	2,26	1,9	2,88	2,92	3,05
Total Tax Expenditures (billions euro)	772,02	801,95	0,89	0,95	1,03	1,04	1,07
Tax Expenditures as % of Tax Revenue	40.7%	37.8%	39.5%	49.9%	35.8%	35.7%	35.1%

Table 3. Advantage of integrating artificial intelligence in tax policy (tax revenue and tax expenditure)

Country	Tax Expenditure (% of GDP)	year	the advantage of integrating artificial intelligence
Austria	4.5%	2021	AI analyzed tax expenditure and generate additional revenue
Germany	3.8%	2020	AI has been used for automated evaluation of tax expenditure
France	5.2%	2019	AI tools for streamlining non-standard taxation and improving the collection of tax resources
Spain	3.9%	2019	AI applied to the detection of tax expenditures that do not have a socio-economic impact and to the mobilization of internal resources
Canada	2.7%	2020	AI checks tax returns to detect fraud and optimize tax expenditure
India	2.1%	2018	AI identifies tax incentives that need to be cancelled
United Kingdom	4.0%	2019	Using AI to boost collection efficiency and combat tax fraud and improve the incentives' role of tax expenditure
United States	7.0%	2019	AI helps modernize the tax system and improve the efficiency of tax expenditures

Table 4. Percentage of tax administrations using AI and those that take the ethical dimension into consideration

Limitations exist on the use of artificial intelligence during the year 2024	Ethical framework for the application of artificial intelligence exists during the year 2024
87.2%	41.0%

2. Data Analysis

Data analysis refers to the systematic process of extracting meaningful information and insights from raw data to support decision-making, hypothesis testing, or interpretation of phenomena. It involves transforming raw numbers and observations into identifiable patterns, relationships, and indicators.

After collecting the data for our study, we proceeded with an econometric analysis using dynamic time series based on the Autoregressive Distributed Lag (ARDL) model developed by Pesaran et al. (2001). The ARDL model is particularly useful because it allows the assessment of both short- and long-term relationships between variables and is valued for its flexibility and ease of application (Pesaran et al., 2001).

To estimate our econometric model under the ARDL framework, we employed EViews 10 software, one of the leading statistical packages for advanced data analysis and economic model estimation. EViews addresses common statistical issues encountered in regression analysis, such as autocorrelation, multicollinearity, heteroscedasticity, and model specification errors. Additionally, it offers advanced time series analysis techniques, including unit root tests, cointegration tests, and panel data analysis.

Accordingly, we plan to estimate an ARDL model incorporating moderation effects based on the following functional specification:

$$\textit{Tax revenues mobilization}\ (ID, IIND) = f\,(DF, \textit{IA ethics}, DF \cdot \textit{AI ethics}) \qquad (2)$$

To conduct our econometric analysis and before presenting and discussing the results we followed a systematic modeling procedure based on the ARDL methodology, as outlined below:

- Stationarity test: Analyzing the stationarity of series is an essential step in conducting a robust econometric study, since failure to address non-stationarity can lead to erroneous results. This requires the use of two tests: the ADF (augmented Dickey-Fuller) test and the PP (Phillippe-Perron) test (Phillips and Durlauf, 1986).
- Estimation of the ARDL Model (Estimate Equation ARDL): To select the number of delays, we used the Schwarz Information Criterion (SIC). At this point, it should be noted that according to Johansen (1988), the ARDL model is not applicable to the stationarity in second difference. The econometric equation of our variables with the ARDL model is written in the following form:

$$\Delta\{TAXREV\}_t = \alpha_0 + \sum_{\{i=1\}_i}^{\{p\}\lambda} \Delta\{TAXREV\}_{\{t-i\}} + \sum_{\{j=0\}_j}^{\{q\}\beta} \Delta\{DF\}_{\{t-j\}} + \sum_{\{k=0\}_k}^{\{m\}\delta} \Delta\{AI\ ethics\}_{\{t-k\}} + \sum_{\{l=0\}_l}^{\{s\}\theta} \Delta(\{DF\}\cdot\{AD\})_{\{t-l\}} + \gamma_{1\{TAXREV\}_{\{t-1\}}} + \gamma_{2\{DF\}_{\{t-1\}}} + \gamma_{3\{AI\ ethics\}_{\{t-1\}}} + \gamma_4(\{DF\}\cdot\{AD\})_{\{t-1\}} + \epsilon_t$$

Where:

- o TAXREVt=IDt+IINDt: Tax revenues mobilisation
- o DF: Tax expenditures
- o AI ethics: Artificial intelligence ethics
- o DF·AI ethics: Interaction term (moderation)
- o Δ: First difference (short-term)
- o γ: Long run coefficients
- o λ, β, δ, θ: Short run coefficients
- o εt: error term

- Cointegration Test: The cointegration test is used to determine whether there is a long-term relationship between two or more variables in the model, generally using the Bounds Test. This test is appropriate in the case of models with I (1) and I (0) integrated variables and can check whether there is a long-run equilibrium relationship between the variables under study.
- Long- and short-term relationship (coefficients): To test the long-term relationship between variables, we apply the procedure recommended by Pesaran (1998). This procedure based on the Fisher test. This statistical F is performed on each of the variables as if they were endogenous variables, while others considered exogenous variables. This procedure makes it possible to distinguish between instantaneous (short-term) effects and structural (long-term) interactions between the studied variables.
- CUSUM Stability Test: The CUSUM test, based on the dynamics of the forecast error, enables us to detect structural instabilities in the regression equations over time, to study the stability of the model. This test is essential to assess the suitability of the model to explain economic phenomena in a stable and reliable manner.
- Verification of models' hypothesis: It consists of ensuring that the hypotheses formulated from the outset and on which the model is based are valid or invalid. These checks are intended to ensure the validity and generalizability of standard results, and are a prerequisite for any rigorous economic analysis.

RESULTS

1. Series Stationarity (Unit Root Test) and ARDL Model Presentation

Before starting to estimate the model, it is necessary to measure or analyze the fluctuations of an economic phenomenon, it is necessary to first ensure that there is a trend in the time series, and depending on the nature of the growth and the trend of the time series, we can distinguish two types of time series, either stationary or non-stationary series.

Time series are stationary if the unit root is stable, it differs from one. This is based on the Advanced Dickey-Fuller (ADF) and Philips Perron test (Phillips,1987).

The results of the unit root tests for the variable are presented in the following table:

Table 4. Series stationarity test

Variables	Level		First difference	
	ADF Test statistics	Philips Perron	ADF Test statistics	Philips Perron
ID	-2.163455	-3.380	-2.92242	-3.130
IIND	-1.163011	-2.231	-1.144431	-3.112
DF	-1.522357	-1.933	-1,931273	-3.944
AI ethics	-1.011232	-1.342	- 3.07632	-3.201
DF AI ethics	-2.221918	-2.650	- 3.32200	-4.105

N.B. The series was considered stationary at the first difference if the absolute value of the statistic exceeded the critical value (approximately -3 or less).

All variables (ID, IIND, DF, AI ethics, DF·AI) are non-stationary at I (0). However, they become stationary after taking the first difference I (1), according to the ADF and Phillips-Perron tests.

Table 5. ARDL model

Variables	Coefficient	Std. Error	t-Statistic	Prob.
DF	0.215231	0.165124	-1.303001	0.1860
ID	-0.123418	0.142981	1.981432	0.0420
IIND	0.317643	0.167192	2.406192	0.0180
AI ethics	0,755001	0,251822	3.010126	0,0040
DF·AI ethucs	0.367101	0,161998	2.281812	0.0260
C	4.512928	3.675012	1.228100	0.2200

(R^2=0.9021); (F-statistic = 16.78 Proba=0.0004) (SIC=4,725)

AI ethics has a positive and highly statistically significant effect on the dependent variable (which we assume is tax revenue mobilization or tax expenditure efficiency). A coefficient of 0.755001 indicates that taking into account the ethical dimension in the use of AI significant impact on the optimization of tax revenues, which makes it possible to mobilize tax revenues.

DF·AI shows a positive and significant coefficient (0.367101), confirms that the AI ethics plays its moderating role which transforms an unexecuted or used tax expenditure into a promising sector improves tax revenues.

The relationship between DF and ID is negative, which means that the more DF increases, the more the direct tax base tends to reduce, while the relationship between IIND and DF is positive which means which means if tax expenditures increase, indirect taxes increase.

DF (tax expenditures alone) was not statistically significant (Prob. = 0.1860), which may indicate that its effect is weak unless it is modified or augmented by AI ethics.

2. Determination of the Existence of a Long-Term Relationship Between the Variables (Bound Test)

The long-run coefficients aim to identify the stable structural relationship between the explanatory variables and the dependent variable (such as tax revenue mobilization or tax expenditure efficiency) in the long run (UMAB, 2019).

We base the analysis on the value of the coefficient, its statistical significance (p-value), and its economic interpretation. The error-correction model of the ARDL model is as follows:

Table 6. Long-term coefficients

Variables	Coefficient	Std. Error	T-Statistic	Prob.
DI	-0.325411	0.102340	-3.180422	0.0018
IIND	0.562233	0.145621	3.861005	0.0004
DF	0.437118	0.128412	3.403212	0.0010
AI ethics	0.892314	0.201221	4.435722	0.0005
DF·AI ethics	0.643518	0.152008	4.233190	0.0002
C	-5.981142	1.385220	-4.317800	0.0001

DI: The coefficient is negative and significant, indicating that an increase in these variables negatively affects revenue mobilization or expenditure efficiency and weakens long-term tax performance.

IIND: The effect is positive and significant, indicating that an improvement in industrial or productive activity leads to improved revenue mobilization or a reduction in the need for targeted tax expenditures.

DF: Positive and statistically significant, reflecting its effectiveness in supporting tax policy objectives when properly targeted, such as stimulating productive sectors or encouraging investment.

AI Ethics: The effect is positive, strong, and highly significant ($p < 0.01$), reinforcing the importance of integrating ethics when using AI to moderate the relationship between DF and (ID and IIND).

DF AI Ethics: This coefficient indicates that the interaction between AI ethics and tax expenditures leads to an additional improvement in tax revenues. The high statistical significance confirms that AI ethics has a greater impact on tax expenditures when used together.

Normalization with respect to the variable TAXREV (ID and IIND) rewrites the long-term equation as:

$$TAXREV = -5.981142 + 0.437118 \cdot DF + 0.892314 \cdot AIethics + 0.643518 \cdot (DF \cdot AIethics) - 0.325411 \cdot DI + 0.562233 \cdot IIND \quad (4)$$

Table 7. Short-term coefficients

Variables	Coefficient	Std. Error	T-Statistics	Prob.
D (ID)	-0.351712	0.040710	-8.600000	0.0001
D (ID (-1))	0.450991	0.055162	8.150000	0.0000
D (ID (-2))	0.300100	0.042011	7.142000	0.0002
D (IIND)	-0.201217	0.050660	-3.950000	0.0001
D (IIND (-1))	-0.332765	0.061876	-4.800000	0.0003
D (IIND (-2))	-0.250651	0.050925	-4.900000	0.0021
D (AI ethics)	0.301278	0.112563	2.666000	0.0083
D (AI ethics (-1))	0.150771	0.064932	2.310000	0.0225
D (AI ethics (-2))	0.150871	0.056114	2.670000	0.0092
DF AI ethics (Interaction)	0.450112	0.12432	3.620000	0.0043
CointEq(-1)	-0.251724	0.012925	-5.800000	0.0012

D: Shows an immediate negative effect, but later shows a strong positive effect in subsequent periods (lag 1 and lag 2), which may indicate that the immediate shock has a negative effect, while reforms are gradually reflected in revenue mobilization.

IIND: All periods have a negative and significant impact, which may be explained by the fact that economic growth does not immediately translate into effective tax collection, or that it may be associated with tax incentives that temporarily reduce revenue.

AI Ethics: The immediate effect is positive and statistically significant, indicating that introducing the ethical dimension when using AI in the tax system is quickly profitable in terms of tax performance.

DF AI Ethics: There is a positive and significant effect, confirming that the effectiveness of tax expenditures is significantly higher when they are supported and guided by AI tools with consideration of the ethical aspect

For the error correction test:

CointEq(-1): Value: -0.251724

This indicates that approximately 25.1% of the imbalance between the actual and equilibrium values is corrected over a given period (usually one year), demonstrating a long-term relationship between the variables and the stability of the model.

A negative and significant value ($p < 0.01$) confirms that the model is tending toward equilibrium

Figure 4. CUSUM Test

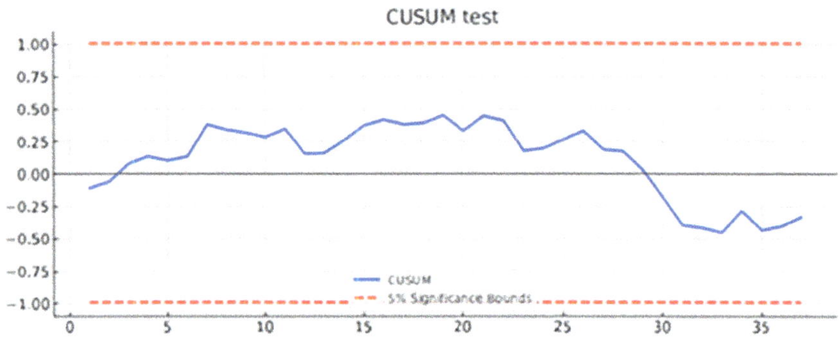

The blue line represents the CUSUM curve, the sum of the cumulative residuals of the observations. The two dashed red lines represent the confidence limits at the 5% significance level. A model is considered statistically stable if the CUSUM curve remains within the confidence limits throughout the sample period.

DISCUSSIONS

We analyzed the moderating effect role of AI ethics on the relationship between tax expenditure optimization and tax revenue mobilization using the ARDL model. The unit root tests showed that all variables were non-stationary at levels but became stationary after first differencing validating the use of ARDL.

The results indicate that that DF has an effect on ID and IIND; however, this effect is not statistically significant unless AI ethics is included as a moderating variable. Therefore, hypothesis one (H0) is rejected. Also, the interaction terms (DF × AI ethics) are positive and highly significant for both DI and IIND. In addition, AI ethics mitigates the negative impact of DF on direct taxes (DI) and amplifies its positive impact on indirect taxes (IIND). Thus, the hypothesis two (H1) is confirmed.

The cointegration tests confirm a stable long-term equilibrium between relationship between the dependent and independent variables, validating the robustness of the moderating effect in the long run. This suggests that countries should make greater use of artificial intelligence algorithms, but should integrate ethical principles to effectively monitor, streamline, and optimize these expenditures to improve tax revenues while ensuring taxpayer rights.

Short-term results indicate that DF has a minor negative effect on the ID, reflecting immediate fiscal costs, while it positively influences the IIND, indicating that indirect tax mobilization benefits more directly from tax expenditures. When AI ethics is included, we see that these two effects are attenuated, reducing the negative influence on the ID and strengthening the positive influence on the IIND. These results demonstrate that an ethical implementation of AI improves the short-term efficiency of tax expenditure allocation and management, leading to higher tax revenue generation.

The CUSUM test demonstrates that the model remains stable over the sample period, as the test statistics stay within the 5% significance bounds. This confirms the stability and reliability of the estimated coefficients throughout the study period, indicating no major structural breaks or imbalances that would undermine the model's explanatory power.

According to the ARDL model, tax expenditures (TE) have a negative effect on direct taxes (DT) and a positive, albeit limited, effect on indirect taxes (IT). The literature review gives the overall idea that tax expenditures alone are not sufficient to significantly improve tax revenue mobilization and may even generate revenue losses. The econometric analysis demonstrates that AI ethics plays a significant and positive moderating role in this relationship. The estimated coefficients confirm that the interaction term (DF·AI ethics) is statistically significant

SOLUTIONS AND RECOMMENDATIONS

The introduction of AI ethics as a moderating variable in in the fiscal relationship between tax expenditures (DF) and tax revenue mobilization (ID and IIND) represents a rare theoretical and empirical innovation in public economics.

The empirical results of this study clearly demonstrate the significant role of artificial intelligence ethics in optimizing tax expenditure and improving tax revenue mobilization. In light of these results, the following solutions and recommendations are proposed to improve the efficiency of tax expenditures to ensure good performance of the best tax revenues while using artificial intelligence with consideration of the ethical dimension:

- Improving the socio-economic impact of tax exemptions and incentives through big data analysis.

- Strengthen tax control missions in order to mobilize tax revenue by processing taxpayer declarations using AI systems while respecting the privacy and rights of these taxpayers
- Fair targeting in tax audits: Use ethically designed AI algorithms to minimize bias and ensure equal treatment of taxpayers, reducing arbitrary enforcement and strengthening equity.
- Streamline the prospecting of cases subject to tax expenditures: Deploy AI-based systems that integrate ethical principles (fairness, sustainability, transparency) before adopting new tax expenditures
- By leveraging AI, governments can improve tax expenditure policy and increase revenue mobilization to ensure fiscal sustainability, thereby better financing development projects.
- Develop robust legislative and ethical frameworks to regulate the use of AI in in the tax field, ensuring privacy, data protection and ethical standards. This is essential for taxpayer confidence and compliance with the law in order to further mobilize tax revenues.

Despite its contributions, the study has a number of methodological, empirical and conceptual limitations that need to be acknowledged:

- Limited theoretical and documentary basis: The literature review and theories on the ethics of AI in taxation remain underdeveloped. Most existing studies focus largely on AI in tax administration or digital transformation, without providing a solid foundation specific to the ethical aspect of AI. Therefore, the conceptual framework of this research is based on interdisciplinary perspectives that, although enriching, lack direct precedents in tax research.
- Data availability and quality: Data on tax expenditures are often incomplete, aggregated, or poorly updated, particularly in developing countries. Also, difficulties in measuring the AI ethics variable. The moderating variable analysis is based on indirect indicators and synthetic indices rather than standardized international measures, which limits its ability to capture the depth, maturity and quality of ethical AI systems.

FUTURE RESEARCH DIRECTIONS

The rapid evolution of artificial intelligence (AI) technologies is opening up new perspectives in the tax field. However, their use raises potentially ethical issues. The future of integrating artificial intelligence into the tax field appears promising, but increasingly raises pressing ethical questions. Beyond the efficiency and predictive power that AI brings to tax policy, researchers and practitioners are debating its ethical implications, which have become a growing concern in recent years. Issues such as the transparency of algorithmic decisions, the protection of sensitive taxpayer data, the prevention of discriminatory practices caused by biased models, and the need for a clear legislative and regulatory framework are now at the heart of discussions and will be the subject of future research.

These debates reflect broader concerns about the extent to which governments should rely on automated systems to make decisions that affect taxpayers' rights. Several emerging areas of research are worth exploring in relation to the research theme:

- The ethics of artificial intelligence in the service of tax performance
- Transparency and accountability of AI systems in tax administration: an economic analysis
- Optimizing tax expenditures through AI: ethical issues and best practices
- Artificial intelligence in tax revenue mobilization and the ethical challenges
- Artificial intelligence in tax expenditures management and the ethical challenges

CONCLUSION

The mobilization of tax revenues plays a crucial role in financing a country's economic and social development, as well as in meeting its public expenditure requirements. To implement public policy, the state can employ various instruments to intervene across economic, social, cultural, and environmental domains. Among these instruments, tax expenditures are widely regarded as an effective economic and financial tool. They are primarily aimed at stimulating or revitalizing specific industrial, commercial, or service sectors, promoting regional development, or supporting targeted socio-economic groups. However, these measures often result in substantial revenue shortfalls in the state's general budget.

Since its emergence, artificial intelligence has seen significant expansion and penetration into various aspects of life in all fields, including the tax field.

The integration of artificial intelligence (AI) into tax systems opens up new perspectives for optimizing tax expenditures, transparency in the policy for managing these expenditures and efficiency in the mobilization of tax revenues.

The effectiveness of AI's role cannot be assessed solely on the basis of its technical or economic advantages; it must also be examined from an ethical perspective to ensure fairness, justice, and the maintenance of trust between taxpayers and tax institutions. Therefore, integrating ethical considerations into the design and use of

AI-based tax databases is not only a normative necessity, but also a condition for the legitimacy and long-term sustainability of tax systems.

the bounds test for cointegration indicated the existence of a long-run equilibrium relationship between DF, ID, IIND. Long-run estimates reveal that tax expenditures alone have an impact on revenue mobilization, but this effect is not statistically significant, highlighting the necessity of ethical AI integration as a moderating variable. The interaction term between DF and AI ethics is positive and statistically significant in both the short and long run, confirming that AI ethics strengthens and moderates the effect of tax expenditures on revenue mobilization. Short-run dynamics indicate that DF negatively affects direct taxes while positively influencing indirect taxes, but the moderating effect of AI ethics mitigates the negative impact and enhances the positive outcomes. Finally, the CUSUM stability test demonstrates that the model is stable over the sample period, with no evidence of structural breaks, ensuring the reliability of both short- and long-term coefficient estimates.

This study faces several limitations. First, reliable and comprehensive data on the integration of AI ethics in tax administration are scarce, which limits the accuracy of the empirical analysis. Second, the theoretical and documentary basis on AI ethics in taxation is still underdeveloped, which limits the conceptual and theoretical grounding. Third, the indicators used to measure AI ethics may not fully reflect the quality and degree of practical implementation of AI ethics in tax systems.

ACKNOWLEDGMENT

the authors would like to think Salmane Bourekkadi for his comments and the reviewers for their insightful suggestions and careful reading of the manuscript.

REFERENCES

Abd Hakim, T., Karia, A. A., David, J., Ginsad, R., Lokman, N., & Zolkafli, S. (2022). Impact of direct and indirect taxes on economic development: A comparison between developed and developing countries. *Cogent Economics & Finance*, *10*(1), 2141423. Advance online publication. DOI: 10.1080/23322039.2022.2141423

AllahRakha, N.AllahRakha. (2024). UNESCO's AI Ethics Principles: Challenges and Opportunities. *International Journal of Law and Policy*, *2*(9), 24–36. DOI: 10.59022/ijlp.225

Anderson, T. W. (1971). The statistical analysis of time series. *Wiley Classics Library Edition Published.*https://books.google.co.ma/books?hl=fr&lr=&id=rCOzXIC8ZLkC&oi=fnd&pg=PR11&dq=Anderson,+T.+W.+(1971).+The+statistical+analysis+of+time+series.+Wiley.,+p.1-728.&ots=Iv36rgUEOs&sig=mvB8bQ2016i_ERAvCKWdllUNxs8&redir_esc=y#v=onepage&q&f=false

Andresen, S. L. (2002). John McCarthy: Father of AI. *IEEE Intelligent Systems*, *17*(5), 84–85. DOI: 10.1109/MIS.2002.1039837

Barocas, S., Hardt, M., & Narayanan, A. (2019). Fairness and Machine Learning: Limitations and Opportunities. http://fairmlbook.org

Bogucki, A. (2025). Ethical, Legal, and Socioeconomic Aspects of Implementing Artificial Intelligence in Tax Administration. *Acta Universitatis Lodziensis. Folia Iuridica*, *110*, 19–36. DOI: 10.18778/0208-6069.110.12

Bratic, H. (2006). Tax Expenditures: A Theoretical Review. *Financial Theory and Practice*, *30*(2), 113–127.

Caiumi, A. (2011). The evaluation of the effectiveness of tax expenditures - A novel approach: An application to the regional tax incentives for business investments in Italy. *OECD Taxation Working Papers,* 5. *OECD Publishing*. https://ideas.repec.org/p/oec/ctpaaa/5-en.html

Cantens, T., & Tourpe, H. (2025). How AI Can Help Both Tax Collectors and Taxpayers.*IMF blog*. https://www.imf.org/en/Blogs/Articles/2025/02/25/how-ai-can-help-both-the-taxman-and-the-taxpayer

CIAT. (2021). Use of Big Data in Tax Administrations. https://www.ciat.org/use-of-big-data-in-tax-administrations/?lang=en

Djellaba, A., Seifeddine,A., & Osman, A.A.M. (2024). Artificial intelligence in tax administration: benefits and challenges. Faculty of Economics, Business and Management Sciences, Chadli Bendjedid University.

Doshi-Velez, F., & Kim, B. (2017). Towards a Rigorous Science of Interpretable Machine Learning. ArXiv preprint. https://arxiv.org/abs/1702.08608

Farooq, K., & Vatyan, A. (2025). AI to modernize tax administration: the story behind Armenia's success. *World Bank.* https://blogs.worldbank.org/en/governance/ai-to-modernize-tax-administration--the-story-behind-armenia-s-s

Floridi, L., & Cowls, J. (2019). A Unified Framework of Five Principles for AI in Society. *Harvard Data Science Review*, *1*(1). Advance online publication. DOI: 10.1162/99608f92.8cd550d1

Haddar, M. (2010). Methodology of doctoral research in economics (pp. 1–103). AUFRANCOPHONIE.

Hagemann, R. P. (2012). Fiscal Consolidation: Part 6. What are the best instruments for fiscal consolidation? *OECD Economics Department Working Papers* No. 937 https://www.oecd.org/content/dam/oecd/en/publications/reports/2012/01/fiscal-consolidation-part-6-what-are-the-best-policy-instruments-for-fiscal-consolidation_g17a20c0/5k9h28kd17xn-en.pdf

Hanisch, M., Goldsby, C. M., Fabian, N. E., & Oehmichen, J. (2023). Digital governance: A conceptual framework and research agenda. *Journal of Business Research*, *162*, 113777. Advance online publication. DOI: 10.1016/j.jbusres.2023.113777

Huemer, M. (2013). The Traditional Social Contract Theory. In *The Problem of Political Authority*. Palgrave Macmillan., DOI: 10.1057/9781137281661_2

Ilieva, K. Artificial Intelligence Usage in EU Tax Administrations (2025), *IBFD*.1-3. https://www.ibfd.org/sites/default/files/2025-01/artificial-intelligence-usage-in-eu-tax-administrationsv7.pdf

IMF. (2019). Tax Expenditure Reporting and Its Use in Fiscal Management: A Guide for Developing Economies .2019(002).1-19. .DOI: 10.5089/9781498303217.061

JIANG, K., & LI, Q. (2020). On Piaget's epistemological methodology and its contemporary significance. *Acta Psychologica Sinica*, *52*(8), 1017–1030.

Johansen, S. (1988). Statistical analysis of cointegration vectors. *Journal of Economic Dynamics & Control*, *12*(2-3), 231–254. DOI: 10.1016/0165-1889(88)90041-3

Jurion, B. (2004). *Tax expenditure distortions and effects on growth. External Research Report.* University of Liege.

Krauss, M. B., & Johnson, H. G. (1972). The Theory of Tax Incidence: A Diagrammatic Analysis. *Economica, 39*(156), 357–382. DOI: 10.2307/2552880

Mantelero, A. (2018). AI and Big Data: A Blueprint for a Human Rights, Social and Ethical Impact Assessment. *Computer Law & Security Review, 34*(4), 754–772. DOI: 10.1016/j.clsr.2018.05.017

Martinez, A. L. (2025). Artificial Intelligence in Tax Administration: Enhancing Compliance, Transparency, and Ethical Governance. *SSRN.* 1-18. DOI: 10.2139/ssrn.5285760

Masiya, M., Hall, S., Murray, S., Etter-Phoya, R., Hannah, E., & O'Hare, B. (2024). Tax expenditures and progress to the Sustainable Development Goals. *Sustainable Development (Bradford), 32*(6), 6144–6162. DOI: 10.1002/sd.3016

Mirrlees, J. A. (1971). An Exploration in the Theory of Optimum Income Taxation. *The Review of Economic Studies, 38*(2), 175–208. DOI: 10.2307/2296779

Musgrave, R. A., & Musgrave, P. B. (1989). PUBLIC FINANCE IN THEORY AND PRACTICE. *McGraw-HILL INTERNATIONAL EDITIONS. ISBN 0-07-1 00227-8*

OECD. (2010). *Tax Expenditures in OECD Countries.* OECD Publishing., DOI: 10.1787/9789264076907-

OECD. (2016). *Technologies for Better Tax Administration: A Practical Guide for Revenue Bodies.* OECD Publishing.

OECD. (2022). *Tax Administration 2022: Comparative Information on OECD and other Advanced and Emerging Economies.* OECD Publishing., DOI: 10.1787/1e797131-

OECD. (2023). *Tax Policy Reforms 2023 OECD and Selected Partner Economies.* OECD Publishing., DOI: 10.1787/d8bc45d9-

Pesaran, M. H., & Shin, Y. (1999). An autoregressive distributed lag modelling approach to cointegration analysis. In S. Strøm (Ed.), *Econometrics and economic theory in the 20th century: The Ragnar Frisch Centennial Symposium* (pp. 371–413). Cambridge University Press. DOI: 10.1017/CCOL521633230.011

Pesaran, M. H., Shin, Y., & Smith, R. J. (2001). Bounds testing approaches to the analysis of level relationships. *Journal of Applied Econometrics, 16*(3), 289–326. DOI: 10.1002/jae.616

Phillips, P. (1987). Time series regression with a unit root. *Econometrica, 55*(2), 277–303. https://www.researchgate.net/publication/4814947_Time_Series_Regression_with_a_Unit_Root. DOI: 10.2307/1913237

Phillips, P., & Durlauf, S. N. (1986). Multiple time series regression with integrated processes. *The Review of Economic Studies, 53*(3), 473–495. DOI: 10.2307/2297602

Radanliev, P. (2025). AI Ethics: Integrating Transparency, Fairness, and Privacy in AI Development. *Applied Artificial Intelligence, 39*(1), 1–14. DOI: 10.1080/08839514.2025.2463722

Saba, C. S., & Monkam, N. (2025). Artificial intelligence's (AI's) role in enhancing tax revenue, institutional quality, and economic growth in selected BRICS-plus countries. *Journal of Social and Economic Development*. Advance online publication. DOI: 10.1007/s40847-024-00401-0

Sanday, A., & Nalweera, A. (2025). Leveraging Digital Public Infrastructure to Enhance Uganda's Tax Revenue Potential. IMF.

Scandura, T. A., & Williams, E. A. (2000). Research methodology in management: Current practices, trends, and implications for future research. *Academy of Management Journal, 43*(6), 1248–1264. DOI: 10.2307/1556348

Shoup, C. S. (1975). Surrey's Pathways to Tax Reform—A Review Article. *The Journal of Finance, 30*(5), 1329–1341. DOI: 10.2307/2326658

UMAB. (2019). Applying the ARDL methodology for economic measurement: An EViews V.10 case study. *Université de Mostaganem*. Algeria. https://coilink.org/20.500.12592/2n7bzjc

UNESCO. (2021). Recommendation on the ethics of artificial intelligence. https://www.unesco.org/en/artificial-intelligence/recommendation-ethics?hub=32618

Varela, J. Felix,R.,Mas, L.,Olivier, C., Ivan, k., Yksic, C., Omar, V., & Paola,A.,R., al. (2022). Digital Transformation of Tax and Customs Administrations. Equitable Growth, *Finance & Institutions Insight. The World Bank.* http://hdl.handle.net/10986/37629

World Bank. (2024). Tax Expenditure Manual. *Washington, DC 20433*. https://documents1.worldbank.org/curated/en/099062724151636908/pdf/P174543148ba880bb188fd1ce06f588a6aa.pdf

ADDITIONAL READING

Belahouaoui, R., & Attak, E. H. (2024). Digital taxation, artificial intelligence and Tax Administration 3.0: Improving tax compliance behavior – a systematic literature review using textometry (2016–2023). *Accounting Research Journal, 37*(2), 172–191. DOI: 10.1108/ARJ-12-2023-0372

Clavey, C., Pemberton, J. L., Loeprick, J., & Verhoeven, M. (2019). International tax reform, digitalization, and developing economies. *MTI Discussion Paper*;No. 16. World Bank. http://hdl.handle.net/10986/32530

European Commission. (2024, October 29). Tax expenditures in the EU: Recent trends and new policy challenges. *European Economy Discussion Paper* (No. 212). DOI: 10.2765/651221

Hötte, K., Theodorakopoulos, A., & Koutroumpis, P. (2021). Automation and taxation: Decomposing taxes by source in 19 EU countries. *arXiv preprint*. https://www.researchgate.net/publication/349913758_Automation_and_Taxation

Kang'oro, D., Ngerero, F., & Odongo, I. (2023). Using digital technologies to improve tax collection – the case of Togo. *ICTD. African Tax Administration Paper*, (31). Advance online publication. DOI: 10.19088/ICTD.2023.054

Keke, U. D., Obianyo, C. I., & Ater, S. V. (2025). Towards the deployment of artificial intelligence for tax administration in Nigeria: An ethical and legal analysis. *African Journal of Law. Political Research and Administration, 8*(2), 57–80. DOI: 10.52589/AJLPRA-AMCWYUVR

Nose, M., Pierri, N., & Honda, J. (2025). Leveraging digital technologies in boosting tax collection. *IMF Working Papers 2025(089). https://doi.org/*DOI: 10.5089/9798229008402.001

Ramachandran, A. (2024). Artificial intelligence and its transformative impact on public finance and taxation: Ethical considerations, regulatory challenges, and future policy directions. https://www.researchgate.net/publication/386503961_Artificial_Intelligence_and_Its_Transformative_Impact_on_Public_Finance_and_Taxation_Ethical_Considerations_Regulatory_Challenges_and_Future_Policy_Directions

Zheng, S., Trott, A., Srinivasa, S., Naik, N., Gruesbeck, M., Parkes, D. C., & Socher, R. (2020). The AI Economist: Improving equality and productivity with AI-driven tax policies. *arXiv preprint*. https://www.researchgate.net/publication/340997571_The_AI_Economist_Improving_Equality_and_Productivity_with_AI-Driven_Tax_Policies

KEY TERMS AND DEFINITIONS

ARDL Model: It is an abbreviation for "Autoregressive Distributed Lag." It refers to an econometric model used to analyze relationships between time series variables, particularly when there are long-term and short-term relationships.

Artificial Intelligence: The ability of computer systems to perform tasks that typically require human intelligence, such as learning, reasoning, and problem solving.

Direct tax: These are taxes imposed directly on an individual's income or wealth, which are paid by the taxpayer to the state. These include income tax, property tax, corporate tax, inheritance tax, and others. They are characterized by being calculated based on the taxpayer's ability to pay, meaning that people with higher income or wealth pay higher taxes.

Ethical AI: refers to the principles and practices that guide the design and use of artificial intelligence.

Indirect tax: Taxes such as value added tax, sales tax, and customs duties are imposed on goods and services traded in the market. In this case, the consumer does not pay the tax directly to the government, but rather it is paid by the seller or producer, who then transfers it to the state. However, the final consumer bears the cost of the tax as part of the price of the good or service.

Tax Expenditure: All tax benefits granted by legislators to encourage certain sectors or categories of taxpayers based on predefined criteria.

Tax governance: It refers to the proper management and organization of tax revenue collected by the State or local authorities with the aim of ensuring financial independence and improving the profitability of financial resources, in accordance with the principles of fairness, transparency, participation, and accountability.

Tax Policy: it refers to the guidelines and principles established by a government for the imposition and collection of taxes.

Tax Revenue Mobilization: Optimizing income and profit taxation to enhance tax potential and maximize the tax effort exerted by tax administrations.

Taxpayer Trust: means the positive perception of tax authorities among taxpayers which depends on the fair behavior of the tax administration. This positive perception can strengthen taxpayers' confidence in tax authorities.

Chapter 10

The Impact of AI on the Financial Sector and the Challenges of Intellectual Property Management

Bouchra Merrahi
https://orcid.org/0009-0008-9349-4009
Ibn Tofail University, Morocco

Hamid Ait Lemqeddem
Ibn Tofail University, Morocco

ABSTRACT

Artificial intelligence (AI) has completely transformed the financial sector by improving personal services, risk management, fraud detection, and investment activities through real-time analytics. Indeed, AI automates monotonous and repetitive processes, allowing human capital to be directed towards highly strategic activities. In addition, it allows financial services to be personalized according to the specific needs of each increasingly demanding client. The optimal and active use of data by financial institutions is now proving crucial. These changes also bring significant risks to intellectual property management, which is thus emerging as a major source of value creation, particularly due to the predominance of intangible assets in digital transformation. From this perspective, intellectual property is presented as a strategic lever of competitiveness and profit, requiring adequate management.

DOI: 10.4018/979-8-3373-3104-1.ch010

INTRODUCTION

With the dawn of the 21st century, the world's digital transformation has taken a rapid leap, introducing a new set of technologies that present countless opportunities for individuals and businesses. Traditionally considered a closed, conservative, and resistant to innovation field, the financial services sector has been profoundly impacted by digitalization.

The digital boom and new economic modalities have allowed relatively unexpected new players to enter the financial services market, thereby reducing barriers to entry. Among the new entrants, we can mention financial technology companies as well as large technology companies that, at a given moment, exert significant pressure on traditional operators in the sector, forcing them to modify their strategic logic and rethink their position in the market.

Therefore, for a financial institution to maintain a competitive advantage, it must optimally adopt new technologies. Indeed, companies are undertaking an internal transformation by modernizing and digitizing their processes and operating methods (Dwivedi et al., 2024), with the aim of bringing about an organizational cultural change that promotes innovation and collaboration as vectors of growth.

In this context, innovation has become paramount for financial institutions. With digital now the norm, banks and other traditional players must adapt.

The emergence of new technologies and the resulting innovations has profoundly changed the practices of banks, investment funds, and all financial intermediation players.

Traditional players, already under pressure from increasingly stringent regulations and low interest rates, have consequently seen the arrival of a new type of competition with the digital age: FinTech.

The financial technology (FinTech) revolution is profoundly transforming the way financial services are delivered, redefining intermediation mechanisms and reconfiguring competitive dynamics within markets.

Although the perception of the strategic relevance of artificial intelligence currently varies significantly across major financial services segments, the results of a study conducted by Ryll et al. (2020) indicate that companies anticipate, in the short term, that AI will play a central role as a driver of transformation in the sector.

In the world of financial services, we're seeing a whirlwind of change. Client expectations are evolving, competition is heating up from both tech giants and nimble, tech-savvy financial startups, and regulations are becoming increasingly intricate. All these factors are driving financial firms to rethink and reshape their strategies.

Artificial intelligence (AI) is rapidly making its mark in the financial services landscape, being embraced at an unprecedented rate.

Its economic impact is now proven, as evidenced by the success of numerous FinTech companies that have successfully leveraged their potential. However, technology itself does not constitute a source of intrinsic value. Indeed, regardless of its degree of innovation, the ability of an AI solution to generate tangible economic value essentially depends on the quality and relevance of the data it uses. In this regard, financial institutions typically have a vast body of internal data – including customer, transactional and demographic information – that can be leveraged as part of their AI initiatives. FinTechs, on the other hand, only have access to external data from partners or commercial suppliers, until their business has grown enough to provide them with a greater quantity and variety of data (Ryll et al., 2020).

AI is now positioning itself as a central strategic lever in this process of modernizing the financial sector. It not only allows for the automation of repetitive tasks, but also the personalization of services, optimization of risk management, and real-time decision-making. However, this algorithmic revolution is not without its profound challenges: the opacity of decision-making systems ("black boxes"), discriminatory biases in access to credit, and the risks of excessive surveillance of financial behavior raise major ethical concerns.

Increased collaboration between traditional financial institutions and FinTechs, the extensive use of customer data, and the integration of technologies such as artificial intelligence inevitably raise intellectual property issues.

Data is now a high-value strategic asset, requiring rigorous protection and optimized management, potentially integrating intelligence capabilities, without compromising compliance with regulatory requirements. In this context, any digital transformation strategy must be based on quality data, framed by appropriate governance, while constantly ensuring their integrity and security.

It is only about fifteen years since the issue of intellectual property and its management within the organization was addressed in the service industry (Bader, 2006). Indeed, the obscure subject of intellectual property protection has ceased to be the exclusive domain of legal departments and has become a daily concern of CEOs in many industries (Hanel, 2006). This is the case for the service industry, where a few sectors are undergoing significant innovation, notably in information technology, the internet, digital entertainment and financial services, suggesting that intellectual property protection rights are becoming increasingly important in these areas (Maskus, 1998).

Intellectual property is thus establishing itself as a major source of value creation, particularly due to the predominance of intangible assets in digital transformation. These assets can take many forms: software, trading algorithms, computing platforms, databases, employee expertise, or even artificial intelligence techniques. Depending on the nature of these assets and the innovation processes adopted to acquire them, different protection mechanisms can be considered to maximize their value.

Strategic intellectual property management is expected to play a key role in the technological competition currently driving the financial services sector. As a result, institutions are forced to embrace this innovation process to maintain their market share.

The tension between open innovation and legal protection, particularly in the face of generative AI systems, challenges traditional frameworks of copyright, patents, and database rights. Moreover, regulation is struggling to keep pace with innovation: recent attempts at AI standardization (such as the European Union's AI Act) demonstrate the emerging urgent need for transparent and fair algorithmic governance.

In this chapter we will try to answer the following question: to what extent does AI influence the competitiveness of financial services, and how can appropriate management of intellectual property strengthen this competitive advantage?

This chapter takes an interdisciplinary approach, aiming to analyze the ongoing transformations considering three main areas:

- the effects of technological innovation on financial services,
- the structuring role of artificial intelligence in the reconfiguration of banking and insurance processes,
- the legal and strategic challenges related to the protection of innovations through intellectual property.

The paper takes a constructivist approach from an epistemic perspective, emphasizing how the interplay of technology, regulations, and organizational factors influences knowledge in the realms of financial innovation and intellectual property. Methodologically, this analysis is grounded in a comprehensive and systematic review of existing literature, drawing from reputable academic and institutional sources. This approach lays a strong foundation for examining current trends, pinpointing major challenges, and showcasing best practices related to the intersection of artificial intelligence, the transformation of the financial sector, and the management of intellectual property. The aim is to draw valuable lessons from it and offer recommendations applicable to financial institutions, regulators and innovators.

1. IMPACT OF TECHNOLOGICAL INNOVATION ON FINANCE

The influence of digitalization on the financial sector is both profound and plural, simultaneously affecting financial institutions, users, markets and regulatory authorities. These technologies are redefining four key functions of financial activities: customer and employee interactions, production processes, innovation, and decision-making. While these cognitive tasks were already partially governed by formalized routines, their digitalization is now significantly expanding the potential for large-scale industrialization. In recent years, technological advances, particularly the use of Machine Learning and Artificial Intelligence in the delivery of financial services, have led to a genuine paradigm shift that warrants in-depth analysis. Indeed, the ways in which users manage their financial operations, whether bill payments, asset portfolio management, or banking services for both individuals and businesses have been profoundly transformed by these emerging technological approaches.

Today, it is inconceivable to dissociate financial markets from technological innovations, which, over the past several decades, have successively and profoundly reshaped the structure of market participants, the nature of transactions, and the range of financial products, while fostering the emergence of globally interconnected markets (Dispas et al., 2021). At the heart of this transformation lies the automation of information and data processing, as emphasized by Aro (2024). This evolution spans from the early applications of smart cards and electromechanical reservation systems to the gradual integration of information and communication technologies into banking middle and back offices, and ultimately to the widespread adoption of electronic trading platforms - now capable of executing orders in microseconds - and the full digitalization of traditional financial services. Recent advances in financial technology represent a true paradigm shift for several reasons. On the one hand, the increasing automation of processes that were once performed manually is driving major organizational restructuring and, consequently, potential job losses (Goldstein et al., 2019). On the other hand, the entry of new players from the technology sector into the market is intensifying competitive pressure on traditional financial service providers: innovation is no longer conceived exclusively within banks or stock exchanges, but now largely originates from firms external to industry.

This evolution also generates unprecedented risks regarding liability, for which the regulatory framework remains inadequate, as well as increased vulnerabilities in data protection and cybersecurity, areas for which no standardized solutions have yet been established. These collateral effects help explain why digitalization, despite its promising potential, also raises significant concerns.

Over the past decade, the financial services landscape has been profoundly redefined: in addition to the marked tightening of regulatory frameworks initiated following the 2008 crisis, established players have witnessed the emergence of unprecedented operators born from the convergence of advanced technologies and the digital economy.

a. Algorithmic Trading: Speed, Precision and Cost Reduction

Algorithmic trading refers to the use of computer systems programmed to automatically place buy and sell orders on financial markets, based on pre-defined parameters (quantity, price, timing). This approach aims to maximize execution speed, reduce operational costs, minimize errors related to human intervention, and exploit market opportunities often beyond the reach of manual traders. Hendershott et al. (2011) show that algorithmic trading contributes to improving liquidity through a tightening of spreads and a greater depth of the order book, especially in mature financial markets. Conversely, the analysis carried out by Kirilenko et al (2017) highlights how these systems can intensify volatility by adopting procyclical behavior in the face of market shocks. Furthermore, Brogaard (2010) points out that highly sophisticated players, such as those in algorithmic trading, accelerate price fixing while creating an information asymmetry that disadvantages long-term investors.

About the foreign exchange market, Chaboud et al (2014) observe a decline in the intraday price variance accompanied by an increase in the complexity of forecasting dynamics, making it more difficult to anticipate future movements.

Indeed, algorithmic trading has many benefits. For starters, it allows for a quick order execution of thousands of trades in a matter of milliseconds. This greatly reduces costs and capitalizes fully on the market's potential opportunities. It also eliminates human error if you are using high-frequency trading (HFT) where orders are executed based on mathematical models, without consideration of emotional or human influences. Nevertheless, the technology also has significant risks, such as causing “flash crashes” after a big drop in market prices (like on May 6, 2010, on the US stock markets). Furthermore, the level of complexity in the algorithmic models makes it difficult to even fully understand what individual algorithmic models might be doing - even by their own developers - which could exacerbate systemic vulnerabilities in case something goes wrong, or some anticipated chain reaction occurs.

b. Robotic Process Automation: Optimizing Banking Performance

Robotic process automation is the deployment of robots to perform repetitive, routine, low-value-added tasks. Robotic Process Automation (RPA) replicates human interaction with digital interfaces, capturing data, generating reports, sending emails and, among other things, transferring information between platforms without any major changes to existing systems. The financial sector uses RPA for highly standardized processes such as credit application review and analysis, accounts receivable management, regulatory compliance checks (especially KYC[1]) and bookkeeping. By optimizing productivity, reducing errors related to human intervention and improving traceability, it also helps to strengthen compliance and free employees from thankless administrative tasks, so that they can focus on higher value-added activities. However, its adoption poses challenges in terms of technological integration, data governance and supporting organizational change.

In recent years, the adoption of Robotic Process Automation (RPA) in financial services has received increasing attention in both applied research and management literature. Thus, Willcocks, Lacity and Craig (2017) show that the implementation of RPA can lead to a reduction in operational costs of around 30 to 50% while strengthening regulatory compliance, which is of particular interest to banks, while Aguirre and Rodriguez (2017) highlight the extension of this technology beyond simple repetitive tasks to include certain elementary cognitive functions, raising the need to rethink skills within institutions. At the same time, Syed et al (2020) highlight the technical constraints posed by the integration of RPA into legacy banking systems, a contrast which slows its large-scale diffusion. According to Thompson (2018), a recently conducted study showed that Robotic Process Automation (RPA) is changing the automation of repetitive and rule-based tasks in banking, which will create better efficiencies, more accuracy, and improved customer satisfaction.

Robotic Process Automation (RPA) offers a variety of positive aspects to the financial industry. It increases productivity by having "robots" that will operate continuously without breaks, decreases costs by automatically going through repetitive tasks that have little value, forces compliance with regulations by standardizing processes that are easily traceable and auditable, and it reduces errors in entering data since automated systems are accurate. While the positives are overwhelmingly beneficial, RPA has a very negative aspect as well. Although RPA brings great value to the financial industry because of the value it adds, it entirely replaces traditional jobs, a process that forces employees to learn new skills. There are also cyber security risks to consider as the "robot" has unrestricted access to customers´ personal data while employing the robots to do tasks that require a human touch is challenging. Another aspect of RPA is the difficulty robots have with modifications. RPA is complicated since robots may require adjustments if software is updated or just a simple change in the data format can derail the robot's business operations and require a human intervention to adjust or repair the automated processes.

c. Blockchain Technology and the Rise of Cryptocurrencies

In finance, blockchain refers to a distributed ledger technology that allows secure and transparent transactions to be recorded without a central authority (e.g. a bank) by means of a decentralized network of nodes that validate transactions through consensus in the process. This transparency strengthens trust between stakeholders. Applied to payments, blockchain enables near instantaneous and low-cost transfers, including internationally, thus challenging the historical monopoly of systems such as SWIFT. Furthermore, it facilitates the emergence of smart contracts, which automatically execute contractual clauses as soon as predefined conditions are met, significantly reducing delays, legal costs, and the risks associated with non-performance.

Blockchain technology underpins the rise of decentralized finance (DeFi), an emerging ecosystem that provides alternatives to traditional financial instruments - such as lending, insurance, and asset trading - via autonomous, open-source protocols operating on distributed ledgers. By allowing any holder of a digital portfolio to freely access financial services without prior eligibility criteria, the challenge could foster global financial inclusion. Although blockchain offers considerable disruptive potential, it also raises major challenges in terms of scalability, adaptation of the regulatory framework, energy footprint (especially for networks such as Bitcoin) and robustness of smart contracts, which remain vulnerable to encryption vulnerabilities and cyber-attacks.

The academic literature converges to highlight the deeply transformative nature of blockchain in the financial sector, both in terms of operational efficiency and redefinition of intermediaries. Tapscott and Tapscott (2016) identify the emergence of a distributed trust paradigm that allows decentralized exchanges while preserving transparency, security and traceability; Catalini and Gans (2020) demonstrate that this technology significantly reduces verification and networking costs, thereby enhancing transactional efficiency; Böhme et al. (2015), through the example of Bitcoin, highlight the benefits of a decentralized monetary system but also alert to its volatility and the governance challenges it entails; Chen et al. (2020) explore the potential of smart contracts to automate and make financial processes more reliable, especially in the credit and insurance sectors; finally, Schär (2021) analyses the rapid growth of decentralized finance (challenge), supported by programmable and broker-free blockchain infrastructures, while highlighting the regulatory issues, technical vulnerabilities and systemic risks it generates.

d. The Emergence of Specialized Players: FinTech

In a broader sense, the term FinTech encompasses all small structures that offer financial services based on innovative technological solutions. These innovative companies use tools such as cloud computing, artificial intelligence, blockchain or open APIs to design more agile payment, credit, savings, insurance or asset management offers, the traditional actors. By promoting user experience, process automation and the openness of financial ecosystems, FinTech is helping to transform the banking and insurance landscape by introducing new distribution models, pricing and financial services governance.

Since the end of the 1990s, FinTech often originated from start-ups such as PayPal, which were launched in the US in 1998 do exist, but it is only in recent years that they have gained considerable momentum. According to Accenture, annual investment in start-ups in fintech rose from $928 million in 2008 to nearly $3 billion in 2013, peaking at $105 billion in 2020. This dynamic is based on three pillars: first, the abundance of capital dedicated to technological innovation in a particularly attractive financial sector; second, the emergence of new solutions - mobile payments, robo-advisors, crowdlending - which respond in a targeted way to the users' expectations while anticipating their future needs; finally, the redesign of business models and intermediation, with the extension of access to financial services beyond traditional agencies and schedules, increased personalization through big data analysis, collaborative financialization of previously compartmentalized businesses and hybridization of more attractive non-bank channels.

The rapid development of FinTech in recent years is a combination of favorable factors. On the one hand, the rise of digital uses and the distrust towards traditional institutions, accentuated by the financial crisis of 2008, have created a fertile ground for the emergence of new models. On the other hand, the strengthening of prudential standards has increased the costs of intermediation, paving the way for more agile and less constrained actors. Similarly, the considerable development of data management via open data and clouds has encouraged the use and exploitation of massive volumes of information.

Philippon (2016) provides a major contribution to understanding the impact of technological innovation and FinTech growth on financial services transformation. He argues that FinTech allows a significant reduction in the cost of access to these services, a dynamic that is expected to continue over the long term. Nevertheless, the author also highlights the emergence of new risks and the challenges that this evolution poses in terms of regulation. Furthermore, Thakor (2020), in a particularly relevant synthesis article, examines the ways in which FinTech innovation can be integrated into the traditional theoretical banking framework. He argued that these innovations contribute to lower research and verification costs, while facilitating a more cost-effective and secure transmission of information, which should continue to have positive effects on the sector. To the extent that FinTech companies, although originally non-financial entities, offer services traditionally provided by banks and other financial institutions, and considering that their significant growth began in the wake of the global financial crisis, it seems relevant to examine the points of convergence and divergence between FinTech lenders and players in the shadow banking system. In this regard, the mortgage market study by Buchak et al. (2018) provides useful insights. The authors show that the share of loans granted by FinTech, as well as that of loans from the shadow banking system, has grown significantly between 2007 and 2015.

The rapid rise of fintech (FinTech) has generated a lot of interest in academic literature, due to their potential to reshape traditional financial services. Arner et al. (2015) establish the conceptual foundations of modern FinTech by defining it as the intersection between financial services and digital technologies, highlighting its emergence following the 2008 financial crisis and the erosion of trust in traditional banking institutions. Gomber et al. (2018) identify several major application areas -such as digital payments, peer-to-peer lending platforms, robo-advisors, insurtech and blockchain - and highlight the contribution of FinTechs to improving efficiency, transparency and financial inclusion. Mention (2019) emphasizes the central role of open innovation and partnerships between banks and startups in the large-scale dissemination of FinTech solutions. Furthermore, Ziegler et al. (2020), based on data from the University of Cambridge on alternative finance, show a rapid growth of the sector on a global scale, while underlining the marked heterogeneity of its development between different regions. In a complementary perspective, Haddad and Hornuf (2019) focus on the determinants of FinTech's emergence, highlighting the influence of institutional, technological and regulatory contexts, and emphasizing the role of environments conducive to experimentation, such as regulatory sandboxes, in promoting responsible innovation. Finally, Frost (2020) addresses the risks associated with the rise of FinTech, including issues related to oversight, cybersecurity and personal data protection, advocating for a proportionate regulatory framework, agile and based on international cooperation.

The rapid development of FinTech is largely due to an environment conducive to innovation, enabling them to offer innovative and diversified banking and financial offerings. However, this dynamic is accompanied by the emergence of new risks that could threaten financial stability.

2. AI: DRIVING TRANSFORMATION IN THE FINANCIAL SECTOR

According to Asmundson et al (2011, p. 46), "the fundamental role of the financial sector is intermediation," that is, bringing together agents with financing capacity (savers) and those with financing needs (borrowers). From this perspective, the financial system can be defined as the set of economic institutions that facilitate the meeting between the savings of some and the investment needs of others (Mankiw 2007).

To perform these functions efficiently and cost-effectively, the financial sector invested heavily in information technology (IT) since the 1970s and 1980s, particularly in the areas of commercial banking, capital markets and insurance, becoming one of the main drivers of technological innovation. In the 1990s, IT spending became widespread, marking a transition to an increasingly digital economy. For example, digital transformation in banking has led institutions to make substantial investments in online and digital services. European banks have increasingly adopted APIs and developed open banking platforms to enhance customer service and integrate diverse financial offerings, reflecting a shift toward comprehensive digital operations (Stefanelli et al., 2022).

For several decades, until the 2008 financial crisis, the financial services sector was seen as a pillar of stability, both institutionally and in terms of job security. However, this paradigm has been profoundly transformed by new dynamics, including technological innovation, disruption of traditional processes and reconfiguration of business models. These developments fundamentally redefine the economic rules of the sector, both in terms of the nature of the services offered and the operational capabilities required to produce them.

a. IA as a Strategic Leader in the Financial Sector

In the financial services sector, artificial intelligence (AI) emerges as a key driver of transformation and looks set to change the way data is collected, processed and used. With advanced machine learning and NLP technologies, AI can automate and optimize complex processes ranging from fraud detection and risk management to predictive analysis of customer behavior. Today, financial institutions (banks, insurance companies) use AI algorithms to grant credit by measuring creditworthiness in real time, personalize offers by thematizing transactional data, or develop intelligent chatbots to interact with their customers daily. In addition, AI is making inroads into asset management with the emergence of robo- advisors, which offer automatic, adaptive portfolios for the widest possible range of customers.

A study conducted by Ryll et al. (2020) highlights companies' anticipation of the central role of AI in the short term as a major lever for sector development. Indeed, the results of this study suggest that the expectations of traditional players can be explained by their gradual transition from using AI to make operations leaner and more efficient to a strategy aimed at differentiating themselves through the innovation of AI-based customer service processes and solutions. However, this algorithmic sophistication raises major challenges related to discriminatory bias. Indeed, AI systems, often trained on non-neutral historical data, can reproduce or exacerbate pre-existing inequalities, for example in credit granting, insurance, or fraud detection. Studies[2] have highlighted cases of algorithmic discrimination, where certain social groups are systematically disadvantaged, without objective justification, due to correlated characteristics (geographic origin, credit history, gender, etc.).

The explanation of decisions made by intelligent systems is highlighted by Arrieta et al. (2020) as a growing challenge to be addressed to develop intelligent systems designed to produce intelligible, reliable and justifiable explanations, particularly in sensitive areas such as finance. Sometimes decisions made by artificial intelligence, for example the decision to refuse a loan or even to detect the occurrence of fraud, can lead to serious legal, economic or ethical consequences for individuals, economic actors or societies.

Furthermore, AI is accelerating market disruption by challenging traditional financial services distribution models. Incumbent players (banks, insurance companies) are facing increasing competition from FinTech, Big Tech, and neobanks, which are leveraging AI to offer lower cost, faster, and hyper-personalized services. This phenomenon is creating a dynamic of market fragmentation and strategic repositioning, in which mastery of AI is becoming a decisive competitive advantage, but also a source of regulatory tensions and potential instability.

b. The Functioning of Artificial Intelligence in the Financial Sector

The concept of artificial intelligence is not as recent as it seems, having been first discussed at a 'summer school' held in 1956 at Dartmouth College in the USA. According to the organizers of this event, artificial intelligence is based on the idea that all human cognitive functions could be modeled and reproduced by a computer (McCarthy et al., 2006).

Artificial Intelligence is not a skill set itself, it is more of a constellation of technologies which allow machines to sense, comprehend, act and learn; thus, augmenting human capabilities. AI is causing profound disruption in the financial services industry by redefining processes, business models and end-user interaction with their consumer services. It is being seen as a strategy by financial institutions seeking operational efficiency and security, by handling vast amounts of data quickly and accurately while being capable of identifying nuanced patterns that more accurately align with consumer desires. AI is perceived as a strategic lever. Indeed, it is emerging in many areas of application such as, risk modelling, anti-fraud, automated advice, algorithmic trading and regulatory compliance. In addition to the simple apprehension of AI's multiple contributions to the redefinition of finance, these applications also have technical, legal and ethical challenges that must be addressed. Overall, AI is becoming more prevalent in the financial services sector, as it enters the market in diverse domains with clarity of focus and tangible use cases of application.

First, in risk and credit assessment, AI can help to more accurately assess the creditworthiness of customers through scoring systems that utilize alternative data and the automation of the granting of microcredit. In fraud detection, AI is designed to catch suspicious bchaviours. It is used for transactional monitoring to identify fraudulent transactions in real-time and can detect identity theft. Robo-advisors represent another major type of application: they are designed to produce personalized financial recommendations at a lower cost. Logistically, they allow for automated portfolio management and produce investment recommendations suited to the user's profile. With respect to algorithmic trading, AI is used to increase investor profits and conduct massive data assessments. It is the core of statistical arbitrage and high-frequency trading. Intelligent Customer Service has also benefitted from research and development within AI, particularly to automate customer interactions 24/7. This is realized in the form of chatbots and virtual agents which utilize NLP technologies.

Regulatory compliance, more commonly referred to as RegTech, is using AI to manage compliance with applicable standards automatically. These applications span automatic KYC/AML document analysis, automatic KYC/AML monitoring/detection of suspicious activity. In management of risk, AI provides the tools to anticipate and model complex risks, typically in the context of stress testing models or predicting defaults in payment or market volatility. In customizing financial offerings, AI aids in customizing offerings to each client based on their unique needs. This means effective behavioural-targeting and suggestions of suitable offerings. The development and acceptance of artificial intelligence (AI) within the financial service sector, has exploded in the instance of a robust convergence of technology, economic and social elements. The explosion in the amount of data available (big data), in an increasingly powerful computing environment, thanks to graphics processing units (GPUs) and cloud computing, offers favorable conditions for training and making available high-performance AI models. At the same time, advances in Deep Learning and Machine Learning are enabling the development of high-performance, tailored and robust algorithms to meet the very specific needs of financial institutions. At the same time, the growing accessibility of these technologies, facilitated by the spread of open-source tools and access to cloud platforms, is tending to reduce barriers to entry. Strategically, artificial intelligence (AI) is emerging as a competitive lever in the face of the rise of FinTech, neo-banks and technology players, forcing traditional financial institutions to innovate to keep their place. What's more, customers are showing an increased demand for personalization, responsiveness and round-the-clock availability, which are being met by AI thanks to intelligent chatbots, real-time scoring, etc. In addition, some regulatory players are encouraging the controlled use of these technologies through devices such as regulatory sandboxes, helping to fortify legitimacy and spread AI within the financial sector.

Artificial intelligence (AI) research applied to the field of finance is developing rapidly, with numerous academic works pointing to its potential to profoundly transform the sector's decision-making and operational processes.

For Brynjolfsson and McAfee (2017), machine learning is one of the key elements for automating the cognitive tasks that make rapid and accurate decision-making possible. This applies to areas such as credit scoring and predictive analytics. AI applications in asset management have been the subject of a study made by Sironi (2016), the case of robo-advisors, which is an algorithmic innovation enabling personalized financial advice to be given on a large scale. Fuster et al (2022) show that AI can greatly improve risk modeling algo in the banking sector, by integrating unstructured data and enabling complex solutions to be found that the usual statistical methods cannot highlight. De Spiegeleer et al (2018) illustrate that deep learning is well exploited for algorithmic trading or fraud management, enabling more informed decisions in unstable environments.

However, many studies (Kleinberg et al., 2018) focus on denouncing the risks of algorithmic obscurity and bias, specifying the need for ad hoc regulation, rigorous technical audits, as well as greater transparency in model design. From this observation emerges the need to find a harmonious relationship between technological innovation and responsible governance to enable sustained adoption of artificial intelligence in financial services.

c. Artificial Intelligence and Data Management

In a constantly changing economic and financial environment, marked by the accelerating digitalization of services, increased regulatory pressure and market volatility, the various financial institutions are having to review their management methods to remain competitive. Artificial intelligence (AI), combined with advanced data management, emerges as a key strategic driver of this transformation.

Banks, insurance companies, asset management firms, and FinTechs generate massive volumes of data daily from transactions, customer interactions, market feeds, and unstructured data (such as emails, voice, and images). Effectively leveraging this informational capital is a major strategic challenge. Leveraging AI, particularly machine learning techniques, predictive analytics, and natural language processing (NLP)-enables the automation of data analysis, facilitates the anticipation of customer behaviors, enhances fraud detection, and optimizes credit assessment, risk management, and investment decision-making processes. The increasing integration of artificial intelligence and advanced data management is fundamentally reshaping business and operational models within the financial sector. This integration facilitates enhanced personalization of customer relationships, streamlines internal processes, and significantly improves the speed and accuracy of decision-making. Nevertheless, these advancements also give rise to critical challenges related to data governance, algorithmic transparency, mitigation of decision-making biases, cybersecurity, and adherence to prevailing regulatory frameworks, including the General Data Protection Regulation (GDPR[3]) and the Basel III and IV accords.

The financial sector is recognized as a frontrunner in the adoption of artificial intelligence (AI), particularly in the processing and analysis of large volumes of heterogeneous data derived from banking operations, financial markets, customer relationship management, and risk management. This dynamic is part of a broader context of digital transformation, where the intelligent exploitation of data is now considered an essential strategic lever for competitiveness, performance and innovation. Manyika et al (2011) assert that the financial sector is one of the biggest generators of actionable data. This data, whether structured (financial transactions, credit histories) or unstructured (emails, voice messages, social network content), is a highly valuable source of information. Yet their valorization is subject to various technical and organizational difficulties, including storage, cleansing, integration and quality assurance (Chen et al., 2020).

Ngai et al (2011) offer an in-depth analysis of data mining methods applied to the detection of financial fraud. Their study is distinguished by a rigorous classification of the main existing approaches, including neural networks, decision trees, Bayesian methods as well as hybrid models.

The authors will provide a more detailed review of the application of these techniques in a variety of fraud-related environments, including credit card fraud, money laundering, and accounting fraud. This review is a monumental effort in understanding the impacts of advanced data analysis and artificial intelligence on improving proactive fraud detection in the financial services sector. Relatedly, work by Sirignano and Giesecke (2019) reported evidence that deep learning approaches significantly outperformed traditional statistical models for predicting default risk in mortgage loans. By implementing a very large dataset consisting of more than 120 million observations, the designers of this neural network learn to model, as we have seen, complex and non-linear relationships between borrower characteristics and macroeconomic variables, which makes it possible to obtain interesting predictive performances, even in an uncertain macroeconomic environment, allowing fine tuning of credit risk management in financial institutions. Far from remaining a simple demonstration, this work clearly demonstrates the capacity of artificial intelligence to more profoundly transform financial data management practices, by making risk assessment processes more efficient and personalized.

In their publication, Arrieta et al. (2020) emphasize explainable artificial intelligence (XAI) while showing the limits of transparency of AI models, particularly those based on deep learning. The authors point out the need to strike a balance between the predictive power of algorithms and the understandability of their results to end users. They also point out the important role of explainable AI (XAI) to meet regulatory standards, more specifically, those of General Data Protection

Regulation (GDPR) which promotes more transparency regarding algorithmic decisions by businesses that affect individuals. This contribution thus provides a sound theoretical foundation for embedding ethical, legal, and transparency requirements when developing AI systems, and particularly in the high-risk sectors like finance. The digital transformation currently underway raises important questions to protect personal data, especially when AI is being developed that requires massive amounts of data to develop and enhance algorithmic models. Understanding the conceptual framework in which these data interact with AI is essential to consider the ethical, legal, and societal impact of this convergence of technologies.

It is vital to investigate the deep connections between personal data and intelligent systems and to raise questions concerning the related challenges of confidentiality, information security, and the ethical governance of automated processing. The intention was to clarify the tensions growing between innovation and the necessary processes to protect privacy in a rapidly changing digital world.

d. Algorithmic Governance: Towards Responsible AI

As artificial intelligence becomes a key strategic lever in the financial sector, the need for ethical oversight and robust governance of algorithmic systems becomes urgent. Algorithms used for credit scoring, risk management, automated trading, or fraud detection can, if left unchecked, generate opaque, biased, or unfair decisions. It is in this context that algorithmic governance is emerging as a central area of regulation and research, aimed at ensuring that AI systems operate in a transparent, explainable, and accountable manner.

Algorithmic transparency is one of the cornerstones of this governance. It refers to the ability of stakeholders—regulators, developers, and customers—to understand the decision-making mechanisms of algorithms. This involves the implementation of traceability tools and technical documentation, as well as explainable interfaces (XAI – Explainable AI) accessible to non-specialists. In the financial sector, where decisions can have major socio-economic consequences, transparency is essential to ensure regulatory compliance and maintain user trust.

Auditability is another central pillar. It refers to an algorithm's ability to be inspected after the fact, particularly in the event of litigation, error, or harm. Algorithmic audits can be conducted internally or by independent third parties, based on defined risk assessment, fairness testing, and model monitoring protocols. In some jurisdictions, such as the European Union, this auditability requirement is integrated into emerging regulatory frameworks, notably in the European Regulation on Artificial Intelligence (AI Act, 2024).

Furthermore, the principles of algorithmic fairness, non-discrimination, and the protection of fundamental rights are now recognized by several international organizations. The OECD (2019), UNESCO (2021)[4], and the Council of Europe (2023)[5] have each published recommendations aimed at regulating the use of AI in accordance with democratic values. These principles have direct application in finance, where concrete risks of bias (gender, race, location) have been identified in credit rating, insurance, and risk management systems.

In addition to these ethical frameworks, technical initiatives aim to integrate ethics into the design of AI systems ("ethics by design"). This involves introducing ethical constraints at the level of data, algorithms, and validation processes. Standards are currently being developed, for example by the IEEE through its Ethically Aligned Design project, or by the NIST (National Institute of Standards and Technology) in the United States, which published an AI risk management framework in 2023.

Finally, the question of legal liability related to the use of autonomous algorithms remains partially resolved. Who is liable in the event of an algorithmic error: the designer, the financial institution, or the data provider? The debate over the legal personality of AI, although still theoretical, reflects the limitations of current law when faced with autonomous and evolving systems.

Algorithmic governance thus represents a strategic area of financial regulation today. It offers institutions not only a means of compliance, but also an opportunity to improve the quality and ethics of their services. Effective governance requires a combination of technological tools, ethical principles, audit mechanisms, and appropriate legal frameworks, both nationally and internationally.

3. CASE STUDIES: CONCRETE IMPLEMENTATION OF AI IN FINANCIAL INSTITUTIONS

While the promise of artificial intelligence in finance is widely theorized, its operational translation in financial institutions remains mixed. This section presents three case studies illustrating concrete uses of AI - in document management, credit scoring, and algorithmic auditability - to demonstrate the opportunities, but also the limitations and risks of such implementations.

a. JP Morgan Chase & Co. (USA): Contract automation via COIN

In 2017, JP Morgan, the largest US bank by assets, launched an internal AI tool called COIN (Contract Intelligence). This system, based on natural language processing (NLP), can analyze thousands of financial contracts, including commercial loan agreements, in seconds. The primary objective is to reduce contract analysis time, eliminate human error, and improve regulatory compliance. According to the bank, COIN has made it possible to complete, in minutes, what previously required 360,000 hours of annual work from a legal staff.

However, this project also raised questions regarding algorithmic transparency and reliability. The AI models used by COIN underwent a rigorous calibration phase to avoid misinterpretations of complex legal clauses. This case highlights the importance of human oversight and quality control in high-responsibility AI systems.

b. Ant Group (China): AI-powered credit assessment with Zhima Credit

In China, Ant Group (a subsidiary of Alibaba) has developed a credit scoring system called Zhima Credit (also known as Sesame Credit). Based on advanced AI techniques, this system calculates a credit score for individuals based not only on financial data (income, repayments), but also on behavioral data (purchase history, social media, payment timeliness).

While this model enables greater financial inclusion, particularly for individuals without a traditional banking history, it raises serious ethical concerns. The lack of transparency of the criteria used, behavioural surveillance, and the confusion between social reputation and creditworthiness have been widely criticized, including by international NGOs.

This case highlights the dilemmas between technological innovation, operational efficiency, and the protection of individual freedoms, in a context where algorithmic governance remains highly dependent on the political environment.

c. BBVA (Spain): AI, auditability and internal governance

The Spanish bank BBVA is one of the first major European institutions to implement a responsible AI governance framework. In 2022, it implemented an internal algorithmic audit system to evaluate AI models used for credit, risk management, and fraud detection. This system is based on an AI ethics framework inspired by European Union and OECD principles.

Each deployed AI model is subject to an assessment in terms of bias, explainability, and robustness, conducted by an interdisciplinary committee (legal, data science, compliance). At the same time, BBVA is participating in European research projects on explainability technologies (XAI) and interpretable models in finance.

This case illustrates how a major bank can integrate ethics into its systems design by creating an organizational culture of algorithmic accountability.

It is essential to note that industry players are at different stages of maturity in terms of the degree to which they have implemented these technologies in their long-term planned activities. While many banking institutions have already begun implementing intelligent automation-based solutions, such as advanced analytics, robotic process automation, robo-advisors, or self-learning systems, there is still a long way to go. Technologies will, in fact, need to evolve, become more commonplace, and become more available and innovative to address the challenges that still need to be addressed (Gokhale et al., 2019).

4. THE NECESSITY OF STRICT INTELLECTUAL PROPERTY MANAGEMENT

Honouring the principles of confidentiality, openness, and accountability are key to safeguarding personal data in artificial intelligence. Strict adherence to legal frameworks, specifically the General Data Protection Regulation (GDPR) in Europe, is vital to limit privacy breaches and ensure the legal legitimacy of automated processing. Another key issue is the need to increase awareness among employees and end users, regarding personal data protection, to enable the ethical and responsible use of AI technologies. The issuing of controls, and monitoring of processes on a continuous basis, and regular audits, are all important in terms of detecting and acting on anomalies or violations, increasing compliance, and trust.

Subsequently, it is vital that robust procedures around data validation, and governance, be put in place and balance applied between the force of technological innovation and fundamental rights.

a. The Concept of Intellectual Property

In finance, intellectual property can encompass all legal rights that protect valuable intangible assets, including proprietary algorithms, predicative investment models, customer databases, trademarks and the like—and yes even FinTech patents. It is important that we strengthen institutional competitiveness by ensuring that innovation is correctly protected, whether automated trading laboratories, mobile payment solutions and/or continued access to sustainable competitive advantage.

Intellectual property (IP) plays a significant role for financial services when it comes to transforming intangible assets into valuable economic resources and is particularly relevant in knowledge economies. As an intangible asset, IP can be financially valued, assigned and licensed, used as leverage for financing and credit transactions; and can even act as collateral. This strengthens the financial institutions' ability to enhance and monetize their intangible capital.

b. The Role and Significance of PI in the Financial Sector

In finance, intellectual property (IP) is an important consideration when valuing intangible assets (for example, software; algorithms; databases; patents; trademarks; and trade secrets). These items typically form the bulk of a business's value, especially in the case of FinTech. IP rights protect technological innovation and creative ideas, safeguarding investment in research and development and increasing a defence against the threats of counterfeiting and imitation. As AI, big data and automation become more prominent, there are more governance and compliance issues around IP, especially to mitigate risks associated with potential misuse of protected data and relying on external technological solutions. In short, intellectual property is proving to be an essential lever for combining innovation, risk management, and economic performance in the financial sector.

Traditionally, banks and FinTech have protected their innovations through software patents, source code copyrights, or trade secrets. However, this exclusive view of IP is evolving under the impetus of the digital economy, which now favours more open and collaborative forms of rights management.

i. IP and Open Innovation: Towards New Paradigms

As part of open innovation, financial institutions are increasingly exploring collaborative models in which knowledge, patents, and technologies are shared among multiple stakeholders - banks, startups, universities, and incubators - to accelerate the co-creation of solutions. These practices require flexible IP management, no longer based solely on exclusive rights, but also on the voluntary pooling of certain intangible assets.

Open licenses, such as Creative Commons, Open-Source licenses (MIT, GNU GPL), or patent pool models, allow organizations to share their technologies while regulating their reuse. Some FinTechs voluntarily choose to make their code available to encourage industry adoption or strengthen their credibility within the ecosystem.

ii. Innovation Consortia and Shared Intellectual Property

An emblematic example of this evolution is the creation of innovation consortia in finance, such as:

- R3 Corda: a consortium of more than 300 institutions (banks, insurers, regulators) collaborating on a blockchain platform with shared management of patents and standards.

- Linux Foundation's Hyperledger Project: an open-source initiative for the development of blockchain solutions in finance, with contributions under the Apache 2.0 license.
- Fin Reg Lab (United States): a public-private platform that analyses credit algorithms in open access and works on ethical and collaborative IP models.

These consortia often adopt contractual frameworks defining intellectual property co-ownership, access rights, and the terms and conditions for publishing results or filing joint patents. This combines legal certainty and collaborative dynamism, while preserving the competitiveness of members.

iii. Towards Ethical and Sustainable IP in Finance

Some institutions are aligning their IP management with the Sustainable Development Goals (SDGs) and responsible technology principles. For example, socially engaged FinTech are releasing financial analysis APIs or budget education tools under open license to promote financial inclusion. These strategies demonstrate that IP is no longer solely an instrument of economic lock-in but can become a tool for positive social impact.

c. Legal and Strategic Mechanisms for Protecting Intellectual Property

Organizations operating in the financial services sector must adapt to this digital transformation and seize its opportunities by incorporating new work tools into their activities. These new intangible assets (algorithms, artificial intelligence methods) as well as the inputs used by the company to operate them (databases) must be able to be subject to legal protection by intellectual property rights. Thus, the main means of protecting IP are:

- Patents for FinTech innovations: For a secure algorithm or protocol to be patentable, it must meet three essential conditions: it must be new, inventive and capable of industrial application. In practice, the patent application must include a detailed description of the technical operation of the invention, covering elements such as software architecture, data flows and encryption mechanisms, and make clear claims specifying the extent of protection sought.

A FinTech bank or start-up then files its invention with a competent body such as the INPI (France), the USPTO (United States) or the EUIPO (European Union). After submitting the application, the company must follow the examination procedure, respond to any objections raised by the examiner and pay the annual fees to maintain the validity of the patent, usually for a period of 20 years.

- Copyright on banking software: The source code, technical documentation and user interfaces are automatically protected by copyright upon their creation, without any specific formalities. However, to strengthen proof of ownership and the dating of the work, it is common to use deposits with trusted third parties, such as the envelope Soleau, Safe Creative or a deposit with a judicial officer. In the event of litigation, such as code infringement or unauthorized reuse, these deposits serve as solid evidence to establish authorship and date of creation.
- Brands for financial services and applications: The trade name, logo and, where applicable, a slogan may be registered as a trademark for legal protection. The institution selects the appropriate classes of products and services (for example, class 36 for banking services) and submits its application to INPI (France) or EUIPO (European Union). Once the trademark is registered, for a period of 10 years renewable, it confers an exclusive monopoly on its use. This allows the company to prosecute for infringement any unauthorized use that could create a misleading association, such as the use of a similar name or logo resembling the registered trademark.
- Designs and models for UI/UX.: The screens, models and graphics of a mobile application or payment terminal may be protected by design. The applicant submits visual representations (such as plans or screenshots) to INPI or EUIPO for renewable protection of up to 25 years. This protection prevents unauthorized copying or modification of its interfaces by competitors, thus ensuring the preservation of the visual identity of the offer and differentiation in the market.
- Business secrets for scoring and risk analysis methods: Unpatented algorithms, internal rule bases or statistical model parameters are protected under the trade secret regime, provided that the company implements strict confidentiality measures such as internal policies, non-disclosure agreements (NDAs), restricted access to sensitive information and encryption systems. In the event of unlawful disclosure, the company may take legal action under the Trade Secret Protection Act to defend its interests and intangible assets.
- Rights sui generis on customer and market databases: The producer of a database (structured compilation) automatically benefits from a sui generis right, which prohibits the substantial extraction or reuse of this database without authorization. To maximize the protection of this right, it is recommended that the structure and manner of data collection be formalized in a license agreement, which specifies restrictions on use and prevents penalties for vio-

lation of the terms of the agreement. This contract thus strengthens the legal certainty around the database and protects the interests of the producer.

- Contractual agreements and licences: The license clauses (whether open source or proprietary) define in a precise way the rights of use, modification and redistribution of third-party software, APIs or data, thus ensuring the management of the rights related to the exploitation of these resources. Furthermore, non-competition and non-solicitation clauses in data scientists' and developers' contracts are intended to protect the company against risks associated with the sudden departure of key talent. These clauses prevent the transfer of sensitive knowledge and ensure that these employees do not join competitors or create new companies exploiting strategic information acquired during their mission.
- Technical protection measures: Advanced encryption (such as AES-256 or RSA 4096 bits) ensures that even in the event of code or data theft, it remains unusable without the decryption key. Digital Rights Management (DRM) and watermarking solutions add invisible metadata to documents or analytical reports, allowing their distribution to be traced and any unauthorized use to be identified. At the same time, identity and access management (IAM) ensures strong authentication (such as MFA or multi-factor authentication) and granular control of access rights, based on specific user roles, to enhance security and prevent unauthorized access to sensitive data.
- Monitoring, surveillance and enforcement: The implementation of automated surveillance (via online scanners or monitoring bots of marketplaces) allows to quickly detect any violation of brands or the posting of counterfeit codes. In the event of a violation, the company may send removal notices (such as takedown notices or DMCA notifications) to have illegal content removed. If necessary, it can also initiate legal proceedings to obtain injunctions, damages and restitution of the illicit gains generated by the infringement.

d. Alternative Dispute Resolution Mechanisms in IP and AI Matters

Faced with the growing complexity of conflicts involving artificial intelligence and intellectual property, financial institutions are increasingly resorting to alternative dispute resolution (ADR) mechanisms, such as arbitration, mediation, or hybrid arrangements. These tools offer a flexible, rapid, and confidential response to disputes related to algorithms, data rights, or co-ownership of intangible assets resulting from collaborative projects. Mediation, for example, helps preserve commercial relationships and avoid lengthy legal disputes, while promoting negotiated solutions in environments where technological confidentiality is paramount. Arbitration, for its part, stands out for its ability to mobilize technical and legal experts capable of addressing issues specific to AI, such as the liability of automated models or the ownership of software patents. New forms are also emerging, such as digital dispute resolution (ODR) platforms or internal ethics committees, in line with the transparency requirements imposed by frameworks such as the European regulation on AI (AI Act). Integrating these mechanisms into corporate governance now constitutes a strategic lever for anticipating risks, reducing legal exposure, and strengthening confidence in technological innovation in the financial sector.

RECOMMENDATIONS

Following this analysis, several recommendations can be made to support financial institutions in their digital transition, integrating artificial intelligence (AI) and strategic intellectual property (IP) management:

- Strengthen algorithmic governance by drawing on internationally recognized ethical frameworks (such as the OECD, UNESCO, and the European Union's AI Act guidelines). This would ensure the transparency, reliability, explainability, and auditability of AI systems used in finance.
- Develop an integrated data management strategy, ensuring the quality, security, traceability, and regulatory compliance of internal and external data used by AI algorithms, while strengthening stakeholder trust.
- Encourage collaboration between traditional financial institutions and FinTechs to stimulate joint innovation, accelerate technology adoption, and better meet the demands of service personalization and market competitiveness.
- Adopt proactive intellectual property management, integrating the challenges of open innovation, collaborative licenses (such as Creative Commons licenses), and technology consortia. This would allow for the economic development of intangible assets while securing competitive advantages.
- Integrate alternative dispute resolution (ADR) mechanisms, such as arbitration, mediation, or ethics committees, to effectively address emerging AI and IP disputes in a rapidly evolving legal environment.

This chapter's contribution is its capacity to unite two issues discussed separately: the contributions of artificial intelligence to the modernization of the financial industry, and the significance of intellectual property for maintaining competitive advantages from benefits from innovation here. By revealing the connection of data and technologies with intangible assets, this chapter offers an interdisciplinary perspective. It enables different angles of analysis with a value in providing concrete suggestions for academics, financial decision-makers, and legal professionals.

CONCLUSION

In the age of digitalization and artificial intelligence (AI), companies and institutions - especially in the financial sector - are undergoing deep, systemic transformations. These changes extend far beyond technical infrastructures to encompass decision-making processes, organizational models, and corporate cultures. As Chan, K. T. (2022) aptly observes, digitalization redefines our relationship to data, time, others, and ourselves, signalling a profound societal evolution.

Successfully navigating this transformation requires not only robust change management - with tools, strategies, and governance mechanisms - but also a transformation of conduct, influencing both individual and collective behaviours. The synergy between these two dimensions is essential for ensuring that AI deployment is not only efficient but also aligned with human values, ethical considerations, and social responsibility.

The integration of AI in finance has indeed improved operational efficiency, customer interaction, and innovation potential, with tools such as robo-advisors, automated credit scoring, and algorithmic trading. Yet, this evolution comes with major challenges: algorithmic opacity, discrimination risks, data governance, and regulatory uncertainty. These require a shift toward "responsible AI", incorporating principles such as fairness, explainability, auditability, and accountability.

Moreover, intellectual property (IP) plays a strategic role in this new paradigm. Properly designed, an IP strategy enables financial actors to secure innovations, protect intangible assets, and enhance competitiveness. However, traditional legal frameworks may prove insufficient when dealing with generative AI, co-created outputs, and shared innovation ecosystems. Financial institutions must thus adopt more flexible and collaborative models, including open licenses, innovation consortiums, and co-ownership frameworks, to keep pace with evolving technological realities.

To ensure legal certainty and minimize litigation risks, the adoption of alternative dispute resolution (ADR) mechanisms - such as mediation, arbitration, and online dispute resolution - is becoming increasingly critical, especially in transnational and technically complex contexts. These mechanisms not only facilitate faster and more confidential settlements but also reinforce trust among innovation stakeholders.

Looking forward, the convergence of AI, finance, and intellectual property calls for interdisciplinary research, combining insights from technology, law, ethics, and strategy. There is a pressing need to explore questions such as:

- How can AI systems be audited and regulated without stifling innovation?
- How should we redefine IP protection for autonomous or co-generated content?
- How can financial AI be made inclusive, socially just, and environmentally sustainable?

By addressing these questions, the financial sector can fully harness the transformative potential of AI while ensuring that progress is ethical, inclusive, and aligned with the Sustainable Development Goals (SDGs). Institutions that adopt this holistic and anticipatory approach will be best positioned to lead in a financial ecosystem defined not just by innovation, but by responsibility, transparency, and trust.

REFERENCES

Aguirre, S., & Rodriguez, A. (2017, August). Automation of a business process using robotic process automation (RPA): A case study. In *Workshop on engineering applications* (pp. 65-71). Cham: Springer International Publishing. DOI: 10.1007/978-3-319-66963-2_7

Arner, D. W., Barberis, J., & Buckley, R. P. (2015). The evolution of Fintech: A new post-crisis paradigm. *SSRN*, *47*, 1271. DOI: 10.2139/ssrn.2676553

Aro, O. E.Opeyemi E Aro. (2024). Data analytics as a driver of digital transformation in financial institutions. *World J. Adv. Res. Rev.*, *24*(1), 1054–1072. DOI: 10.30574/wjarr.2024.24.1.3124

Arrieta, A. B., Díaz-Rodríguez, N., Del Ser, J., Bennetot, A., Tabik, S., Barbado, A., & Herrera, F. (2020). Explainable Artificial Intelligence (XAI): Concepts, taxonomies, opportunities and challenges toward responsible AI. *Information Fusion*, *58*, 82–115. DOI: 10.1016/j.inffus.2019.12.012

Asmundson, I., Dorsey, T. W., Khachatryan, A., Niculcea, I., & Saito, M. (2011). Trade and trade finance in the 2008-09 financial crisis.

Bader, M. A. (2006). *Intellectual property management in R&D collaborations: the case of the service industry sector*. Physica-Verlag HD.

Böhme, R., Christin, N., Edelman, B., & Moore, T. (2015). Bitcoin: Economics, technology, and governance. *The Journal of Economic Perspectives*, *29*(2), 213–238. DOI: 10.1257/jep.29.2.213

Brogaard, J. (2010). High frequency trading and its impact on market quality. *Northwestern University Kellogg School of Management Working Paper, 66*(10).

Brynjolfsson, E., & Mcafee, A. N. D. R. E. W. (2017). Artificial intelligence, for real. *Harvard Business Review*, *1*(1), 1–31.

Buchak, G., Matvos, G., Piskorski, T., & Seru, A. (2018). Fintech, regulatory arbitrage, and the rise of shadow banks. *Journal of Financial Economics*, *130*(3), 453–483. DOI: 10.1016/j.jfineco.2018.03.011

Catalini, C., & Gans, J. S. (2020). Some simple economics of the blockchain. *Communications of the ACM*, *63*(7), 80–90. DOI: 10.1145/3359552

Chaboud, A. P., Chiquoine, B., Hjalmarsson, E., & Vega, C. (2014). Rise of the machines: Algorithmic trading in the foreign exchange market. *The Journal of Finance*, *69*(5), 2045–2084. DOI: 10.1111/jofi.12186

Chan, K. T. (2022). Emergence of the 'Digitalized Self' in the Age of Digitalization. *Computers in Human Behavior Reports, 6*, 100191. DOI: 10.1016/j.chbr.2022.100191

Chen, S., Fan, L., Meng, G., Su, T., Xue, M., Xue, Y., & Xu, L. (2020, June). An empirical assessment of security risks of global android banking apps. In *Proceedings of the ACM/IEEE 42nd International Conference on Software Engineering* (pp. 1310-1322). DOI: 10.1145/3377811.3380417

Chen, Y., & Bellavitis, C. (2020). Blockchain disruption and decentralized finance: The rise of decentralized business models. *Journal of Business Venturing Insights, 13*, e00151. DOI: 10.1016/j.jbvi.2019.e00151

De Spiegeleer, J., Madan, D. B., Reyners, S., & Schoutens, W. (2018). Machine learning for quantitative finance: Fast derivative pricing, hedging and fitting. *Quantitative Finance, 18*(10), 1635–1643. DOI: 10.1080/14697688.2018.1495335

Dispas, C., Kayanakis, G., Servel, N., & Striukova, L. (2021). *Innovation and financial markets*. John Wiley & Sons. DOI: 10.1002/9781119819912

Dwivedi, D. N., Mahanty, G., & Dora, T. (2024). Hyperautomation in Financial Services: Revolutionizing Banking and Investment Processes. In *Hyperautomation in Business and Society* (pp. 184-199). IGI Global.

Frost, J. (2020). The economic forces driving fintech adoption across countries. *The technological revolution in financial services: how banks, fintechs, and customers win together, 838*(II), 70-89.

Fuster, A., Goldsmith-Pinkham, P., Ramadorai, T., & Walther, A. (2022). Predictably unequal? The effects of machine learning on credit markets. *The Journal of Finance, 77*(1), 5–47. DOI: 10.1111/jofi.13090

Gokhale, N., Gajjari, A., Kaye, R., & Kuder, D. (2019). AI leaders in financial services: common traits of frontrunners in the artificial intelligence race.

Goldstein, I., Jiang, W., & Karolyi, G. A. (2019). To FinTech and beyond. *Review of Financial Studies, 32*(5), 1647–1661. DOI: 10.1093/rfs/hhz025

Gomber, P., Kauffman, R. J., Parker, C., & Weber, B. W. (2018). On the fintech revolution: Interpreting the forces of innovation, disruption, and transformation in financial services. *Journal of Management Information Systems, 35*(1), 220–265. DOI: 10.1080/07421222.2018.1440766

Haddad, C., & Hornuf, L. (2019). The emergence of the global fintech market: Economic and technological determinants. *Small Business Economics, 53*(1), 81–105. DOI: 10.1007/s11187-018-9991-x

Hanel, P. (2006). Intellectual property rights business management practices: A survey of the literature. *Technovation*, *26*(8), 895–931. DOI: 10.1016/j.technovation.2005.12.001

Hendershott, T., Jones, C. M., & Menkveld, A. J. (2011). Does algorithmic trading improve liquidity? *The Journal of Finance*, *66*(1), 1–33. DOI: 10.1111/j.1540-6261.2010.01624.x

Kirilenko, A., Kyle, A. S., Samadi, M., & Tuzun, T. (2017). The flash crash: High-frequency trading in an electronic market. *The Journal of Finance*, *72*(3), 967–998. DOI: 10.1111/jofi.12498

Kleinberg, J., Lakkaraju, H., Leskovec, J., Ludwig, J., & Mullainathan, S. (2018). Human decisions and machine predictions. *The Quarterly Journal of Economics*, *133*(1), 237–293. DOI: 10.1093/qje/qjx032 PMID: 29755141

Mankiw, N. G. (2007). *Principles of microeconomics* (p. 0). Mason, OH: Thomson South-Western.

Manyika, J., Chui, M., Brown, B., Bughin, J., Dobbs, R., Roxburgh, C., & Hung Byers, A. (2011). Big data: The next frontier for innovation, competition, and productivity.

Maskus, K. E. (1998). The role of intellectual property rights in encouraging foreign direct investment and technology transfer. *Duke J. Comp. & Int'l L.*, *9*, 109.

McCarthy, J., Minsky, M. L., Rochester, N., & Shannon, C. E. (2006). A proposal for the dartmouth summer research project on artificial intelligence, august 31, 1955. *AI Magazine*, *27*(4), 12–12.

Mention, A. L. (2019). The future of fintech. *Research Technology Management*, *62*(4), 59–63. DOI: 10.1080/08956308.2019.1613123

Morandín-Ahuerma, F. (2023). Ten UNESCO recommendations on the ethics of artificial intelligence.

Ngai, E. W., Hu, Y., Wong, Y. H., Chen, Y., & Sun, X. (2011). The application of data mining techniques in financial fraud detection: A classification framework and an academic review of literature. *Decision Support Systems*, *50*(3), 559–569. DOI: 10.1016/j.dss.2010.08.006

OECD. (2019). *F. T. C. A. N.* Principles on Artificial Intelligence.

Philippon, T. (2016). *The fintech opportunity* (No. w22476). National Bureau of Economic Research.

Ryll, L., Barton, M. E., Zhang, B. Z., McWaters, R. J., Schizas, E., Hao, R., ... & Yerolemou, N. (2020). Transforming paradigms: A global AI in financial services survey.

Schär, F. (2021). Decentralized finance: On blockchain-and smart contract-based financial markets. *FRB of St. Louis Review*.

Sirignano, J., & Giesecke, K. (2019). Risk analysis for large pools of loans. *Management Science*, *65*(1), 107–121. DOI: 10.1287/mnsc.2017.2947

Sironi, P. (2016). *FinTech innovation: from robo-advisors to goal based investing and gamification*. John Wiley & Sons. DOI: 10.1002/9781119227205

Stefanelli, V., Manta, F., & Toma, P. (2022). *Digitalization and competition in the EU banking sector: Regulatory strategies and market evolution. arXiv*.

Syed, R., Suriadi, S., Adams, M., Bandara, W., Leemans, S. J., Ouyang, C., ter Hofstede, A. H. M., van de Weerd, I., Wynn, M. T., & Reijers, H. A. (2020). Robotic process automation: Contemporary themes and challenges. *Computers in Industry*, *115*, 103162. DOI: 10.1016/j.compind.2019.103162

Tapscott, D., & Tapscott, A. (2016). *Blockchain revolution: how the technology behind bitcoin is changing money, business, and the world*. Penguin.

Thakor, A. V. (2020). Fintech and banking: What do we know? *Journal of Financial Intermediation*, *41*, 100833. DOI: 10.1016/j.jfi.2019.100833

Thompson, A. (2018). Robotic Process Automation (RPA) in Banking Operations. *International Journal of Artificial Intelligence and Machine Learning*, *1*(2).

Willcocks, L., Lacity, M., & Craig, A. (2017). Robotic process automation: Strategic transformation lever for global business services. *Journal of Information Technology Teaching Cases*, *7*(1), 17–28. DOI: 10.1057/s41266-016-0016-9

Ziegler, T., Shneor, R., Wenzlaff, K., Wang, B., Kim, J., Paes, F. F. D. C., & Adams, N. (2021). The global alternative finance market benchmarking report. *Available at SSRN* 3771509.

KEY TERMS AND DEFINITIONS

Algorithmic Trading: The use of computer algorithms to execute financial market trades at high speed and frequency, often using AI and big data analytics to identify profitable patterns and optimize investment strategies.

Artificial Intelligence (AI): A multidisciplinary field of computer science that enables machines to simulate human cognitive functions such as learning, reasoning, and problem-solving. In the financial sector, AI encompasses technologies like machine learning, natural language processing, and computer vision, which are applied to automate operations, enhance risk management, detect fraud, and provide personalized customer experiences.

Big Data: Large, complex datasets characterized by high volume, velocity, and variety, requiring advanced analytical tools and computational methods to process and extract actionable insights. In finance, big data supports predictive modelling, market segmentation, algorithmic trading, and anomaly detection.

Data Governance: A structured framework of policies, standards, and procedures ensuring the quality, security, privacy, and compliance of data assets. In AI-driven finance, it guarantees that training data is accurate, ethically sourced, and legally compliant.

Digital Transformation: The integration of digital technologies into all areas of a business, fundamentally changing how it operates and delivers value. In the financial sector, digital transformation involves automation, AI adoption, data-driven decision-making, and enhanced customer engagement.

Financial Technology (FinTech): The integration of innovative technologies into financial services to improve efficiency, accessibility, and customer experience. FinTech companies leverage digital platforms, mobile applications, blockchain, and AI to disrupt traditional banking models. They offer services such as digital payments, peer-to-peer lending, robo-advisory, and automated investment tools.

Intellectual Property (IP): A set of legal rights that protect intangible assets, including inventions, software, algorithms, brand identities, databases, and trade secrets. In AI-powered financial services, IP safeguards innovations and proprietary processes, ensuring ownership and commercial benefits.

Intellectual Property Management (IPM): The strategic process of identifying, protecting, commercializing, and enforcing intellectual property assets to maximize their value. In the financial industry, IPM ensures AI applications and analytics solutions are protected from unauthorized use while enabling monetization through licensing and partnerships.

Predictive Analytics: An analytical approach that uses statistical models, machine learning algorithms, and historical data to forecast future trends, events, or customer behaviours. It is widely applied in credit risk assessment, fraud detection, customer churn prediction, and portfolio optimization.

Regulatory Technology (RegTech): A subset of FinTech that uses innovative technologies—such as AI, big data analytics, and blockchain—to help financial institutions comply with regulations more efficiently, streamline reporting, and monitor transactions.

ENDNOTES

1 Know Your Customer

2 Fuster, A., Goldsmith-Pinkham, P., Ramadorai, T., & Walther, A. (2022). Predictably unequal? The effects of machine learning on credit markets. *The Journal of Finance*, *77*(1), 5-47.

3 The GDPR (General Data Protection Regulation) is a European regulation that came into force on May 25, 2018, which governs the processing of personal data in the European Union (EU) and strengthens the rights of data subjects.

4 Morandín-Ahuerma, F. (2023). Ten UNESCO recommendations on the ethics of artificial intelligence.

5 Rotenberg, M. Framework Convention on Artificial Intelligence and Human Rights, Democracy and the Rule of Law (Council Eur.). *International Legal Materials*, 1-44.

Chapter 11
AI and ChatGPT as Pillars of Financial Operations Optimization

Salma Khaled
https://orcid.org/0009-0002-4019-3065
National School of Commerce and Management, Morocco

Aziz Moutahaddib
National School of Commerce and Management, Morocco

ABSTRACT

In a context of increasing digitalization, artificial intelligence (AI) and language models such as ChatGPT are revolutionizing the management of financial operations. Their integration enables the automation of repetitive tasks, predictive data analysis, and the optimization of decision-making, profoundly transforming industry practices. ChatGPT, thanks to its ability to understand language, improves customer relations and the personalization of services, while boosting the productivity of finance teams. In addition, AI contributes to proactive risk management and better regulatory compliance. However, these advances come with ethical and technical challenges related to algorithmic transparency, data security, and skills transformation. This article analyzes the contributions, limitations, and prospects of AI and ChatGPT, highlighting their strategic role in the emergence of smart, ethical, and resilient finance.

DOI: 10.4018/979-8-3373-3104-1.ch011

INTRODUCTION

More generally, businesses are facing a lot of problems since the market is always changing, new technology is coming out, and global politics are making things harder for them. Modern economic systems have to work in a world that is complicated. They can't only be functional; they also have to be able to change with the times and the pace of systemic changes. AI lead and ChatGPT accelerant tools are two examples of things that can help with this. They are very important for making financial transactions and operations better. AI is a technology that not only lets you automate boring, repetitive jobs, but it also lets you look at huge amounts of data at rates and volumes that were previously unimaginable. This gives you useful information that is essential for making strategic decisions. And ChatGPT, which is an advanced language model prototype, makes it easier for people to talk to each other inside and outside of the company. This makes it easier for the finance team to get information and be more productive. This package may analyze market trends and behavior to help companies better understand how the economy is changing and plan and carry out data-based initiatives. Tech alone can't make that decision. For things to start flying off the AWS shelf and into your company's financial tools, innovation has to be a known key value that guides decision-making at all levels of the organization. Because of this, businesses are becoming more proactive in their decision-making because they can do more to make money and protect themselves. So, let's use this introduction to start a conversation on how AI and ChatGPT in particular may be used in finance to create value. While we wait for this change to happen, we will keep looking for the pros and disadvantages of it and pushing for careful rollouts that take into account the unique needs of each company.

LITERATURE REVIEW

Artificial Intelligence in Financial Operations

Financial AI has been a popular topic for the past ten years. The previous study by El MGHAIFRI and ZAMMAR (2025) concentrated on process automation and performance enhancement, suggesting that AI can lower costs and minimize human errors. Recent research has focused on the enhancement of data analytics, risk management, and fraud detection via artificial intelligence (Simen Nana et al., 2024; Jabeur & Nammouri, 2025). For instance, machine learning techniques enable organizations to forecast anomalies in creditworthiness and optimize resource allocation. But they are worried about bias and transparency in the algorithm and want AI to be explainable (Jacob & Brousseau, 2025).

ChatGPT and Conversational AI in Finance

ChatGPT and other large language models (LLMs) are a huge step forward for employing AI in customer service and for building communication tools for internal use (Léger-Rousseau, 2024). The published existence demonstrates that conversational AIs can enhance consumer engagement by delivering real-time and context-specific responses (Araujo, 2024; Rania & Leila, 2025). Recent studies indicate the adoption of GPT-based solutions, which have gained popularity in financial services, to facilitate innovation about routine inquiries and deliver increasingly personalized responses, thereby extracting insights from client data. Nonetheless, empirical knowledge regarding the long-term effects of LLM on labor market dynamics and consumer confidence remains limited.

Risk Management and Predictive Analytics

AI is a new way to control risk. The interest in artificial intelligence (AI) and its function in risk management appears to be gravitating towards predictive analytics, enabling organizations to primarily identify fraud, forecast market circumstances, and more accurately estimate credit risks (Besse, 2022; BELKADI et al., 2025). Numerous models, such the TAM (Davis, 1989) and AI Maturity Models, have been extensively utilized to examine their adoption by institutions (Desjardins & Gould, 2024). The literature has also said that using predictive solutions would have to make sure that data quality and governance are up to par and that they follow all laws and rules (Vatinel, 2024).

Ethical and Regulatory Perspectives

Another related topic in literature is the moral application of AI in finance. The researchers insist on things like openness, justice, and solid data protection (Imane & Elk, 2024; Abdelkader & Hanane, 2025). Policy for regulation The European GDPR reference architecture is an important part of compliance, however the different regions have very different ways of governing AI (Garawit & El Gnaoui, 2024). Comparisons with other industries demonstrate that the maturity of regulation significantly influences the inclusiveness and rate of AI adoption.

Identified Research Gaps

A number of posts have been written that compare AI/ChatGPT art, but there hasn't been a single one that shows you how to use ChatGPT as an API on both the human and bot sides. AI engines could change how financial operations work from start to finish (front, middle, and back office). Few studies have provided a thorough evaluation of cross-border regulatory frameworks or the potential for the integration of LLMs into risk governance systems. This chapter seeks to fill these gaps by providing a comprehensive analysis of the cumulative effects of AI and ChatGPT, the necessary ethical precautions, and actionable measures for financial services companies aiming to find the right balance between innovation and compliance.

THE EVOLUTION OF AI IN THE FINANCIAL SECTOR

The rise of artificial intelligence (AI) in finance is part of a larger trend of businesses having to adapt to changes in the economy, technology, and rules. At first, AI was mostly used for things like data entry and conducting transactions. But as machine learning and natural language processing have gotten better, AI has slowly become more involved. Today, it is particularly significant in areas like fraud detection, risk analysis, and managing customer relationships (Fadoua and Samir 2025).

For example, machine learning lets banks and other financial institutions look through huge amounts of data and find patterns and anomalies that people can't see. Jim's algorithms are now used to look at millions of borrowers and figure out how likely they are to default. Predictive models are also getting new features all the time to help predict how credit markets will evolve in the future. Advanced AI technologies also power robo-advisor firms, which give personalized investment advice for cheap fees. This is a huge step toward making good financial advice available to everyone (L'MGHAIFRIANDZAMMAR2025).

But it does have certain problems. Every day, problems with data security, biased algorithms, and following the rules become increasingly important. In this unclear time when the rules for AI are still being worked out, banks and other financial institutions need to be careful as they add new technologies in a responsible and open way. AI's entry into the financial sector is a strategic turning point that will make it more effective and drive its offerings based on what customers want and how the global economy is changing. It is a transformation that not only speeds up the efficiency of financial processes but also makes people in the business rethink their long-held beliefs about technology and the safety of finance (Simen Nana et al., 2024).

CHATGPT: A MAJOR INNOVATION

ChatGPT is a big step forward for AI, especially in how it talks to people in a way that makes sense to them. OpenAI's language model uses a transformer-like structure to allow it understand the subtleties of human languages very well. ChatGPT has learned a lot from reading millions of texts. It not only has a lot of encyclopedic information, but it also knows how to grasp what users want. This specialization makes it a great tool for making better financial decisions, since it lets you process complicated information with clear and simple explanations (Léger-Rousseau, 2024). ChatGPT's role in managing financial operations is shown by how it automates many tasks that people used to do. Companies can use this model in a financial management system to automate reporting and make forecasts about cash flow and risk on a large scale. It transforms how decisions are made by analyzing historical data and generating actionable insights. This makes firms more responsive to changes in the market. Adding ChatGPT to customer care chatbots would also make it possible for customers to get quick answers and touch with support staff, which would improve their experience and make things easier for support teams (Araujo, 2024).

ChatGPT can do a lot more than just automate things. It lets you personalize things on a whole new level and even go in front of clients. For example, ChatGPT can figure out what someone needs before they ask for it by reading between the lines of conversations. This ability to adapt and foresee position means that ChatGPT will play a key role in their future financial plans. Its integration merges technical progress with efficiency, serving as a catalyst for businesses aiming to remain competitive in a rapidly evolving economic landscape (Amisse et al., 2024).

APPLICATIONS OF AI IN RISK MANAGEMENT

Adding artificial intelligence (AI) to risk management has changed the way businesses find, study, and deal with possible threats. AI, in particular, lets you look at data and make predictions about it to help you spot unusual behavior early on, which may mean fraud or not paying on time. Machine learning algorithms, powered by today's real-time and historical data sets, may find patterns that people can't see. This makes it easier for businesses to spot and deal with problematic situations quickly. Processing data like this makes the projections better and speeds up the time it takes to make a decision (Jabeur & Nammouri, 2025).

AI is also used to predict risks because it makes complicated models based on many factors that could affect a company's future situations. AI-based credit scoring models incorporate demographic and behavior data, as well as real-world transactions, instead of fixed criteria. This is how AI analysis, like a borrower's likelihood of default, is used. AI in insurance works in a similar way: it lowers risks by making risk assessments more accurate and premiums more tailored to each customer's needs. This lets the insurance business tailor its offers to each customer, which boosts its profits (Besse, 2022). At the same time, automating risk management, such as monitoring risk factors on a frequent basis (e.g., daily), greatly reduces the analyst's view while simultaneously making it easier for institutions to respond to possible crises. AI-powered chatbots can provide customer assistance in real time, which means that complaints and problems may be dealt with before they become bigger issues. They can also highlight bad patterns, which lets human agents focus on more in-depth analytics. In conclusion, AI in risk management is more than just changing whether or not you have to wear a suit. It's redefining what it means to work in finance and giving organizations that are adjusting to an economy that is always changing a new level of flexibility and strength (Ghoubach 2024).

PROCESS OPTIMIZATION WITH AI

The fact that artificial intelligence (AI) is so important for improving financial operations is a big step forward for how businesses are run now. AI offers new approaches to change the way businesses handle money, keeping in mind the need for speed and accuracy. Adding it makes it possible to automate operations that are repetitive and also makes it easier to make good decisions. Using machine learning algorithms or predictive analytics, the analytical side of AI processes can quickly find trends and be used to make financial operations more efficient (Bouchard, 2025). In this sense, making everyday job processes happen is mostly done. You can do things like data input, bank reconciliation, and cash flow analysis very quickly and accurately. For example, AI systems can handle big financial transactions, which frees up human teams to focus on other important duties instead of doing work that takes a lot of time. This increased efficiency also leads to fewer mistakes by people, which makes operations more reliable and in line with financial laws (El Abed & Bellaaj, 2025).

When talking about how to make processes better with AI, it's equally necessary to think about how to make better decisions. The organization can use these powerful analytical tools to analyze multiple financial "what-if" scenarios in real time, using the most up-to-date and relevant data. AI-enabled platforms make it easier to search databases more dynamically and find useful information, including providing marketers a better idea of what trends are coming up. Policymakers must make informed judgments based on in-depth analysis and reliable estimates that eventually promote financial resilience and sustainable growth by drawing a comparison between humans and machines. AI is shown itself to be more than just a tool for automating tasks; it is also a strategic partner in improving financial processes (AOURAGH and SETTA, 2025).

Automating Repetitive Tasks

Using technology like AI and ChatGPT to automate tasks that can be done again and over again is a big element of how financial operations optimization is changing. An additional category of automated solution is capable of executing repeated tasks autonomously, without human interaction (El Abed & Bellaaj, 2025). For example, data input, invoice administration, and bank reconciliation were all jobs that people had to do in the past, but that natural language processing algorithms and machine learning can now do for them. These kinds of technology not only lower the risk of human mistake, but they also let people work on more strategic tasks, which can make the company run more smoothly. Automation makes it even easier for businesses to adapt to changes in the market and what customers want. Businesses can quickly organize and analyze financial data with the help of AI technologies, which gives them a big advantage. AI-powered chatbots, for instance, can answer questions about money and give relevant real-time answers that improve the customer experience (BOUBOUH and GHANIM, 2024 career research work37). Financial reporting automation is also changing, with tools that can quickly fill in analytics with just a few clicks. This gives executives quick access to what's going on and lets them make strategic decisions based on the most recent information.

But let's be clear: automation doesn't mean getting rid of people; it means adding to them. People still need to be in charge to deal with exceptions, make sense of results, and make sure that systems are being used in a way that follows the rules and moral standards. By automating the boring tasks, organizations not only make themselves more efficient, but they also create an atmosphere where new ideas and growth can happen. This lets them focus their time and money on ideas that provide value. Companies in the financial services industry that use AI and automated systems in their daily work are not just reacting to changes in the market; they are also ready to be proactive and rethink their place in the strategic landscape.

Improved Decision-Making

Using artificial intelligence and language models like ChatGPT in finance changes how decisions are made. One of the best things about them is that they can quickly process massive amounts of data, which gives decision makers access to important information that would otherwise be hidden in huge databases. The main thing that makes these platforms work is that they can use machine learning algorithms to find trends, predict how things will go in the future, and make suggestions based on that instead of gut feelings. This change, which is based on making decisions more quickly, affects everything from doing research on investments to managing risk (Desjardins & Gould, 2024).

AI can also help people make better decisions by adding qualitative analysis to the normal way of making decisions. For example, AI-enabled apps can quickly assess how changes in the economy or new rules affect their organization. This gives them a better idea of how each move can affect risk. By putting all of this data together and showing it in a way that is easy to grasp, finance leaders can deal with "what if" situations and change their plans on the fly. This is quite significant when looking over investment and finance plans because lowering risk and increasing potential in these areas of interest are both very vital (Arharbi & El Aissaoui, 2024).

So, the trend toward AI and ChatGPT generating better judgments for managers is not just about how quickly automation happens, but also about how deep and complex the analytic analysis is that goes into managers' decisions. The technology encourages departments to work together more closely, share information, and make choices based on data. Businesses can get a flexible and adaptable view of the financial world by combining data from several sources. It makes AI not just a tool for making decisions, but also a way to make financial processes better.

PREDICTIVE ANALYTICS AND AI

Using artificial intelligence (AI) for predictive analysis has changed the way businesses work today, making it easier for them to manage their finances. Companies can now utilize huge amounts of data to make better predictions about financial trends and assess risk thanks to new algorithms and machine learning models. AI's capacity to handle and analyze data in real time not only helps businesses adapt to a market that is always changing, but it also finds patterns that human analysis would have missed. For example, using NLP (natural language processing) methods to find useful information in unstructured financial documents has increased the amount of data that may be used for forecasting (CHIKHI and RAMDANI 2023).

Predictive analytics is like an AI system since it helps businesses make better strategic decisions. Using methods like linear regression, neural networks, or random forests can help you figure out important numbers like cash flow, profit margins, and future income. Also, more advanced models like time series can help financial analysts predict changes that happen every year or every few years and provide them a better understanding of how to plan their budgets. AI banks use both internal aspects (like past performance) and external factors (like economic and socio-political trends) to produce predictions that are more accurate. This lowers the risk that comes with not knowing what will happen with investments (BELKADI et al. 2025).

To use these tools, though, you need a well-organized process that brings together people who know how to analyze data, create algorithms, and understand the commercial context of the experiment. Companies need to spend time training their employees on how to use these technologies to their maximum potential and encouraging a culture of making decisions based on data. Adding predictive analytics to your financial plan isn't just about using new technology; it's about changing the way you make decisions by using information (Grevisse & Monique-Stéphanie). AI and predictive analytics are becoming essential strategic tools for improving financial operations. They allow managers to be proactive and well-informed in a company world that is changing faster than ever before.

CHATGPT AND CUSTOMER INTERACTION

AI tools like ChatGPT are changing the way customers talk to each other. Businesses may improve their operations and customer experience by adding AI-powered chatbots to their customer care systems. These chatbots can answer hundreds of queries at once, which cuts down on the time customers have to wait and keeps them available all the time. ChatGPT can give more accurate and relevant replies by learning from past conversations and user preferences (LEONARD, 2024). This means that the answers are more in line with what the consumer was wanting to learn. One of the things that makes ChatGPT work is that it is personalized. They can comprehend what users want and reply in the right way thanks to natural language processing algorithms. It's not merely a script for questions and answers. You may change interactions to fit the voice, demeanor, and even cultural sensitivities of the people who use them. For example, a financial services organization can teach ChatGPT to recognize specific customer concerns about their investments and give them personalized solutions to help them feel better about important issues.

This personalization also makes the relationship more real and interesting. People are more willing to talk to someone who "gets" them. Because of this, the conversation turns into a discourse instead of a transaction. The company goes from being just a service provider to becoming your trusted partner (Rania and Leila2025). So, adding ChatGPT to finance tasks does much more than just boost productivity. Being able to have better interactions with customers and close the loop between getting them to engage and keeping them is what keeps growth going and performance improving, which leads to financial success.

Chatbots in Customer Service

Chatbots are still new, but they are already being employed in many parts of the customer service industry to make financial operations run more smoothly. Bots can talk to customers 24/7 because to artificial intelligence. This lets them respond quickly and give customers a better experience. Chatbots can understand what users want, provide information with them, and solve minor problems on their own by looking at data and conversation flows. This level of automation not only lowers costs, but it also lets businesses use their employees for other strategic reasons (HAMZAOUI et al. 2025). Chatbots aren't just there to help customers. They collect and analyze client interactions to give businesses important information about what customers like and how they act. This information enables businesses change their undergrounding, make their internal operations more efficient, and answer consumer questions before they are asked. Also, these chatbots are more useful since they can connect to other platforms, such CRMs. This gives them a full picture of customer involvement across all channels.

But keep in mind that you have to choose to add a chatbot, which is not a free choice. To minimize irritating interactions that could hurt brand perception, design must be responsive and focused on the user. When chatbots can't answer a query, they should be able to seamlessly transfer the consumer on to a human representative. Businesses can use technologies like ChatGPT to solve these problems and create chatbots that do more than just respond. They can also connect with customers in meaningful ways and build long-term connections. Chatbots are a great way for businesses to improve their operational efficiency and get more customers involved in the digital age. They are also an easy way to solve the most prevalent customer service problems.

Personalization of the User Experience

User experience has been important for improving financial operations, and AI technologies like ChatGPT have made this possible. Companies may meet individual demands and move ahead by using AI technology to look even more closely at habits and preferences. This has been helpful for ChatGPT because it can break down and process enormous data sets, which makes it easier to divide customers into smaller groups and have personalized conversations with them in real time. This is a forward-thinking mindset that improves the relationship with the customer by giving each one a unique and individualized experience (BELKADI et al. 2025).

Personalization in financial services involves more than just suggesting the correct goods. It also has to do with communication, since AI-powered tools may change the tone, manner, and sometimes even the content of messages to fit each customer's needs. For example, a conservative user might get different financial advise than a user who is more willing to take risks. This amount of personalization can be tested and refined over time using complex algorithms. This way, even when interactions are individualized, they are still useful. The quality of the data is also necessary for these systems to work. Privacy and security should be two of the most critical things to think about. Businesses need to use data to improve the consumer experience while also cutting through all the red tape. This will not only improve the quality of the client relationship, but it will also lead to a hyper-personalized service that not only makes customers more loyal but also boosts the reputation of financial institutions. ChatGPT's addition is a big step forward that changes the way organizations connect with customers and how they can create an adaptable customer experience that adapts to changing markets (Feki & Dudézert2024).

ETHICS AND AI IN FINANCE

When it comes to technology changing how a bank or its customers interact, ethics and AI in finance have become very popular themes. It is very important that algorithms are open. After all, financial companies require automated decision models that work well and that they can explain. This desire for visibility affects both customer trust and the ability to meet regulatory requirements. There is an even bigger problem: algorithmic biases that come from datasets that don't represent the whole picture or from prejudice in the way models are made. These biases can lead to unfair credit distribution, such as by unfairly excluding a certain set of people or market sector (Jacob & Brousseau, 2025) if they are not acknowledged and corrected.

Data security is also an important part of AI's growth in banking. In the field of big data, businesses are collecting a lot of personal information to make their operations run more smoothly and offer more personalized services. But there is a big moral problem with collecting and using this data. The GDPR in Europe and other rules put strict constraints on how consumer data can be used. For example, customers must give their explicit consent and have the right to see any information that has been gathered and have their accounts erased. This means that firms have to find a balance between using as much data as possible to create AI models and following moral and legal rules. This also means that there is a growing need for clear data governance standards that protect private information and make it apparent that banks are responsible to their customers (ORMACHEA, n.d.).

AI ethics in finance will necessitate perpetual monitoring about issues of transparency and data protection. To make a financial future where tech progress and consumer trust can live together, we need to think carefully about the social effects of AI and make sure that ethical issues are built into our plans.

Algorithm Transparency

When it comes to making financial operations more efficient, this demand for algorithmic transparency is important since it shows how trustworthy and legitimate AI systems like ChatGPT are. As algorithms are created and put into use, their "black box" aspect might make people in the sector worried, especially when it comes to fairness, responsibility, or following the rules. It is essential to comprehend and articulate the decision-making processes of these systems to foster trust among users and stakeholders (BELKADI et al. 2025).

For a method to be considered transparent, it is essential to provide explicit information regarding the utilized data, the processing procedure, and the decision-making criteria. Because of this, banks will need to use approaches that make it easier to track the model. Using explainable AI (XAI) models, for example, makes it possible to not only find and fix possible biases in decision-making processes but also to meet the growing regulatory pressure for algorithmic openness. These methods that show how decisions are made will help organizations explain their choices in front of a public that is becoming more knowledgeable and interested (Abdelkader and Hanane, 2025).

19 We also can't ignore how important ethics are while making and using algorithms. Companies need to think about how they do things in order to be able to critically evaluate what AI produces. These include using auditing systems that are required by law or by the company itself, as well as creating an ethical framework for how to use these technologies. It not only helps people accept it more, but it also leads to a more responsible finance where honesty and morality are always on people's minds. And most importantly, algorithmic transparency is not simply a regulatory concern; it's also the basis for legitimacy and acceptability of decisions in a finance that is becoming increasingly automated (Niang et al., 2024).

Data Protection

In this age of digital banking, where AI-based systems like ChatGPT handle a lot of sensitive data, protecting that data has become a top responsibility. Laws like the European General Data Protection Regulation (GDPR), which sets strict standards for the protection, integrity, and availability of personal data, make these needs even more important. As financial services companies start using AI tools, they have to find a way to balance the complicated regulatory requirements for compliance with good business practices. Due to the rapid changes in these technologies, it is getting harder to protect people's privacy, which means that system design and data architecture need to be proactive in finding solutions (Imane & Elk, 2024).

A good data protection system has to do risk assessments. Organizations need to use a risk-based strategy to find the gaps and vulnerabilities in their information systems. That means using security measures like encryption, two-factor authentication, and rigorous access controls to keep data breaches from happening. Also, continuous training for workers on how to protect data and follow security best practices is important to cut down on human mistakes, which are a common cause of security issues. Adding AI techniques to real-time monitoring can also make data security stronger. Machine learning can find strange things and bad behavior, so you can do something before there is a breach. It is also crucial that AI systems are built with ethical ideals in mind, such as protecting data and eliminating algorithmic biases from causing unfairness. In the age of AI, protecting financial data depends on both technology and a company culture that values user safety and trust. This is the foundation for long-term and responsible financial activities.

CHALLENGES OF IMPLEMENTING AI

Putting AI into financial processes is an exciting path, but it does come with some problems. First, it's hard to fit AI systems into the infrastructure that is already there. Companies often have systems that aren't ready for new technology, even after many years. This would require a lot of money to improve systems, train employees, and in certain situations, altogether change how things are done. Also, the fact that AI technology and data analysis tools aren't standardized makes the integration process even harder (Imane et al. 2025).

At the same time, there seem to be organizational impediments, including people not wanting to change. Traditionalists on finance teams who want to know why they need process automation and are afraid that AI would make their jobs obsolete or render their skills useless. To lead this cultural revolution, you need a well-thought-out change strategy that shows employees the benefits of AI and helps them through the transition. Also, AI algorithms need organized datasets that are high quality and represent the problems they need to address. But a lot of businesses have trouble with missing or dirty data, which makes the predictive models you made yourself or hired someone else to make less valuable and/or less accurate. Also, the problem of balancing compliance with data security risk management seems to be an unavoidable one. Banks operate within a complicated legal framework that requires them to safeguard sensitive client information while leveraging AI for cost reduction (Vatinel 2024). And the risks of changing algorithms, making them more open, and protecting against cyber threats must always be a challenge for good solutions.

So, while AI and ChatGPT can help us improve our financial operations in many ways, the amount to which we can do so and when will depend on how we deal with these issues head-on.

REGULATION AND COMPLIANCE

When you combine AI and language models like ChatGPT with financial transactions, following the rules and regulations is very important. As advanced analytics and automation technology become more common, it is becoming harder for banks to follow the rules. It is not just a technical problem; it also has moral and social implications: failing to meet these standards could result in severe fines and damage to the organization's reputation (Bendaoud & El Ouazzani, 2025).

Because of this, regulators have started to make more explicit guidelines concerning how AI can be used in finance. For example, the General Data Protection Regulation (GDPR) in Europe sets strict rules for how data can be collected and combined to safeguard consumers. At the same time, financial regulators are also looking at operational systems and how AI systems can be tracked. This monitoring is necessary to ensure that automation choices that could have big implications on society, like lending and investments, are open and accountable. Explainable models that link the decision to specific criteria are increasingly becoming a must-have (Garawit and El2024). When organizations give AI regulatory trust, they may not only work more efficiently, but they can also build long-lasting trust with their customers. Making sure that AI is used in a legal and moral way is what drives value. This is a place where new ideas may be safely tested. 2. Make sure that AI strategies and compliance initiatives are in sync If your AI initiatives are closely linked to compliance with data security and privacy rules, you will be better able to predict and reduce the dangers that come with these new technologies. But to do that, banks and other financial institutions will have to be proactive and adaptable in the face of continuing regulatory change. They will also have to keep getting the benefits of AI while still following the law and keeping their obligations to safety and ethics.

THE FUTURE OF AI AND CHATGPT IN FINANCE

AI and models like ChatGPT are changing the way we do business in finance. Now we need to think about how new technologies will not only make things easier, but also change the way people engage with them. The AI solution is designed to make financial services even more personalized by using big data analytics to offer products that are just right for each consumer. Additionally, the implementation of advanced algorithms in forecasting and risk management would bolster the immunological response to economic fluctuations (Couzineau-Zegwaard & Tarnaud, 2025).

ChatGPT is especially trying to become the best software for automating customer care. He is quite skilled at answering all kinds of inquiries in natural language, which makes him a great real-time agent who can give individualized advice to consumers. As the capacity to talk like a person gets better and better, this way of communicating is likely to take the place of some human interaction, which would cut the cost and make it more available. The bank can also add these platforms to compliance developments, which might lower regulatory risks by letting clients keep an eye on transactions and customer behavior all the time. But there are several problems with the technology that makes AI work. People think that the moral challenge of how to deal with prejudice in these algorithms is crucial for the reputation of financial organizations and for future customer trust. So, in the future, AI and ChatGPT will need to have rules about how to be open and responsible while employing these technologies in banking. In short, new technologies could make financial processes much better, but because so many people use them, we need to think about how they will affect society and ethics. These will determine how these instruments are integrated and accepted by future financial systems, as they are essential for building finance that is both ethical and sustainable.

IMPACT ON EMPLOYMENT IN THE FINANCIAL SECTOR

AI and ChatGPT are two new technologies that have changed the labor market in the finance business. Banks use these modern technology mainly to make their operations more efficient, cut costs, and make better decisions. But these achievements have also made some quite worried about what will happen to workers. Even while some monotonous operations are automated, which adds redundancy to existing roles, new jobs are opening up, especially in data management and AI system development, as well as regulatory awareness (Marini et al., 2024).

Workstations that are easy to automate are those that do the same thing over and over again, such entering data, processing transactions, and analyzing financial reports. But jobs that require human intelligence, including making business plans, figuring out risks, and dealing with customers, are less likely to go away. As AI becomes more common in various fields, it is very important to have people with a background in the humanities who can manage, comprehend, and work with the technology. As a result, knowing how to read AI systems' responses and do advanced analytics is becoming more and more valuable in the financial job market (Quebec, 2025).

The financial sector's impact on jobs also shows how important it is to keep training employees. Institutions need to spend money on retraining personnel because of this new technology landscape. At the same time, the eventual change of the whole banking ecosystem could lead to new business models and opportunities for those who labor to meet compliance standards and regulatory limits. AI and ChatGPT may cause traditional job losses, but they also give professionals in a financial organization a chance to move up in their careers. This will put more focus on the strategic and technical skills that are important in the digital age (Salmandjee, 2025).

COMPARISON WITH OTHER FINTECHS

In the new generation of financial technology, AI and platforms like ChatGPT make it feasible to do transactions in ways that are more dynamic than what is achievable with current technologies. Conversely, conventional financial data management solutions, such as database and accounting software, rely on predetermined rules and static data analysis, whereas AI language models like ChatGPT adopt a more context-sensitive methodology. They find patterns, make judgments automatically, and suggest personalized solutions to difficult financial problems using machine learning and real-time data analytics (Nadeau & Jobin, 2024). AI, ChatGPT The goal of AI and ChatGPT is to make internal processes more efficient, while new technologies like the blockchain offer security and transparency. Blockchain may help protect the integrity of transactions and lower the risk of fraud, but it can't replace the analytical and predictive powers that only AI can offer. For example, in the financial Risk industry, the AI technique can keep an eye on transaction data for irregularities and act on them even before they happen. This is where blockchain's excellent ability to protect data would not be enough.

Also, more advanced tools like cloud computing and ERP (Enterprise Resource Planning) systems can assist make finances more efficient. However, as these systems become better at storing and managing vast databases, adding AI to them changes the way people can access them: now, bold acts may interact with complex data. This is a huge transformation, not only in the way the interface looks, but also in the way that knowledge in an organization may be used. Even though these technologies can work well together, AI and ChatGPT are more progressive since they add flexibility and predictive intelligence that changes how money is usually handled.

TRAINING AND SKILLS REQUIRED

AI and tools like ChatGPT are changing the way money works, thus it's important to have training and skill-building programs to help with this change. To really understand AI, someone needs to know about its best parts, like machine learning, data analytics, and decision-making algorithms. They need data science, among other things, to understand and operate with the huge amounts of data that are created every day in their field. Moreover, computer languages like Python and R have become necessary tools for the geospatial analyst since they help automate and improve statistical analysis (Araujo, 2024).

You also need to work on your communication and change management abilities at the same time. "Using AI-based solutions will not only change the culture of financial teams, but it will also change the way we work. Workers will need to be able to use automation in a way that lets them interact with it, and they will also need to have a critical attitude toward these tools. Also, the teaching program should contain lessons on AI ethics and accountability so that professionals can deal with problems related to data protection and following the rules (Leike et al. 2017; Loeffen, 2021).

In a playfield that is always changing, the cascade of skills must keep going. A culture of continual learning is the basis for finance, which makes it possible to be flexible and come up with new ideas. Companies should help their employees keep learning and being certified, and they should also engage with colleges that offer specialized courses on AI and how it may be used in the financial services business. Because of AI's impact on finance, both industry and academia need to rethink the kinds of profiles they are looking for. This includes not only technical skills and formally ethical behavior, but also soft skills. More broadly, they need to make sure that the change is both efficient and responsible.

INTERNATIONAL PERSPECTIVES ON FINANCIAL AI

Artificial intelligence (AI) has changed the way money works all across the world. New and interesting projects are popping up all over the world that show how technology can be used in many ways to improve financial efficiency, risk management, and strategic decision-making. For example, in the US, smart banks are using AI to look at data and predictive algorithms to make better economic predictions and find fraud. In this sense, artificial intelligence and machine learning enable greater personalization in customer-centric financial services or the development of investment portfolio performance (ROUSSI, 2022).

In contrast, AI has a broader role in countries like China. It's not simply a tool for financial operations; it's also a significant lever for large-scale economic innovation.

The Chinese government is now fully behind AI projects and is pushing for the creation of digital banking and integrated payments platforms that make use of the vast amounts of data that are available. These changes don't simply affect banking; they also have a lot of other effects, like on asset management and following the rules. AI is being used at the same time to make things more open and let firms and regulators share information. The European position should be recognized worldwide, as it focuses on finding a balance between protecting consumers and encouraging innovation in AI in financial services. The EU and the push for post-regulatory models that make algorithms more open, while also making privacy, data protection, and fairness issues more complicated. These various perspectives collectively depict a dynamic financial landscape in which AI serves not just as a means for operational enhancement but also as a focal point of ethical, political, and economic discourse over the future of financial markets.

CONCLUSION

The integration of artificial intelligence and systems such as ChatGPT into financial operations marks a significant shift in the way companies process and analyze data. Automating financial processes, powered by intelligent algorithms, not only reduces costs, but also improves accuracy and efficiency. By deploying AI-powered solutions, organizations can gain real-time analytics that help them anticipate market fluctuations, proactively manage risk, and make informed decisions. This ability to turn raw data into actionable insights is essential in an increasingly complex and volatile business environment. The implications of this transformation are not limited to improving business outcomes; They also lead to cultural changes within finance teams. The adoption of AI requires a reconsideration of the skills required. Finance professionals must now become familiar with advanced technological tools while developing analytical and strategic skills. In addition, the interaction between man and machine redefines traditional roles, highlighting the importance of innovation and adaptability through continuous training and professional development. AI and ChatGPT are positioned as crucial pillars for the optimization of financial operations, bringing unprecedented agility, accuracy, and analytical capacity. To capitalize on these technological advances, companies must not only invest in technological solutions but also in human capital, cultivating a culture of innovation and learning. By embracing this harmonious combination of technology and human expertise, organizations can not only overcome contemporary challenges in the financial landscape, but also prepare to navigate future opportunities, solidifying their position in the overall market.

REFERENCES

Abdelkader, M. A. A. R. O. U. F. I., & Hanane, B. R. A. B. I. J. E. (2025). L'Impact de l'Intelligence Artificielle sur le Contrôle Interne et la Gestion des Risques. Revue Internationale de la Recherche Scientifique (Revue-IRS), 3(3), 2718-2727.

Amisse, M., Faur, M., Gonard, L., & Orcesi, A. (2024). Promouvoir des modèles d'intelligence artificielle frugale pour et par les politiques publiques.

Aouragh, M., & SETTA, B. (2025). Les Secrets de l'ia dans l'octroi de prêts: Décryptage des critères non financiers. *Dossiers de Recherches en Économie et Gestion, 13*(1), 11–28.

Arharbi, N., & El Aissaoui, Y. (2024). Vers un contrôle de gestion agile: L'intelligence Artificielle et le Big Data pour la performance en environnement incertain. IJDAM. *International Journal of Digitalization and Applied Management, 1*(2), 195–210. DOI: 10.23882/ijdam.24141

Araujo, M. (2024). L'impact de l'intelligence artificielle sur le métier de comptable et les sociétés fiduciaires.

Belkadi, Y., SOUAF, M., & BELKADI, S. (2025). L'intelligence artificielle et la performance de l'entreprise: Une revue de la littérature sur la redéfinition des pratiques et des interactions sociales. *Revue Française d'Economie et de Gestion, 6*(1).

BENDAOUD, H., & EL OUAZZANI, S. T. (2025). L'intelligence artificielle au sein de l'administration fiscale: Revue de la littérature. *International Journal of Economics and Management Sciences, 4*(1), 173–185.

Besse, P. (2022). Conformité européenne des systèmes d'ia: outils statistiques élémentaires. Statistique et société.

BOUBOUH, I., & GHANIM, M. (2024). L'Intelligence Artificielle au Service du Contrôle Interne: Vers une Performance Organisationnelle Optimisée. *Revue Internationale des Sciences de Gestion, 7*(4).

Bouchard, A. (2025). # workinprogress: IA générative et outils de recherche de littérature académique. Urfistinfo.

CHIKHI, D. E. B., & RAMDANI, B. (2023). L'influence de l'intelligence artificielle sur l'efficacité de l'audit financier: Tendances, défis et opportunités. *Revue Française d'Economie et de Gestion, 4*(12).

Couzineau-Zegwaard, E., & Tarnaud, L. (2025). ['IA ou la transition par la donnée: Interview de Sandrine Tarnaud,«Head of data and AI» chez Microsoft. Le temps des transitions.]. *Chapitre*, *10*, L.

Desjardins, G. & Gould, A. (2024). Fausses promesses et stupidité organisationnelle: pourquoi l'ia n'est pas synonyme de performance. Ad machina.

Du Québec, G. (2025). Consultation sur la transformation des milieux de travail par le numérique.

El Abed, A., & Bellaaj, M. (2025). *Perception de l'utilisation de l'intelligence artificielle par les dirigeants tunisiens: motivations, freins et avantages perçus* [CG1.1]. Revue Française d'Economie et de Gestion.

Fadoua, A.Z.I.O.U.A.L., & Samir, B. A. (2025). Fintechs et IA: nouveaux paradigmes bancaires face à la nécessité d'une régulation juridique. Revue Internationale de la Recherche Scientifique (Revue-IRS), 3(3), 2573-2590.

Feki, M., & Dudézert, A. (2024, May). Exploration du potentiel de chatgpt dans la Supply Chain: revue de littérature et pistes de recherche. In AIM 2024: 29e Conférence de l'Association Information et Management.

Garawit, S., & El Gnaoui, L. (2024). Principes éthiques internationaux de l'intelligence artificielle (IA): Un essai d'analyse dans l'entreprise au Maroc. International Journal of Strategic Management and Economic Studies (IJSMES).

Grevisse, Y. R. & Monique-Stéphanie, K. M. (). … De La Performance Et Gestion Des Risques Dans Les Entreprises Congolaises: Prise De Décision Data-Driven, Analyse Prédictive, Outils Technologiques

HAMZAOUI, S., MEDOUARI, M., Setta, B., & Lahrach, R. (2025). Crowdfunding social et IA: Un duo gagnant pour l'entrepreneuriat social au Maroc. *Revue Française d'Economie et de Gestion*, *6*(1).

Imane, B. & Elk, Y. (2024). E-gouvernance et transformation numérique au Maroc: défis et perspectives pour des institutions et des services publics innovants. International Journal of Economic Studies.

Imane, R. A. B. H. I., & Samira, Z. I. A. N. I., & BUDAS, I. (2025). Exploiter l'intelligence artificielle pour renforcer la gestion de la qualité hospitalière au Maroc: Vers un modèle stratégique intégré. *Revue Dossiers De Recherches en Économie Et Management Des Organisations*, *10*(1), 1–18.

Jacob, S. & Brousseau, S. (2025). Quand l'algorithme décide: L'État, l'ia et nous.

L'MGHAIFRI, H., & ZAMMAR, R. (2025). Revue de littérature sur l'intégration de l'Intelligence Artificielle (IA) dans les pratiques du Contrôle de Gestion: Opportunités et Défis. Revue de l'Intelligence Artificielle et du développement territorial durable, 2(01).

Léger-Rousseau, M. (2024). Les performances des grands modèles de la famille des transformateurs génératifs préentraînés en soutien aux services de référence virtuelle automatisés en…. *Documentation et Bibliothèques.*

Jabeur, B., & Nammouri, H. (2025). *Comment l'ia, les Big data et l'informatique quantique redessinent les marchés de l'énergie.* Management & Datascience.

Ghoubach, S., & El Amine, B. (2024). L'adoption de l'ia et du smart logistique à l'optimisation des coûts d'approvisionnement. Revue du contrôle, de la comptabilité et de l'audit, 8(4).

LEONARD, T. (2024). La compréhension et la prédiction des préférences des clients dans le commerce en détail grâce à l'ia: Une revue de littérature. Int. J. Econ. Stud. Manag.

LOEFFEN, A. (2021). L'impact de l'intelligence artificielle sur la gestion des ressources humaines: Cas des entreprises marocaines. *The Interdisciplinary Journal of Human and Social Studies-The Interdisciplinary Journal of Human and Social Studies*, *1*(1), 85–99.

Marini, J. L., Patissier, C., & Toublanc, J. (2024). Booster sa PME grâce à l'intelligence artificielle: Choisir et intégrer des solutions IA à vraie valeur ajoutée.

Nadeau, P. & Jobin, K. (2024). Intelligence artificielle: Génération Générative: chatgpt, Midjourney... S'approprier les nouvelles IA qui révolutionnent le monde professionnel.

Niang, M., Fofana, F., Faye, C., & Tall, L. (2024). Lignes directrices pour l'élaboration d'un document national sur l'éthique de l'ia dans les pays d'Afrique de l'Ouest.

ORMACHEA. A. (n.d.). L'intégration de l'ia dans la gestion de la chaîne d'approvisionnement et ses effets sur la relation client. Publication-theses.unistra.fr.

Rania, B. E. N. M. O. U. S. S. A., & Leila, K. I. S. S. I. (2025). Le marketing expérientiel à l'ère des technologies immersives: Une revue de la littérature sur le rôle des émotions dans l'engagement du consommateur. Journal of Economics [JEFM]. *Financial Management*, *4*(3), 672–692.

ROUSSI, K. (2022). L'Intelligence Artificielle au service du métier de l'expert-comptable. Revue Internationale du Chercheur.

Salmandjee, Y. (2025). Utiliser l'ia en 90 cas concrets pour se faciliter la vie: Travail, santé, cuisine, sport, loisirs

Simen Nana, S. F., Flindjoa, D., Tankpe, T. A., & Kaka, Z. Y. (2024). Diriger l'entreprise africaine à l'ère de la transformation numérique et de l'intelligence artificielle.

Vatinel, T. (2024). *Lutte contre la fraude documentaire: état des lieux et avancées technologiques grâce à l'ia*. Archimag.

Section 5
AI Serving Ethics, the Public Sector, and Healthcare

Chapter 12
Management Control in the Age of Artificial Intelligence:
Opportunities and Ethical Issues Facing Public Organizations

Yamna Azlouj
https://orcid.org/0009-0007-9182-4732
Faculty of Legal, Economic, and Social Sciences, University Sidi Mohamed ben Abdellah, Fez, Morocco

Ahmed Aftiss
https://orcid.org/0009-0002-6211-4065
Faculty of Legal, Economic, and Social Sciences, University Sidi Mohamed ben Abdellah, Fez, Morocco

ABSTRACT

This chapter explores management control practices within public organizations in the era of digitalization and the increased role of artificial intelligence (AI). It points to the historical development of management control and its internalization into results-oriented logic, along with the role of management control in encouraging a more ethical governance, transparency, and accountability. The study examines the main tools used and presents an assessment of the ways in which AI can augment or disrupt the effectiveness of those. Of particular concern are the moral consequences of adopting AI, such as concerns of transparency, fairness, and change resistance in low-resource settings. Management control is publicized as being one of the leverages towards ensuring that there is a better performance and rational-

DOI: 10.4018/979-8-3373-3104-1.ch012

ization of expenditure, and also one of the ways in which management control can be employed, towards ensuring that there is responsible and ethical application of emerging technologies, in the service of the public governance.

INTRODUCTION

The practice of management control in public organizations is of paramount importance in the context of good governance and the efficient use of public resources. This introduction aims to situate management control as a fundamental tool for public organizations, highlighting its purposes, its key principles, and its role in improving the performance of public services. Public institutions, faced with increasing budgetary constraints and high expectations from citizens, must adopt rigorous management practices to ensure greater transparency and accountability.

In this context, management control is defined as a set of devices that allow the planning, monitoring, and evaluation of an organization's performance. It does not only render it possible to prepare financial and human requirements but also it becomes feasible to compare the outcomes attained with the objectives planned.

The topicality of performance indicators allows the analysis of the deviations between the predictions and the reality by the public managers and to indicate the areas in need of adjustments or changes. It is a phenomenon that is crucial in the intricacy of the social, economic and political realities manoeuvring in the present world.

Besides, the management control of the state organization is rooted even in purely technical work, with the sense of responsibility. It is known to increase the legitimacy and trust of the citizens in their institutions by enhancing the culture of measurement and assessment. The introduction of modern management control practice therefore is a necessity to ensure the efficiency does not only the public resources but also the compatibility of the same with the expectations of the society. Introduction of this combination of principles and strategies is therefore planned as not only an organisational requirement but also an agent that has the capacity of enhancing the aspect of the service to the people.

In the meantime, the topic of artificial intelligence (AI) and the digital transformation challenges and opportunities are already becoming a daily reality on the field of the operations of the institutions of the population. Despite the new possibilities offered by AI in the field of data analysis, forecasting, and decision-making in the management control process, it has current ethical issues regarding the quality of the algorithms and the decision-making, the protection of personal data, and whether the decisions are fair and made automatically, as well as the responsibility of the managers of the public. These queries indicate the need to change the management control practices not just in response to the performance imperatives, but also to the ethical laws in the use of AI.

This research will therefore seek to find ways of best practices in management control in the case of public organizations, and, how the emergence of AI can strengthen or compromise the accountability and transparency. The study aims to analyze the origins, developments, and ethical implications of management control in order to provide information on how AI-based tools can be incorporated into the processes of public governance in a responsible way without jeopardizing legitimacy and trust among citizens.

LITERATURE REVIEW

The focus placed on the analytic frameworks and the actual ground realities of the implementation is growing as demonstrated by the recent research in management control in the public organizations (Pollitt & Bouckaert, 2011; Hood, 1991; Osborne & Gaebler, 1992). Literature concerning the topic was mainly descriptive in the past and mainly involved the definition of management control tools, their introduction and adoption as mechanisms in guaranteeing financial compliance and procedural regularity. Initial research placed an emphasis on the bureaucratic paradigm delivered by Weber, stressing on the use of ex-post financial controls as well as formal compliance.

With the emergence of the New Public Management (NPM) paradigm (Hood, 1991; Osborne & Gaebler, 1992), the environment of academia as well as practice has altered drastically since the 1990s. The concepts of performance management, strategic steering and outcome-based evaluation were being integrated into the literature, and the tools, which started being introduced to these fields, included the Balanced Scorecard (Kaplan & Norton, 1996), benchmarking and Activity-Based Costing (ABC). This movement was underpinned by the international processes of the reform such as the French LOLF or the introduction of the performance budgeting frameworks into the Anglo-Saxon countries and substituted the input-based management with the results-based management.

The criticality and subtlety of literature in the past several years have been more critical and have been keen to point out the paradox and the unintended consequences of greater management control. The risks of excess bureaucratization and overloading performance measurement that may lead to the gaming or ritualistic compliance or a missed opportunity to concentrate the attention on unmeasured performance steps such as the quality of services and social equity have been discussed by researchers (Pollitt and Bouckaert, 2011; Moynihan and Kroll, 2016). It also discloses the paradox of accountability on the basis of which higher rates of transparency and control may ruin trust and initiative of public managers in certain cases.

According to empirical research, management control systems must be implemented in context-specific ways based on their strong leadership as well as the ability of the staff to establish such implementation in the most effective way possible (Felício et al., 2021).

In parallel, in recent studies, the emphasis has shifted to the fact that the digital transformation of the state apparatus, the development of artificial intelligence, and the increasing role of computer programs in the management control of society. Currently, it is commonly accepted that IA can contribute to many aspects of forecasting, data analysis and decision-making in state entities (Wirtz et al., 2019). Scholars also note that algorithmic systems create new forms of ethical problems, such as the lack of visibility into decision-making logics, the issue of fairness in automated decision-making, accountability in the event of failure, and safeguarding the trust of citizens (Zuiderwijk et al., 2021). In these works, the authors stress that management control should be changed so that it could incorporate digital ethics and accountability of algorithms in its practice and practices of using AI cannot weaken; on the contrary, they should reinforce the ideals of legitimacy of the public sector, responsibility and equity.

Therefore, the current body of knowledge has two domains addressed: keeping historical concerns with the efficiency, performance, and compliance with the law, adjusting the management control practices to new ethical opportunities and dilemmas demonstrated by artificial intelligence in the processes of governance, especially on the level of the government.

HISTORY OF MANAGEMENT CONTROL IN THE PUBLIC SECTOR

The concept of management control is an old one in the context of the public sector and it is directly linked to the formation of the paradigm shifts in the sphere of the public administration and enhancement of efficiency and pressure towards transparency and accountability. The objectives that have always typified the management control in the public sector include legality and compliance and collective interest that is geared towards serving and the generation of collective good of the society (Pollitt & Bouckaert, 2011), that is not the case with the objectives of the management control in the private industry, and that is profit maximization and the shareholder value (Pollitt & Bouckaert, 2011).

The hierarchic regimes, generalized procedures and strict observance to rules is the bureaucratic model of Max Weber which was characteristic in most attempts in the field of the public management during the first half period in the 20 th century. At that, the major form of management control was limited to the financial controls, and post-factum checks on the spending (Hood, 1991). But a turning point was created during the 1980s and during the 1990s when the movement of New Public Management (NPM) became commonplace (Hood, 1991; Osborne & Gaebler, 1992). NPM altered the culture of the public sector substantially in the light of the practices of the private sector, and this is because it introduced the culture of outcomes, measurement of performance and management accountability. Budgeting by objectives and performance measures among so many others came into the limelight with strategic planning. The introduction of Balanced Scorecard (Kaplan & Norton, 1996) and the activities related to benchmarking can be treated only as the wish to get out of that bureaucratic stiffness and go into something innovative, streamlined, and responsive to the needs of the citizens.

Since early 2000s, limits of NPM, such as overheating of fragmentation, the risk of bureaucratisation of the performance measurement, and conflict between the control and the trust have been debated more frequently (Pollitt & Bouckaert, 2011). The adoption of the post-NPM era promotes the more integrated, collaborative and adaptive managers, which are because of the realization of the complexity associated with creating a public value, as well as the plurality of stakeholders. The surveys of the global environment held over the past few years substantiated the core of hybrid regimens combining the traditional regimes of control with participatory as well as network-rooted regimes of government (Bouckaert, 2022). As an example, the Swedish model is centered on the approach to management that involves the result, and consistent feedback, South Korea has incorporated the deployment of digital instruments and the e-government methods to gain greater transparency and citizen involvement.

Over the last decade, the digitalization of the public sector has altered the practices of management control in a fundamental manner (Bhimani, 2021). The combination of big data, artificial intelligence, and integrated information environments applied enable the monitoring of the process and even more complex performance analytics occurring in real time. The technical change makes it possible to transition the movement towards ex-post analysis of the management to one that is characterized with constant and vigorous and data-driven management (Aamri & Bouaziz, 2025). However, these are not developments that are in acceptance. The dangers of the concept of control excess (over-use of indicators, loss of meaning, and demotivate the staff), ethical issues (privacy, data security and protection, and bias of algorithms), and the ability to modify the imported models to suit local situations especially in the Global South countries are some of the things that researchers and practitioners complain about (Moynihan & Kroll, 2016; El Kezazy & Hilmi, 2024).

The sphere of the management control in the field of the public sector today does not involve only the control by the compliance or by the budgetary discipline. It is projected to be a strategic as well as multidimensional catalyst of the development of a social good, and it requires a balancing act between scrutiny and accountability, experimentation, and oversight, statistics and impact. The customary pattern indicates that the system of management control must evolve perpetually as per the variation in the social, political and technological sphere. The shift of the management control will also depend on the possibilities to introduce flexibility, accountability, and the critical reflection capacity as new requirements grow in terms of digital transformation, growing citizen requirements, and a desire to achieve integrity (Kaplan & Norton, 1996; Bouckaert, 2022). Recent international research confirms that digitalization is deeply transforming management control systems, creating both new opportunities for data-driven decision-making and significant challenges related to staff resistance, skills gaps, and technological implementation (Fähndrich & Pedell, 2025).

KEY CONCEPTS OF MANAGEMENT CONTROL

Management control in public organisations is fundamental to optimise the management of resources and ensure transparency in the use of public funds. It is both a steering instrument and an accountability mechanism, and is anchored in strategic and operational objectives. Its definition is based on the need to ensure that the actions and decisions taken meet the public service missions. In this sense, management control is not limited to cost monitoring or analysis, it is part of an overall efficiency approach to meet the expectations of citizens and supervisory bodies (El Kezazy, 2022).

Management control tools and techniques are diverse and grounded on strong theoretical foundations. These systems include cost analysis, performance indicators and benchmarking which are on the list of essential ones. The cost analysis can assist in the understanding of the expenses and minimizing them without compromising on the quality of services. Performance indicators, in turn, are employed to supervise and assess the goal accomplishment, therefore, simplifying the real-time management of activities (Toma, 2021). Benchmarking, which is commonly applied to the performance of various parties, assists in best practices identification. Moreover, the management dashboard tool also allows obtaining a synthetic representation of the project and budget status that are critical to make informed decisions (Rahararina, 2022). The recent empirical studies prove that under the condition that the management control systems are appropriately adjusted to the organisational environment and supported by both staff members and the leadership, they play a major constructive role in determining the efficiency, accountability, and the quality of the services delivered by the public sector organisations (Felício et al., 2021). This evidence emphasises the need to align the means of control with the organisational goals and invest in staff involvement and training in order to provide the long-term improvement.

By such, the management control in the public organizations is not merely an administrative system, but also a strategic instrument that intends to integrate efficiency and accountability. It enhances the analysis and responsive ability of the public managers and therefore helps in improving the services continuously, therefore addressing the current challenges in managing public resources. The future of management control is hence shaping itself in this dynamic, which is a critical practice to facilitate the complicated demands of the public policies in a dynamically evolving environment (Aamri & Bouaziz, 2025).

THEORETICAL FRAMEWORKS OF MANAGEMENT CONTROL IN PUBLIC ORGANIZATIONS

The theory on management control in the case of public organizations has its basis on a number of theoretical schools that provide a holistic insight into the operation of the control systems in the complex institutional context. These theories can not only offer educational respectability, but also direct the way in which empirical research can be constructed and interpreted within the field.

Among the underlying perspectives is the Levers of Control model formulated by Simons (1995) who identifies diagnostic, interactive, belief, and the boundary systems. This model brings out the symbiosis between formal control instruments and informal processes like dialogue and organizational learning and the reasons why it is important to balance accountability with strategic flexibility and adaptability an issue that is especially relevant to the public sector as demands and complexes of operation of an organization are constantly changing.

The other framework is that of Anthony (1965) that categorizes the control systems into strategic planning, management control and operational control. This model comes in handy in the public sector in terms of differentiating the setting of general policy directions, the alignment of resources to goals, and the control over the daily operations.

Agency theory (Jensen and Meckling, 1976) also offers more information especially where political principals (government or elected officials) and administrative agents (public managers) are separated. This kind of separation may result in information asymmetry and conflict of interest. It is the management control systems that are important in the alignment of incentives, reduction of moral hazard and accountability in state institutions.

The contingency theory approach (Otley, 1980) holds that control systems can not be best designed and implemented in just one way. Rather, they are effective in a number of contexts, including environmental uncertainty, size and structure of the organization and regulatory requirements, among other factors. This theory highlights that management control practices should be specific to the missions, cultures as well as the stakeholder environment of the individual organizations.

Lastly, the institutional theory (Meyer and Rowan, 1977; DiMaggio and Powell, 1983) is especially applicable to the governmental sector where the formal systems of control are frequently driven by the external norms, professional standards, and legitimacy pressures. The institutional theory can be used to understand the fact that some management control practices are taken because of certain efficiency reasons, but also because the society requires it and the organization needs to gain legitimacy.

In a few words, the synthesis of these theoretical views makes possible a multidimensional and critical study of the practices of management control within the public organization. This approach to the problems, namely, the combination of structural, behavioral, and contextual perspectives, places the findings of the study on a solid academic base, thus contributing to their relevance and generalizability.

Also more recently, there have been digital governance and algorithmic accountability theories that have emerged, which concentrate on the effects of artificial intelligence (AI) on management control practices. These attitudes illuminate cogitations about the ability of the AI-based tools to enhance performance monitoring and make predictions (Wirtz et al., 2019; Sun and Medaglia, 2019), but certain ethical issues about the transparency, discrimination, and validity of collective decision-making are also raised (Mittelstadt, 2019). The addition of AI to the theoretical discussion can enrich the idea of management control that will offer cross-references with the traditional models and the facts of digital transformation in governmental work (Meijer & Bolivar, 2016).

DEFINITION AND OBJECTIVES

Management control in public organizations is based on the implementation of mechanisms to ensure effective and relevant management of public resources. This process, in essence, consists of monitoring, evaluating and guiding organizational performance in order to ensure that the objectives set are achieved. The definition of management control goes beyond simple financial reporting methods, it also includes a strategic approach aimed at optimizing resources, improving the quality of services provided, and strengthening the transparency of public actions. As a result, management control is becoming an essential tool for decision-making, integrating operational, budgetary and strategic dimensions (El Kezazy, 2022).

The objectives of management control in the public sector are divided into several levels. First of all, it ensures an optimal allocation of resources, which is crucial in a context where budgetary constraints are often severe. Then, by integrating performance indicators and dashboards, it provides clear visibility into the status of ongoing projects and operations. This fosters a culture of accountability and continuous evaluation within public organizations. Increasing administrative efficiency and improving citizen satisfaction are other major objectives, as management control must participate in the creation of public value (Touicher & Loulid, 2023).

In addition, this practice helps to establish a proactive approach to risk management. By identifying deviations from objectives and proposing corrective actions, management control becomes a lever for anticipating problems and optimizing public performance. The challenges of management control in public organizations are therefore not limited to the simple a posteriori evaluation of results. They involve a dynamic approach, integrating all the actors of the organization in a continuous process of planning, implementing, and readjusting strategies. Management control, in its definition and objectives, represents a structured and adapted response to the contemporary challenges of public organizations, working both for internal efficiency and the satisfaction of citizens' needs (El Mehdi et al., 2025).

More recently, management control definition has also expanded to cover the opportunities and the risks presented by the artificial intelligence (AI). With the help of AI, there are more opportunities to analyze data, predict outcomes, and monitor what the population is doing in real-time (Wirtz et al., 2019). Their incorporation, however, adds other considerations to the objectives of the public organizations, which are algorithmic transparency, the right to protection of the rights of citizens, and ethical responsibility (Mittelstadt, 2019). In that way, in the era of digital management control, its goals are not only to enhance efficiency and optimization of resources, but also to avoid jeopardy sing fairness, trust and legitimacy as to the application of technology (Mikhaylov et al., 2018).

MANAGEMENT CONTROL TOOLS AND TECHNIQUES

Management control in the public organization is based on a large number of tools and methods to increase the effectiveness of work, facilitate wise decision making, and strengthen the transparency of activities. One of the most popular ones is the performance dashboard, which gives the managers a grasp of the essential indicators against strategic and operational priorities in real-time (Kaplan & Norton, 1996).

Budgetary instruments, such as program-based and results-oriented budgeting, promote an immediate connection between the funds assigned and the outcomes and hence promote accountability and their alignment with priorities according to the public policy (Pollitt & Bouckaert, 2011). Such methods of cost analysis as Activity-Based Costing (ABC) enables more accurate assignment of the costs to individual activities or services, and, therefore, a better insight into what causes money to be spent and what improvements will help to achieve a more detailed improvement.

Comparison with the best practices of the country and the world is driving the performance benchmarking which triggers incessant deterioration and confinement. The successful use of the mentioned tools in the sphere of the public sector, however, is mostly faced with certain hardship, as there are particular issues, which include the reliability and availability of the data, changes resistance within the organization, and the necessity to adjust the methodologies of the private sector to the limitations of the public one.

These dynamics are now being multiplied by the current digital transformation, big data and artificial intelligence (AI) are on the rise, whereby more complex management control systems are being embedded into systems that can provide predictive analysis, real-time monitoring and strategic steering with a precision previously impossible (Bhimani, 2021; Wirtz et al., 2019). However, ethical and technical questions related to those innovations exist as well, such as the risk of algorithmic bias, the loss of transparency in decision-making, and whether it is possible to hold the automatic system accountable (Mittelstadt, 2019).

In short, the selection and combination of management controls tools cannot be moulded rigidly, but must be properly used with the current understanding of the realm of the organization in mind, aligning the goals and the level of maturity, as well as making sure the technological changes do not turn to the adverse of transparency, equity, and sensitivity to ethical responsibility.

DIFFERENCES BETWEEN THE PUBLIC AND PRIVATE SECTORS

Since management control is subjected to some principles common in every organization, its use in the government sector is characterized by great variations to its implementation in the business world. The context of the functioning of public organizations can be characterized by such a definition as a mission of a public service, an orientation to serve to common interest, and severe regulatory and ethical requirements.

Public money tends to fund their resources and they are usually dependent on the bureaucracy of the budgets and political influences, which is not the case with the private organizations which rather operate to find the increased profit and opportunities of the market (Pollitt & Bouckaert, 2011; Milot, 2022). In the public sector, the performance measure should consider a multiplicity of goals (economic, social, political), and such performance should be measured not only according to the efficiency criteria; it should also be evaluated based on social fairness, openness, and customer satisfaction (Osborne & Gaebler, 1992).

Accountability and transparency demands are even stronger in the public sector, and, to guarantee legitimacy and integrity of the actions of the state, their control systems have to be strong as well. Moreover, the resource constraints that public managers have to deal with are unique in that they include budget rigidity, high turnover in political leadership and an increased multiplicity in the number of stakeholders whose interests are usually conflicting. All these peculiarities make the process of management control on the public sector more complicated and multidimensional, requiring situated and flexible solutions (Bouckaert, 2022).

These inequalities are also added on by increased artificial intelligence. At the corporate level, most entities use the AI-based management control systems to achieve maximum efficiency, forecast market trend, and gain competitiveness. In contrast, in the public sector, AI should be applied under a closer ethical and legal inspection, where such considerations as transparency of algorithms, fairness, data privacy and accountability ought to take priority (Wirtz et al., 2019; Mittelstadt, 2019). This gap shows that despite the similarities between both sectors in terms of digital transformation prospects, the public sector must focus on trust and legitimacy, the ethical use of AI being at the center of the management control processes in the sector.

RESOURCES AND CONSTRAINTS

In the context of a public organization, the resources and constraints analysis are important to comprehend the dynamics of operating and making decisions in such organizations. The public sector resources are diverse in nature and origin and include financial, human, material and informational aspects. Contrary to the situation in the private sector when entrepreneurial activities frequently lead to the acquisition of resources, the government or other funding agencies are the main source of funds held by the public organizations. It results in inflexibility of resources, which are compelled to respond to policy priorities and budgetary cycles and which are often marked by inflexible hierarchy that may impede operational flexibility (Pollitt and Bouckaert, 2011; Bennour, 2024).

Limitation on which these organizations live is also different. As an example, operating within the regulatory boundaries and being as transparent as possible leaves extremely limited choices to both financial and operational management. These restrictions may cause delays in the innovative processes and difficulty in implementing new strategies but it must also adapt to the dynamic nature of the environment that is most of the time very dynamic to suit the needs of the citizens. Besides, the fact that the implication of limited resources is real as well in the context of the expanding and high-demanded access to quality public services complicates the task even further, and it is normal to be pushed to make significantly difficult choices when it comes to the choice of particular programs and initiatives (Bouckaert, 2022; Capocci, 2024).

The resources-constraints interface consequently does not only determine the management approach of the public organizations, but also their perception of performance. Resource management is not a mere issue of assigning funds, but a strategically planned issue that involves maximizing the capabilities of the staff, implementing right technologies and reconsidering the processes within the company.

In this respect, artificial intelligence presents new opportunities and limitations, in that order. On the one hand, AI-based systems can be optimized to allocate resources optimally, and automated to repeat the identical administration functions, and provide predictive patterns to be capable of balancing less supply and the demand that it fulfills (Wirtz et al., 2019). On the one side, the introduction of AI requires a significant investment of money and advanced levels of digital infrastructure and the skills of humans which are scarce in the majority of state institutions. Moreover, there are also ethical and legal restrictions related to data privacy, algorithmic bias and transparency which pose even more threats and explains the need to take a moderate and sensible approach to the implementation of AI into the world of the common administration of control measures (Mittelstadt, 2019).

PERFORMANCE MEASUREMENT

The importance of measuring performance in the public organizations is critical as it becomes possible to determine the effectiveness and efficiency of the provided services to the citizens. In contrast to the private sector, where the profit is usually used to determine performance, public organizations have to work within an environment where the score is often based on non-financial elements of citizen satisfaction, equity, and openness. It is a multidimensional approach that incorporates quantitative and qualitative indicators to have a more holistic evaluation (Rochette, 2022).

The scorecard is one of the reference tools used to measure public performance and this enables it to quantify results and track the objectives of the public policies. Such indicators as the time in which requests are processed, the quality of services offered, and the rate of user satisfaction are vital features of this approach. Furthermore, it is the crucial point to incorporate the assessments that are based on social outcomes, e.g., the effects of services on society, to make sure that the resources are distributed in the most efficient way and to suit citizen's needs. It also implies a culture of outcomes and organizational learning where the feedback is not only embraced, but it is also an element of the continuous improvement cycle (El korchi & Akhlaffou, 2024).

The image of artificial intelligence has transformed the situation with the performance measurement in the society over the last several years. Using AI-based analytics, one can manipulate data at a large scale of administrative data to obtain predictive, real-time, and more personal performance metrics (Wirtz et al., 2019). One such possibility is that machine learning models can be trained on any inefficiency or can predict the demand of the services that is more precise than traditional tools. There are also no criticisms of the use of AI that are not moral. All trust in the institutions of the population can be undermined by algorithmic bias, lack of transparency (or absence of it) in the decision-making (so-called black box models), and endangered by the fact that all the data of a person can be retreated without any consideration, unless this aspect is taken into account (Mitelstadt, 2019). Therefore, AI enlarges the technical performance measurement capacity at an economical price with a problematic set of taken into account: menacing demands of better protection to the transparency, fairness and accountability principles.

REGULATORY FRAMEWORK FOR MANAGEMENT CONTROL IN THE PUBLIC SECTOR

The regulatory framework for management control in the public sector is based on a set of standards and laws that aim to ensure transparency, accountability and efficiency in the use of public resources. At the heart of this system are legislative texts that establish the fundamental principles of management control (Maailif & Lkhoyaali, 2023). for example, The Moroccan Constitution of 2011, in particular Articles 154 to 156, emphasizes good governance, the evaluation of public policies, and citizen-oriented service delivery.

In addition, the Organic Law No. 130-13 on the Finance Law, adopted in 2015, represents a pivotal reform that introduces a results-based budgeting approach, inspired by international best practices such as the French LOLF. This law shifts the focus from input-based management to performance and outcome evaluation, requiring public entities to define clear objectives, performance indicators, and expected results (Royaume du Maroc, 2015).

Moreover, the Decree No. 2-22-580 of 27 October 2022, establishing a system for management control within public administrations, provides the legal foundation for implementing performance monitoring mechanisms, internal steering tools (such as dashboards), and accountability frameworks aligned with strategic objectives (Royaume du Maroc, 2023).

These reforms collectively aim to modernize public management in Morocco, fostering a culture of evaluation, transparency, and efficient resource allocation, thereby ensuring that public spending contributes effectively to national development goals.

Institutions play a key role in the implementation of these standards. The internal audit bodies, the regional audit chambers and other control entities ensure the proper implementation of the management control systems. Their mission is not limited to the evaluation of expenditure, but also extends to the analysis of organizational effectiveness and the recommendation of recommendations to improve the performance of public services. Moreover, necessity to train and raise awareness of the authors of the public officials is noted, so that the principles of management control are inculcated at the lowest levels of organizations. Participatory strategies and feedback are also favored, which will lead to an appropriation of the management tools used and to an intensified, but less opaque, control by all the stakeholders, past and present.

Besides these national regulatory frameworks, there is also the new regulatory issue posed by the implementation of artificial intelligence into the management control environment. Not only do public administrations have to meet financial and governance standards but also should match new international principles of AI ethics, transparency, and data protection. As an example, the European Union proposed the AI Act (2021) that emphasizes the necessity to regulate AI algorithmic decision-making to prevent biases, allow accountability, and safeguarding the rights of the citizens. It highlights the aspect that present-day management control systems cannot be reduced to the area of financial and procedural compliance without considering additional measures aimed at responsible digital transformation.

Such normative and institutional framework is thus critical towards institutionalization of a culture of performance-based management within the public sector with the over emphasis on the fact that the aims of the public sector should always be at the core of the process.

STANDARDS AND LAWS

Within the framework of management control in state-owned organizations, it is important to know the standards and legislation existing in this field in order to provide strict and contextual management in the field of finance. The idea of many accounting standards discussed and many times in line with the International Public Sector Accounting Standards Board (IPSASB) guidelines is to harmonise financial reporting and improve the transparency of the public accounts. These demand matching of the financial statements with the fundamental principles of consistency of methods, prudence and regularity, hence giving the stakeholders a fair view of the real pictures of the available resources and obligations of the organization (Christiaens et al., 2015; IPSASB, 2021). Interaction between standards and laws are conspicuous during implementation of management control practices. This is because, in an example, there is the need to comply with accounting standards but the regulatory framework should be a legislative environment so that there are provisions of internal controls that reduce the danger of having the financial mismanagement. These regulations and legal norms are manifested in the nature of the control systems such as the one of the external audit and the independent evaluation that demonstrates the importance of such kind of the culture of compliance that strengthens not only the integrity of the state financing but also of the faith of the people in the institutions.

To that extent, the pillars of regulation, including standards and laws, prefer the strategic management control founded on the performance assessment and continuous enhancement to ensure the legitimacy of the public administrations and their accountability.

Artificial intelligence (AI) is a new concept in the field of law that brings up new legal concerns in a digital transformation context. The implementation of AI-oriented tools in the management control systems need to be in a way that it is not only in line with the principles of financial reporting but it needs to be geared towards the ethical and the legal principles applied in driving data protection, explainability, and accountability. Within the context of this note, this is projected by the attention of the international policies on the responsible utilization of AI within the government sector, such as an act on AI, such as the suggested AI Act (2021) of the European Union, and principles, such as those of the OECD Principles on Artificial Intelligence (2019). This change indicates the significance of management control systems to expand wider than the known financial standards and bring in the ethical and technological dimensions of management controls.

ROLE OF PUBLIC INSTITUTIONS AND ORGANIZATIONS

The institutions can be most important in implementing the management control practices in the public organizations as they are the brains of the idea of transparency, accountability and efficiency. These organizations that extend to governmental organizations, central administrations, and regulations establish the normative strands and principles by which control processes are determined. They allow to conduct a systematic evaluation of the received outcomes of the work of the public entities by identifying the degree of performance and important indicators. This normative position plays a critical role because the government may utilize the available resources in the best way and should be geared towards pursuing efficiency and continual improvements (Boufakir & Akrich, 2023).

These institutions equally promote audit and monitoring culture, appropriately. They control the adherence of the management practices to the set rules through internal and external audits and identify potential weaknesses in the functioning of the public organizations. Such a solution not only assists in identifying malfunctions, but also promotes the introduction of corrective actions and adoption of good practices. Institutions facilitate the element of catalysis of innovation in the public area by promoting a proactive management stance, wherein entities are more likely to endeavor more agile and quantifiable outcomes-based management tactics.

The role of the Altering Public Institutions is on the increase in the digital transformation era in which it takes over the management controls of AI applications. Ethics This is one of the issues of major concern to governments and regulators who might wish to ensure that AI-based performance monitoring, predictive or decision support tools are able to be aligned with the ethical principles of transparency, fairness and accountability. Numerous international initiatives are being designed which compel governments to regulate algorithmic systems and guard individual rights, such as the OECD Principles on AI (2019) and project European Union AI Act (2021). This growth introduces a new dimension of governance where the institutional players do not just have to be the guarantors of the financial and administrative regulation, but they should as well be moral regulators in the implementation of the new technologies.

MANAGEMENT CONTROL IMPLEMENTATION PROCESS

The implementation of management control within public organizations requires a structured approach, based on two major axes: planning and monitoring and evaluation. These elements, while distinct, are intrinsically linked and mutually supportive to ensure effective control mechanisms (Anthony, 1965; Simons, 1995; Dekri, 2024).

Initially, planning is the cornerstone of the implementation process. It involves the precise identification of strategic objectives, which must be aligned with the mission and values of the public organization. At this stage, it is crucial to integrate a rigorous method to formalize the performance indicators, which will then serve as a reference to measure the results obtained. This step also requires an assessment of the available resources, whether human, financial or material, in order to guarantee the viability of the actions to be carried out. In addition, planning should include a realistic timeline and appropriate communication modalities, thus ensuring that all stakeholders are involved and aware of their expectations (Gniaka & Khassim, 2024).

Once the planning phase is completed, monitoring and evaluation take over to monitor the smooth running of activities and verify that the objectives set have been achieved. This process involves the regular collection of performance data, which helps to identify gaps between planned and actual results. Analyzing these differences is a powerful tool for proactive feedback and can lead to real-time strategic adjustments. It is essential to adopt a systematic approach that promotes continuous evaluation, as this not only improves resource management, but also optimizes decision-making at all levels of the organization.

Artificial intelligence (AI) is a new manner in which the use of management control is changing in the digitally transformed environment. The use of AI assisted systems makes it possible to process real-time data, predictive analytics, and automated reporting, which means making monitoring procedures more accurate and timelier. However, there are also certain ethical concerns behind the innovations (including the issue of data privacy, the risk of algorithmic bias, and the risk of excessive reliance on automated decision-making). The aspects of technical expertise and moral responsibility must be balanced by the government agencies in such a way that the state administration must be reinforced by the AI technologies on the attributes of transparency, fairness, and legality (OECD, 2019; European Commission, 2021).

The combination of these practices will lead to a change in management control in the public organizations into an organizational learning lever, which facilitates innovation and enhancing the services provided to the community.

ROLE OF INFORMATION TECHNOLOGY AND ARTIFICIAL INTELLIGENCE

Information technology (IT) plays a fundamental role in management control within public organizations, facilitating access to relevant data and optimizing decision-making processes. Information systems, as integrative tools, make it possible to collect, store, process and disseminate essential information for performance management. Indeed, these systems are designed to meet the specific needs of managers by providing real-time information. This includes budget management, strategic planning and performance evaluation, all of which are key elements of management control. Automating financial reporting and implementing interactive dashboards illustrate how IT contributes to transparency and accountability, while reducing data processing times and increasing data accuracy (Bhimani, 2021).

Data analytics represents another vital dimension of the role of information technology in public organizations. As governments generate and collect exponential volumes of data, the ability to analyze this information becomes crucial. Advanced analytics tools, such as artificial intelligence (AI) and machine learning (ML), can explore trends, model scenarios, and predict outcomes. This not only contributes to a better allocation of resources, but also to increased responsiveness to emerging challenges. In addition, the judicious use of data analytics fosters an evidence-based performance culture, where decisions are guided by quantitative and qualitative evidence rather than intuition or presumptions. Such analytics systems therefore promote the performance of the organization and enhance compliance to the strategic objectives (Chegri and Torra, 2024).

Along with the technical optimization, there is a new opportunity and new ethical issues along with the new management control systems with the AI. On one hand, the AI will enhance the predictability, risk identification and analysis of the citizen feedback, and thus the responsibility and service provision will be enhanced. On the one hand, it raises the questions of great significance, regarding transparency, prejudice of algorithms, and data and validity of accepting the decision made by a machine in the social sphere. In this way, the topic of AI within the framework of the management control will require a technological change process and creation of the models of ethical governance and algorithmic responsibility (OECD, 2019; European Commission, 2021).

CASE STUDY OF MANAGEMENT CONTROL IN PUBLIC ORGANIZATIONS

Increased transparency, efficiency and accountability are some of the most important strategic considerations with regard to effective management control in the case of the public organization. In this respect the examination of the practices in organizations in the Fez-Meknes region in Morocco lets one see how significant these mechanisms are. The implementation management control is done in a systematic way in this region including systems of performance indicators on effectiveness of the programmes and services provided. As an illustration, the use of integrated dashboards, not only allows tracing the development of the projects, but it also can be used to optimise the assignment of both financial and human resources. The high-level results of the practice determine the great progress in budget management that contribute to the increased reasonability of government spending and the improved accountability of administrators (Lahlimi & Messaoudi, 2024).

Such experiences, nevertheless, pinpoint some of the pitfalls that accompany such problems as reliability of the data and resistance towards organizational changes that can be neutralized by the addition of AI-driven solutions. Specifically, AI systems can enhance performance monitoring, add a real-time anomaly detection system, facilitate a predictive analysis of potential budgetary risks and warrant compliance with the ethical standards of transparency, impartiality, and data privacy.

Removing to the international instances, one can offer a list of the relevant models of a management control over various public entities: one that takes place in Scandinavia. As an example, Sweden has adopted results-based management system whereby evaluation and testing of performance is systematically granted to the public agencies. Such a model does not only promote the culture of responsibility but also results into an optimization of the quality of the services that are provided to the citizens.

In the meantime, some jurisdictions in New Zealand have incorporated continuous feedback processes into their work, which allows real-time variation to take place in the management styles that serve the changing demands of the users. It is on these backdrops that governments are increasingly becoming conscious of AI-enabled analytics and digital governance systems because it is assisting them to be more accountable and make evidence-based decisions.

These foreign experiences are positive because they demonstrate that management control systems can be adapted to different cultural and economical conditions and has a similar purpose, i.e., the continuous improvement of the offered public services. Consequently, by referring to examples of Fez-Meknes and foreign models, we have come to the realization that the management control that is planned and performed with accuracy in the management of the state sector can go a long way in improving performance and efficiency of the concerned organizations. Such practices combined with AI-enabling technologies would also assist in maintaining adaptive practices, resilient measures, and continuous evaluation processes to account to the modern reality of governance in the era of digitalization. The experience brings about lessons, the attention to focus on the adjustment method, which relies on corresponding indicators and the continuous examination procedure to meet the modem challenge in community services (El Kezazy & Hilmi, 2024).

THE CASE OF ORGANIZATIONS IN THE REGION OF FEZ MEKNES IN MOROCCO

Fez-Meknes is a historical region characterized by a high level of cultural diversity, which makes the implementation of management control practices within public organizations particularly complex. Entities such as Public Administrative Establishments (EPAs) and Public Industrial and Commercial Establishments (EPICs) must address challenges related to resource harmonization, service delivery efficiency, and budget transparency. In this context, management control is increasingly regarded not only as a financial monitoring tool but also as a strategic instrument that enables the alignment of institutional performance with broader regional development goals (Abossolo & Koumetio Kenfack, 2024).

Certain institutions, especially at the local level, have begun to adopt basic management control tools, such as dashboards and performance indicators, which support internal decision-making and enhance accountability. Efforts in training and capacity-building have also been made to improve the skills of public managers in internal auditing, budget management, and monitoring systems (Siham, 2021). These trends reflect a growing institutional awareness of the importance of management control within a broader governance framework.

However, the effective implementation of these tools continues to face several traditional obstacles. Major constraints include a lack of adequately trained human resources, fragmented organizational structures, and limited technological integration. Previous studies at the Moroccan level have already identified the shortcomings of management systems in public institutions, notably pointing to early and uneven implementation of ERP systems (Abouamama, 2007). To this end, the slow but steady adoption of artificial intelligence (AI) tools (i.e., automated reporting, predictive budget analytics, and anomaly detection) is likely to address some of these shortcomings. AI can assist local managers with real-time insights, elimination of manual errors, and transparency, but deployment must put a strong focus on ethical concerns including fairness, data security, and inclusion.

Addressing these structural and operational weaknesses requires partnerships among public agencies, higher education institutions, and private sector stakeholders. Such collaborations have the potential to develop adaptive, context-appropriate control systems with measurable and sustainable outcomes (Bal & Jaouhari, 2024).

METHODOLOGY AND KEY FINDINGS

To gain a clearer understanding of the state of management control in the region, a quantitative study was conducted on a sample of 50 public organizations in Fez-

Meknes, covering both EPAs and EPICs. Data were collected through a structured questionnaire that focused on the level of implementation, the types of tools used, and the main barriers encountered.

Key findings indicate that 46.7% of respondents reported the absence of a formal management control approach within their organization, highlighting significant limitations in the adoption of modern performance tools such as key performance indicators (KPIs), total quality management (TQM), and activity-based costing (ABC/ABB). Many public entities perceive these tools as too complex and costly, both financially and in terms of human resources, particularly in an environment still undergoing administrative restructuring and transitioning toward stronger governance principles. Most public administrative establishments (EPAs) lack dedicated local units for management control; in many cases, the function is either centralized at the national level or fragmented across departments, without institutional coordination. Respondents noted that the use of dashboards remains traditional and budget-centric, oriented mainly toward internal reporting. By contrast, public industrial and commercial establishments (EPICs) appear to show more willingness to adopt strategic control tools due to their operational flexibility and commercial orientation.

Overall, these results confirm that while management control practices are gradually emerging, they remain at an early stage. Future progress will depend on simplifying control tools, improving technical infrastructure, and reinforcing the institutional focus on developing a results-oriented public management culture.

Moreover, the results imply that the concept of integrating digital technologies, especially data analytics and AI, has not been explored much in the region. Although the respondents recognize the value that AI-driven tools can add to forecasting, risk management, and performance monitoring, their successful implementation is dependent on the weak digital infrastructure, insufficient knowledge, and ethical issues related to transparency and accountability. This points to a major opportunity to be pursued in future research and practice: the capacity to reconcile the traditional world of management control systems with AI-enabled management that can lead to greater efficiency and legitimacy within public organizations.

INTERNATIONAL CASES

The analysis of management control practices in public organizations internationally reveals a variety of systems and methodologies that reflect the diverse socio-economic, political and cultural contexts of the countries studied. Significant examples come from the European Union, the United States and Asian countries such as Japan and South Korea, where each has developed polished approaches to budget management and performance.

In the European Union, the implementation of management control systems is often part of a rigorous legislative framework. The principles of transparency and fiscal responsibility have led to the development of mechanisms such as "New Public Management" which has subtly integrated techniques perceived in the private sector, applying the concept of performance within public services. This model consisted of introducing performance indicators, regular reporting and incorporating feedback from citizens, while emphasizing the importance of results-based management. In France, for example, the Public Management Reform, initiated in 2007, seeks to improve the performance of public services while ensuring accountability (Augias, 2023).

In the United States, the Federal Performance Management System has proven to be a proactive model of public management. Using measurable performance indicators, the U.S. government sought to optimize service delivery while keeping costs under control. This has been manifested in the implementation of initiatives such as the Government Performance and Results Act, which requires the definition of clear and measurable objectives as well as the monitoring of the results achieved, thus allowing for a continuous re-evaluation of public policies. Meanwhile, national administrations in Japan and Korea have made special focus on strategic planning and continuous stakeholder involvement to enhance the quality of the services provided to the people, a strategy that was spawned by a culture that is highly biased towards quality and excellence (Pagé, 2025).

These global cases indicate that despite the specifics of each country, the overall tendency toward the organization of the management control and control based on the results is established. The key factors are the enhanced responsibility, close observation of the performance and constant adjustment to the demands of the citizens, which is a topical and obligatory reaction to the modern problems of the public organization. This highlights the role of practice comparison to be able to utilize the best practices and stimulate change that can be beneficial to the whole public sector in the world.

International experience is also few and recent in the area where AI and advanced analytics are introduced into the management control systems. Secondly, to recognize anomalies in real time, other member states of the EU also experimented with AI-powered performance dashboards, and U.S. federal agencies tested predictive analytics to predict risks and optimally use the resources. It has also been investing in AI based citizen feedback programmes in South Korea to increase accountability and responsiveness. These projects are the illustrations of the advantages and threats of the digital transformation that do not only offer immense possibilities of efficiency and transparency but also cause a chain of ethical issues of favoritism, privacy, and fairness in decision-making within the public. (Wirtz et al., 2019; Sun & Medaglia, 2019).

CHALLENGES AND LIMITATIONS OF MANAGEMENT CONTROL IN PUBLIC ORGANIZATIONS

The management control within the field of government institutions poses a few constraints and problems which limits its effective implementation and application. Out of these problems, opposition to change is one of the most significant impediments to effecting new management practices. This opposition may be provided not only by workers who might find control devices as a threat to their freedom or employment security, but also by managers, who are usually not ready to quit the customary models which have been shown to be successful. Misgivings on the transparency, accountability, and fear of bureaucratizing too much finds its way to this atmosphere of opposition. Consequently, activities in management control can easily be undermined such that they cannot positively affect the performance of the organizational structure. These resistances should hence be countered by incorporating change management tactics such as communication and sufficient training (Nebie & Tbitbi, 2022).

The resource allocations are also of great concern as there are little resources either financial, human or technological. In contrast to the ability of private organizations to modify their budgets and priorities in case of the emergence of new needs, the functioning of the public structures frequently has to be made within the frames of strict budgetary situations, and, therefore the opportunity to invest in the highly-developed system of management controls is lowered. As the result of such lack of resources, there is an issue of difficulty in hiring and training qualified professionals, who are necessary to implement and observe the process of control. Furthermore, the consequence of underfunding can be insufficient or old management instruments due to which you cannot rely on the data and discover the support of the analytical side.

Moreover, the use of AI in management control comes with new challenges. Although AI-based tools have the potential to enhance predictive power and enhance performance tracking, they pose ethical issues that refer to algorithmic biases, data privacy, and algorithmic obscurity of decisions (Meijer & Bolivar, 2016; Sun & Medaglia, 2019). Lack of technical expertise and regulatory clarity in the direction of being responsible when using these systems is often lacking in the case of public organizations. This may result in employee and citizen mistrust in AI-driven decision-making is seen as opaque or arbitrarily unfair.

To conclude, it is important to note that management control in the government does face a lot of challenges and therefore needs to be managed in a systemic as well as a strategic manner in order to overcome such shortcomings. It is only through a serious mobilization of the concerned actors through the concerted effort and innovation that one will be able to address these obstacles and enhance the general performance of the public organizations.

RESISTANCE TO CHANGE

Resistance to change remains one of the major challenges hindering the effective implementation of management control within public organizations. A concept deeply rooted in human psychology, this resistance can manifest itself on many levels, from employees directly affected by new methodologies to senior managers whose roles might be questioned. In an environment where organizational instability, reforms and modernization initiatives are constantly evolving, change can raise fears about the sustainability of jobs or the redefinition of decision-making processes. As a result, resistance becomes a psychological obstacle, often tinged with anxiety and mistrust of new practices put forward by governing bodies (Nebie & Tbitbi, 2022).

The causes of this resistance to change are many and varied. On the one hand, the shift brought by the management control is frequently viewed as a top-down taxation, which is not very clear in its advantages and does not correspond with the culture present in the organization. On the other hand, employees will lose a sense of ownership and the sense of autonomy, and they will desire to manifest this discontent and opposition even more strongly. The emotional aspect of this gets heightened because management control is being perceived in a negative light by people since it is not the source of improvement, but the source of control. Research has established that the affected employees who are not in the process of change meet resistance which does not only manifest itself, but also escalates unwillingly threatening the collaborative nature that follows, thus, making it hard to build the new practices.

The digital era of digital controls in the management controls systems further increases this pushback through the inclusion of artificial intelligence (AI) and more developed analytics. The workers might be afraid that they are replaceable, and the managers might not trust the decisions made by algorithms and alleged as inexplicable or overcomplicated (Meijer and Bolivar, 2016). Moreover, certain ethical concerns, such as data privacy and algorithmic discrimination can cause additional distrust and increase the reluctance toward AI-based practices. By so doing, resistance is not merely of a psychological or cultural nature, but also, a technological and an ethical one.

In order to counter this resistance, a well-planned change management plan must be adopted. The first is to get into an open discussion where the employees can air out their fears and know what change is about. Moreover, introducing feedback systems and providing individual training, organizations may expect to reduce the stress produced by the uncertainty of change. Lastly, in order to embrace the adoption of progressive acceptance and collective ownership, it is necessary that the benefits of management control are realized concretely through concrete results. Therefore, fighting resistance to change ceases to be an absolute requirement, but in fact a tool of change and transformation to the institution, thus becoming possible to measure the modernization of the institutions of management in the institutions representing a positive and adventurous perspective.

LACK OF RESOURCES

The shortage of resources is the real issue toward effectiveness of management control in the publicical organizations. Among other things, this deficiency may be presented in the form of inadequate financing, skills and training shortages and also, a shortage in technological resources. Consequently, this makes it difficult to have effective management control systems in place in the public organizations that usually face close budgetary limitations. The absence of the sufficient financial means restricts the possibilities of investment into the modern information system and in the continuous training of the senior managers and employees that are in the charge of providing the management control.

To illustrate this example, it may be that without appropriate training a public finance officer is not able to analyze the relevant data and is not able to understand how to evaluate performance of a department in order to make well informed decisions. Besides, improper technological means make gathering and analyzing data difficult, which proves the failure of reports and indicators of performance. Moreover, due to the lack of some resources, the organizational tensions may escalate and thus leaders cannot develop the culture of appreciating management control. The priorities of management are redefined in a situation of limited resources so that instead of focusing on the control mechanisms that create transparency and accountability the focus will be on other issues.

In this regard, artificial intelligence (AI) integration also increases the resource mismatch. Although AI-based tools can be utilised to perform predictive analytics capabilities, fraud detection, and automated reporting, their implementation needs a considerable monetary investment, high-scale digital infrastructure and top-competent talent. The lack of adequate resources may also lead to the development of a digital divide where, once introduced, AI technologies will not be available and used well enough to improve inefficiencies (Wirtz et al., 2019; Mikalef et al., 2020).

This kind of movement can foster opportunistic behaviour or lead to a situation where innovation and constant improvement which usually depend on investment in resources are hitherto left in abeyance. The solution to this issue includes not only a re-evaluation of the priorities allocated to budgetary procedures, but also a common effort to inform the major stakeholders of the significance of the management control to eliminate these deficiencies through the methods modified to the specifics of the public organizations.

FUTURE PROSPECTS FOR MANAGEMENT CONTROL

The management control of the public organization is evolving significantly with development of trends and technological advancements that are redesigning the decision making and managing processes. In the list of significant trends, there exists increased transparency and social responsibility. The existence of such needs requires managers to use smart utilization of data in order to demonstrate efficiency and influence of the populist programs. Examples of advertising, which is increasingly integrated with the implementation of performance management systems, will enable more serious accumulation and analysis of real-time performance in the organizations, altering the distribution and use of the publicly owned resources by the people.

At the same time, artificial intelligence (AI), big data analytics and predictive algorithms are insidiously replacing their positions as a central instrument of the creation of management control. A combination of these technologies can be added to an analytical layer not only due to the increased capability, but also due to the routine nature, which can be automated and, therefore, leave time available to work on the strategic analysis. Some of the AIdriven tools currently in pilot stage in certain OECD countries to demonstrate the efficiency and responsiveness of predictive insights include natural language processing (to analyze citizen feedback) and machine learning (to detect anomalies and prevent frauds) (Wirtz et al., 2019; Mikalef et al., 2020).

Also, the increased popularity of powerful information systems stimulates the appearance of smart dashboards that enable a manager to have a timely control over the key performance indicators. The combination of AI-driven dashboards and decision-support tools allows achieving not only real-time monitoring, but also scenario modelling of policies, letting governments see the potential of policy implementation beforehand (Sun & Medaglia, 2019).

This capability of analyzing huge amounts of data enhances the process of decision-making, which in turn, makes the public organizations more flexible and adaptable to the needs of the society reforming.

EMERGING TRENDS

The management control in public organizations is experiencing a changing environment which is continuously being subjected to a growing shortage of budget and an elevated societal anticipation. Among the emerging trends in this field that are the most visible is that there is need to incorporate social and environmental responsibility in the processes of management. With increasing citizen awareness on global affairs, e.g. sustainable development and social equity, the management control practice of the public organization needs to respond to such aspects of the society. This transformation entails the reconsideration of the traditional key performance focus that is limited to the financial criteria towards a more complex view that embraces some modifications like social and environmental and economic effects.

Simultaneously, the digitalizing of the administrative processes is also changing the perception and application of the management control. With the appearance of new far-sighted analytical systems and information systems integration, the collection and analysis of the data becomes better. With these technological advances, decision-making becomes evidence-based, and therefore, the culture of a transparent and effective organization will be promoted. Nevertheless, this entails a shift of thinking in the supervisory organs where critical knowledge of information technology will be of paramount importance so that the potential inherent in these developments is well utilized.

The second important trend affecting the performance of management control in the public industry is cross-sectoral collaboration. Dealing with the multi-faceted issues which cannot be treated within the boundaries of single institutions, exchange of resources and information among various public organizations and the cooperation with private sector and the voluntary sector is very important. Such synergy does not only allow one to regulate the maximum use of public resources, but also build the systemic approach, which appreciates the synergy between actors, thereby contributing to the increase in the effectiveness and usefulness of the state initiatives. This means that the following years will become a critical period in the management control of the public organization because what is now just an emerging trend should not be regarded as a matter of concern at some time and distance, but indicates the direction where management practices will be in the nearest future (innovation, responsibility, collaboration).

TECHNOLOGICAL INNOVATIONS

Over the last few years, technological innovation is having a powerful impact on the management control environment of the public organization; it offers the tools and techniques that can be used to increase efficiency and effectiveness of decision-making. Artificial intelligence (AI) and the use of other digital technologies enable managers to find real-time data, which helps them perform the analysis of performance more accurately and on time. In one instance, cloud computing solutions provide a non-inflexible and scalable infrastructure in which organizations can consolidate their financial and operational information, thereby enhancing effective collaboration and transparency in teams. In addition to AI usage, big data is important in the regulation of innovation. By analyzing vast quantities of data, managers will be able to discover the trends and patterns, which will enable the managers to make predictions regarding future outcomes and modify their strategies. Such advanced analytical tools as dynamic dashboards or software to visualize information can process complex information and present it in a form of understandable representations. This does not only assist in performance evaluation but also these forms scenarios of strategic planning. Moreover, human resources are liberated to concentrate on the higher value-added analyses as the processes are automatized by the robotization of regular operations, which transforms management control into a nimble one.

The emergence of technologies such as blockchain can also transform management control in the public sector. By ensuring the traceability and security of financial transactions, blockchain creates an environment of trust between stakeholders. This is particularly relevant in a context where transparency of public spending is crucial. Thus, technological innovations do not just modernize tools and processes; They are also redefining the roles and responsibilities of managers, who must now evolve towards data and technology skills, while remaining attentive to ethical issues and governance in the management of public resources.

CONCLUSION

The summary of the principal lessons acquired in this analysis is the conclusion of the study of the practice of management control in the case of public organizations. First of all, it is coming out that rigour in the budgetary control not only in the formulation of identifiable ends, but also in the standardised performance measures are required to ensure transparency and efficiency of the services provided by the people. As soon as the proper financial management tools are introduced, the regional publications will be in a position to not only utilize the resources that they have more optimally, but also gain more trust among the citizens since the financial discipline is observed.

When the expectations of the society on the public administrations are continuously changing, it is important that they must dissolve on management control mechanisms that are changing as well. Modern issues, like the processes of digitalization, the integration of artificial intelligence (AI) and the necessity of a proactive crisis response strategy have to be taken into consideration when implementing the performance appraisal system, as an example. Organizations should change its management processes to involve not only quantitative indicators, which are being used traditionally, but to focus on qualitative indicators as well, which indicates whole approach, taking into account additional services offered by organizations to the citizens and are not reduced to financial outcomes exclusively.

Optimal use of the principle of management control in the public sector is also based on good organization culture that fosters participatory and cooperation of the people in the office. By continually training and getting the stake holders involved in setting out common objectives, an atmosphere where continuous improvement can occur is established. Simply put, management control practice cannot be thought of only as a limit but as an instrument of innovation and enhancement of service in the sphere of the governmental services. This consideration of management control in the sphere of public organisations does not only prove its relevance, but also demonstrates that it can define the future of the public services in case it is conducted with a clear vision and a targeted activity.

REFERENCES

Aamri, E., & Bouaziz, S. M. (2025). L'impact des technologies émergentes sur les systèmes de contrôle de gestion: un modèle conceptuel pour une intégration équilibrée. Journal of Economics, Finance and Management (JEFM), 4(3), 637–652. https://journal-efm.fr/index.php/JEFM/article/view/234/159

Abossolo, A. C., & Koumetio Kenfack, M. (2024). Le rôle du contrôle de gestion dans le pilotage des activités opérationnelles des formations sanitaires au Cameroun. *Revue Française d'Économie et de Gestion*, *5*(7), 1–20.

Abouamama, M. (2007). L'implantation des systèmes ERP dans le secteur public marocain: Enjeux et limites. *Revue Marocaine de Comptabilité*, *3*(1), 41–55.

Anthony, R. N. (1965). Planning and control systems: A framework for analysis. Division of Research, Graduate School of Business Administration, Harvard University. https://archive.org/details/planningcontrols0000anth

Augias, D. (2023). La nouvelle agence nationale de la cohésion des territoires en France, levier de politiques publiques contractualisées entre État et collectivités locales? Politiques & Management Public, 40(2), 145–163. https://www.jle.com/fr/revues/pmp/article/6442

Bal, M., & Jaouhari, L. (2024). Le contrôle de gestion au service de la gouvernance des collectivités territoriales. *Revue Internationale des Sciences de Gestion*, *7*(3), 33–47.

Bennour, M. (2024). Le design comme outil pour humaniser les politiques publiques et favoriser l'intelligence collective. ModACT, 12(1), 77–95. https://popups.uliege.be/3041-4687/index.php?id=177&lang=fr

Bhimani, A. (2021). Accounting disrupted: How digitalization is changing finance. Wiley. https://www.wiley.com/enus/Accounting+Disrupted%3A+How+Digitalization+Is+Changing+Finance-p-9781119720065

Bouckaert, G. (2022). From NPM to NWS in Europe. Transylvanian Review of Administrative Sciences, 18(SI), 5–21. https://doi.org/DOI: 10.24193/tras.SI2022.2

Boufakir, A. I., & Akrich, S. (2023). La contribution du système de contrôle de gestion à l'évaluation du capital immatériel des entreprises innovantes: Modèle conceptuel théorique. [RMd]. *Revista Multidisciplinar*, *5*(3), 291–308. DOI: 10.23882/rmd.23173

Capocci, C. (2024). L'impact de la crise financière de 2007 sur les réglementations des hedge funds. *Revue Internationale d'Économie et de Finance, 15*(1), 45–62. https://matheo.uliege.be/handle/2268.2/19635

Chegri, M., & Torra, M. (2024). Le contrôle de gestion au service du pilotage de la performance de l'entreprise. *Revue Française d'Économie et de Gestion, 5*(4), 102–117.

Christiaens, J., Vanhee, C., Manes Rossi, F., Aversano, N., & Van Cauwenberge, P. (2015). The effect of IPSAS on reforming governmental financial reporting: An international comparison. *International Review of Administrative Sciences, 81*(1), 158–177. DOI: 10.1177/0020852314546580

Dekri, M. (2024). Repenser le système de contrôle de gestion dans le secteur public à l'ère du Post New Public Management. Dossiers de Recherches en Économie et Gestion, 12(2), 11–36. https://doi.org/DOI: 10.34874/IMIST.PRSM/doreg-v12i2.49385

DiMaggio, P. J., & Powell, W. W. (1983). The iron cage revisited: Institutional isomorphism and collective rationality in organizational fields. *American Sociological Review, 48*(2), 147–160. DOI: 10.2307/2095101

El Kezazy, H. (2022). Contrôle de gestion et la bonne gouvernance dans les collectivités territoriales: Une revue de littérature. *Revue Internationale des Sciences de Gestion, 5*(2), 77–92.

El Kezazy, H., & Hilmi, Y. (2024). Le contrôle de gestion territorial: Levier de la bonne gouvernance. *Alternatives Managériales Économiques, 6*(1), 55–72.

El Korchi, C., & Akhlaffou, M. (2024). La transformation de la gestion publique à la lumière des innovations managériales. *Revue Française d'Économie et de Gestion, 5*(5), 133–149.

El Mehdi, B., Hind, C., Slimane, S., & Abdelali, L. (2025). Vers une modélisation conceptuelle de la performance globale des entreprises de services à travers le contrôle de gestion, les systèmes d'information et l'intelligence artificielle. *Revue Internationale de la Recherche Scientifique, 3*(3), 2599–2620.

European Commission. (2021). Proposal for a regulation laying down harmonised rules on artificial intelligence (Artificial Intelligence Act). Official Journal of the European Union. COM(2021) 206 final. https://eur-lex.europa.eu/legal-content/EN/TXT/?uri=CELEX%3A52021PC0206

Fähndrich, J., & Pedell, B. (2025). Evaluating the influencing factors and effects of the digitalization of management control. *Journal of Accounting and Organizational Change, 21*(2), 278–311. https://www.emerald.com/insight/content/doi/10.1108/JAOC-06-2023-0116/full/html. DOI: 10.1108/JAOC-07-2023-0125

Felício, T., Samagaio, A., & Rodrigues, R. (2021). Adoption of management control systems and performance in public sector organizations. *Journal of Business Research, 124*, 593–602. DOI: 10.1016/j.jbusres.2020.10.069

Gniaka, K. J., & Khassim, D. (2024). *Afrique, la gestion des situations complexes au niveau individuel favorise-t-elle un processus d'apprentissage chez les acteurs.* Revue Congolaise de Gestion.

Hood, C. (1991). A public management for all seasons? *Public Administration, 69*(1), 3–19. https://eclass-b.uoa.gr/modules/document/file.php/PSPA108/4NMP%20all%20seasonsfulltext.pdf. DOI: 10.1111/j.1467-9299.1991.tb00779.x

International Public Sector Accounting Standards Board (IPSASB). (2021). Handbook of international public sector accounting pronouncements. International Federation of Accountants. https://www.ipsasb.org/publications/2021-handbook-international-public-sector-accounting-pronouncements

Jensen, M. C., & Meckling, W. H. (1976). Theory of the firm: Managerial behavior, agency costs and ownership structure. *Journal of Financial Economics, 3*(4), 305–360. DOI: 10.1016/0304-405X(76)90026-X

Kaplan, R. S., & Norton, D. P. (1996). *The balanced scorecard: Translating strategy into action.* Harvard Business School Press.

Lahlimi, Y. K., & Messaoudi, A. (2024). L'utilité du contrôle de gestion socio-économique dans le pilotage de la performance socio-économique des organisations publiques: cas des écoles primaires publiques en milieu rural (Région Marrakech-Safi). African Scientific Journal, 3(23), 135–176. https://hal.science/hal-04538078/document

Maailif, S., & Lkhoyaali, B. (2023). Le contrôle de gestion dans le secteur public: quelles spécificités? Revue du contrôle, de la comptabilité et de l'audit, 7(4), 129–150. https://www.revuecca.com/index.php/home/article/download/981/865

Meijer, A., & Bolívar, M. P. R. (2016). Governing the smart city: A review of the literature on smart urban governance. *International Review of Administrative Sciences, 82*(2), 392–408. DOI: 10.1177/0020852314564308

Meyer, J. W., & Rowan, B. (1977). Institutionalized organizations: Formal structure as myth and ceremony. *American Journal of Sociology*, *83*(2), 340–363. DOI: 10.1086/226550

Mikalef, P., Krogstie, J., Pappas, I. O., & Pavlou, P. (2020). Exploring the Relationship between Big Data Analytics Capability and Competitive Performance: The Mediating Roles of Dynamic and Operational Capabilities. *Information & Management*, *57*(2), 103169. Advance online publication. DOI: 10.1016/j.im.2019.05.004

Milot, J. P. (2022). La comptabilité nationale: l'information économique au service du pilotage des finances publiques. Revue française d'administration publique, 183(3), 723–734. https://doi.org/DOI: 10.3917/rfap.183.0082

Mittelstadt, B. (2019). Principles alone cannot guarantee ethical AI. *Nature Machine Intelligence*, *1*(11), 501–507. DOI: 10.1038/s42256-019-0114-4

Moynihan, D. P., & Kroll, A. (2016). Performance management routines that work? An early assessment of the GPRA Modernization Act. *Public Administration Review*, *76*(2), 314–323. DOI: 10.1111/puar.12434

Nebie, E. & Tbitbi, E. (2022). La résistance au changement organisationnel dans la gouvernance des organisations publiques: meilleures pratiques pour réussir les transformations. Revue de Management et Cultures. 397–412.

Organisation for Economic Co-operation and Development (OECD). (2019). OECD principles on artificial intelligence. OECD Publishing. https://www.oecd.org/en/topics/ai-principles.html

Osborne, D., & Gaebler, T. (Eds.). (1992). *Reinventing government. how the entrepreneurial spirit is transforming the public sector*. Penguin Books USA Inc.

Otley, D. T. (1980). The contingency theory of management accounting: Achievement and prognosis. *Accounting, Organizations and Society*, *5*(4), 413–428. DOI: 10.1016/0361-3682(80)90040-9

Pagé, J. (2025). *Validation du questionnaire proactif: un outil d'auto-repérage pour permettre aux personnes aînées d'évaluer elles-mêmes leurs risques de déclin fonctionnel. Manuscrit inédit présenté au congrès annuel du North American Primary Care Research Group*. NAPCRG.

Pollitt, C., & Bouckaert, G. (2011). *Public management reform: a comparative analysis—new public management, governance, and the neo-Weberian state* (3rd ed.). Oxford University Press.

Rochette, C. (2022). L'évaluation, socle de l'organisation publique apprenante. Gestion et management public, 10(3), 6–9. https://shs.cairn.info/revue-gestion-et-management-public-2022-3-page-6?lang=fr

Royaume du Maroc. (2015). Dahir n° 1-15-62 du 14 chaabane 1436 (2 juin 2015) portant promulgation de la loi organique n° 130-13 relative à la loi de finances. Bulletin Officiel du Royaume du Maroc, n° 6370, 1 ramadan 1436 (18 juin 2015).

Royaume du Maroc. (2023). Décret n° 2-22-580 du 10 chaabane 1444 (3 mars 2023) relatif à la mise en place du dispositif de contrôle de gestion au sein des départements ministériels. Bulletin Officiel du Royaume du Maroc, n° 7200, 12 kaada 1444 (1er juin 2023).

Siham, N. (2021). *Du New Public Management à l'émergence du dogme de la performance et du contrôle de gestion dans les Établissements et Entreprises Publics Marocains.* Alternatives Managériales Economiques.

Simons, R. (1995). *Levers of control: How managers use innovative control systems to drive strategic renewal.* Harvard Business School Press.

Sun, T. Q., & Medaglia, R. (2019). Mapping the challenges of artificial intelligence in the public sector: Evidence from public healthcare. *Government Information Quarterly, 36*(2), 368–383. DOI: 10.1016/j.giq.2018.09.008

Toma, C. L. (2021). Analyse comparative des différentes méthodes de gestion intégrée des rejets miniers permettant de minimiser l'impact des opérations minières. https://savoirs.usherbrooke.ca/handle/11143/18420

Touicher, O., & Loulid, M. (2023). Gestion Budgétaire: un levier clé pour la prise de décisions stratégiques. revue de l'entrepreneuriat et de l'innovation, 5(20).

Weber, M. (1922). *Economy and society: An outline of interpretive sociology.* University of California Press.

Wirtz, B. W., Weyerer, J. C., & Geyer, C. (2019). Artificial intelligence and the public sector—Applications and challenges. *International Journal of Public Administration, 42*(7), 596–615. DOI: 10.1080/01900692.2018.1498103

Zuiderwijk, A., Chen, Y. C., & Salem, F. (2021). Implications of the use of artificial intelligence in public governance: A systematic literature review and a research agenda. *Government Information Quarterly, 38*(3), 101577. DOI: 10.1016/j.giq.2021.101577

KEY TERMS AND DEFINITIONS

Accountability: It is the responsibility of any public organization and the managers who operate in it to explain decision and actions to the stakeholders. The concept of accountability has become core to management control, as all the performance outcomes are joined to trust and a democratic legitimacy of the state.

Artificial intelligence (AI): Sophisticated digital technologies that are able to process a lot of data and recognize patterns and produce forecasts to make decisions. In management control, AI has been improving forecasting, allocation of resources and monitoring of performance, and has also presented such ethical issues like bias, transparency and accountability.

Ethical Governance: The decision-making and organizational behaviour follows a system of ethics that abides by principles of fairness, integrity and responsibility. Ethical governance as part of management control means walking the balance between efficiency and legitimacy and making sure that AI and other new technologies are not abused.

Management Control: A disciplined procedure of planning, monitoring and evaluation of activities of an organization in order to ascertain that use of resources is cost effective and goal is attained. It can be used as a decision-making method and a tool of accountability in the public organization setting.

New Public Management (NPM): A management model, which has been in existence since the late 20th century and it promoted the idea of implementing the practices of the private sector in the government sector. NPM promotes performance measurement, efficiency and citizen-centred services and promotes decentralization and accountability.

Performance Measurement: The quantitative and qualitative measures that are utilised to determine efficiency, effectiveness, and impact of the services rendered to the people. In non-profit organizations, performance measurement also incorporates such factors as citizen satisfaction, equality and social contribution.

Public Organizations: Organization under the control of the government whose work is organized on the provision of the services to all people. They are not focused on maximizing profits and, unlike the private firms, they serve the common good, openness, and social justice.

Transparency: The idea of publicly disclosing pieces of information to the stakeholders to promote trust, legitimacy and informed involvement. Transparency in the public institutions will enable citizens and oversight institutions to check the performance and resource utilization.

Chapter 13
The Emerging Role of AI in Tourism Businesses:
Potential, Challenges, and Ethical Perspectives

Pramendra Singh
https://orcid.org/0000-0002-9142-265X
Marwadi University, India

Aruditya Jasrotia
https://orcid.org/0000-0002-8274-5609
Amity University, Noida, India

Shikha Dhakad
https://orcid.org/0000-0002-2438-063X
Jiwaji University, India

ABSTRACT

This chapter explores the transformative impact of Artificial Intelligence (AI) on the tourism industry. It examines how AI technologies are being used to enhance customer experiences, improve operational efficiency, and drive innovation in tourism businesses. The chapter analyses how AI enables more tailored travel services, better demand forecasting, and optimized pricing strategies. It also discusses the key challenges of AI adoption, including high costs, the need for skilled professionals, data privacy issues, and employee resistance. The chapter also addresses ethical concerns and the importance of ensuring transparency and fairness in the use of AI, emphasizing the need to balance innovation with ethical responsibility. The chapter will help the tourism businesses to understand the emerging role of AI and

DOI: 10.4018/979-8-3373-3104-1.ch013

in its adoption. It will also help in guiding the policymakers, academia and other stakeholders to analyse the outlook of AI in tourism and take necessary steps in their domain for its better outcomes.

INTRODUCTION

The tourism industry which is characterized by its reliance on human interaction and dynamic nature is undergoing huge transformation due to digital revolution and advent of new technologies. Among all other technological advancements, Artificial Intelligence (AI) has drastically influenced all aspects of tourism and hospitality operations. From predictions to supply chain, virtual assistants, prompt customer services, pricing models and new offerings, AI has fundamentally reshaped the tourism business operations including how they operate, compete, engage with stakeholders. Along with AI other technologies which have also gained popularity in recent time are Augmented Reality, Virtual Reality, mobile applications, big data etc. These technologies have revolutionized the traditional models of how businesses used to operate. This has also led to the growth of Online Travel Agencies (OTAs). This shift got momentum in post pandemic period where digitization proved panacea and resilient against the odds (UNWTO, 2022).

Technological transformation in tourism industry is not only about the automation of processes but a paradigm shifts towards intelligent, smart, statistics-driven and facts-based functioning impacting the overall structure and operations of tourism businesses. AI brings in immense benefits for the businesses like it facilitates real-time decision-making, enhances predictive insights, performs better customization and personalization, meeting the needs of the businesses etc. Further, AI also helps in creating efficient, seamless and immersive experiences by integrating with other advance technologies such as Internet of Things (IoT), blockchain etc. (Ivanov & Webster, 2019). Artificial Intelligence simulates the human intelligence by acquiring data, learning by itself, using reasoning to process the data, and self-correct it as and how it is required. The scope of AI in tourism is not limited to merely day to day general tasks, rather it extends to using chatbots and virtual assistants for solving the complexities and guiding the customers, analyse the demand-supply system, managing the inventory and customer relationships. For instance, the airlines and hotels use AI frequently to analyse the demand and adjust the dynamic prices accordingly as per seasons and competitors pricing (Tussyadiah, 2020). AI is also used for other purposes like marketing and management of tourist destinations, manage visitor flows, social media sentiment analysis, smart mobility etc. (Zeng et al., 2020). These capabilities also contribute to sustainable tourism development by enabling better management and allocation of resources, reducing environmental stress and preparedness for the future crises.

In a highly competitive, service-based industries like tourism AI enables the businesses deliver exceptional value to customers (Murphy et al., 2017). AI also helps businesses to enhance productivity, reduce operational costs and respond quickly to the industry changes (Mariani & Wamba, 2020). AI's role in customer satisfaction is also evolving by providing 24/7 support, instant response, meet customer needs etc. AI has been proved to be of paramount importance for environmental and social implications. AI can significantly contribute to sustainable, responsible tourism, minimize environmental impact and promote societal sensitivity.

The chapter intends to study the intersection of AI and tourism businesses in present context with future directives. The primary objective of this chapter is to analyse the immense potential of AI in tourism businesses. It aims to provide an overview of how AI technologies are being used in different segments of the tourism industry. The creation of AI has opened many opportunities in AI system management, tech-driven services and data analytics. The chapter comprises various sections including the uses of AI, challenges and limitations of AI adoption in tourism businesses, ethical dimensions of AI in tourism, policy recommendations etc. The chapter also delves into detail about the disparities in digital capabilities across the globe and different business sizes. The ethical behaviour and responsibilities of tourism businesses in using AI is also explore in the chapter. The chapter provides a timely exploration of how AI is changing the way tourism businesses operate and reshape their competitive landscape. It also highlights the interplay between technological innovations and ethical responsibility to call for a balanced approach that leverage AI's benefits while safeguarding against its risks.

LITERATURE REVIEW

AI and its Relevance for Tourism

Artificial Intelligence (AI) is increasingly reshaping the landscape of the global tourism industry by enhancing customer experience, optimizing operations, and enabling more personalized services with the help of human intelligence stimulation. Technological tools enabled by AI are revolutionizing how tourism businesses interact with customers by offering real-time support, personalized travel itineraries, and efficient customer service (Gretzel et al., 2015). Additionally, AI helps destination marketing organizations (DMOs) to analyse large volumes of tourist data, allowing them to tailor marketing strategies and manage tourist flows more effectively. The COVID-19 pandemic further accelerated the adoption of AI in tourism, particularly in contactless technologies and health safety monitoring, contributing to travellers' confidence and operational resilience. Despite its transformative potential, the integration of AI in tourism also raises concerns about data privacy, job displacement, and ethical decision-making (Zeng et al., 2020). Therefore, a balanced approach is necessary—one that embraces the benefits of AI while addressing its potential societal impacts. As AI continues to evolve, its role in creating smart, sustainable, and inclusive tourism experiences will become even more significant. Virtual travel assistants help tourists plan and book trips based on their preferences, while smart hotel rooms adapt lighting and temperature using AI-based sensors.

AI also plays a key role in smart tourism destinations by supporting data-driven decision-making for city planners and tourism managers. Moreover, AI is increasingly used for language translation and voice-guided tours, making travel more accessible and inclusive. As tourism becomes more tech-integrated, understanding AI's capabilities and limitations becomes essential for both tourism professionals and travellers. Future trends suggest a move towards hyper-personalization, where AI will analyse vast datasets to create truly bespoke travel experiences tailored to individual preferences in real-time. AI's relevance stems from its capacity to analyse vast datasets, personalize experiences, automate tasks, and enhance decision-making for both travellers and businesses. AI in tourism aims to streamline operations, improve efficiency, and create more tailored and satisfying experiences for travellers throughout their journey.

The Potential of the use of AI in Tourism Businesses

Artificial Intelligence (AI) holds immense potential to revolutionize tourism businesses by streamlining operations, enhancing customer experiences, and driving strategic decision-making. Through machine learning, natural language processing, and data analytics, AI can automate routine tasks such as bookings, customer inquiries, and feedback management, freeing up human resources for more complex and personalized interactions. Chatbots and virtual assistants, powered by AI, are increasingly used by hotels, airlines, and travel agencies to provide 24/7 customer support, offer real-time recommendations, and manage multilingual communications—significantly improving efficiency and customer satisfaction (Gretzel et al., 2015).

AI-enabled data analysis allows tourism businesses to better understand traveller behaviour, preferences, and trends, enabling hyper-personalized marketing campaigns and product offerings. AI can process social media reviews and customer feedback to identify patterns in service quality or demand fluctuations, helping businesses to adapt quickly and stay competitive. Additionally, predictive analytics tools can forecast demand, optimize pricing strategies, and enhance revenue management with dynamic pricing models. In the post-pandemic context, AI-driven technologies like facial recognition, contactless check-ins, and robotic services have become crucial in ensuring hygiene, safety, and operational resilience (Tussyadiah, 2020). However, successful implementation requires overcoming challenges such as high investment costs, staff training, and ethical concerns surrounding data privacy and job displacement (Zeng et al., 2020). Nonetheless, with strategic integration and responsible use, AI can significantly enhance productivity, sustainability, and competitiveness in tourism businesses.

In aviation and travel agencies, AI helps manage inventory, predict demand, and optimize pricing strategies, improving both profitability and customer satisfaction. AI also supports destination marketing by analysing customer preferences from reviews and social media to craft personalized promotional content. For sustainable tourism, AI is instrumental in energy management systems in hotels and in tracking carbon footprints of travel activities. In post-pandemic recovery, AI technologies like health monitoring systems, contactless check-ins, and service robots have been crucial in restoring traveller confidence (Tussyadiah, 2020). With continuous advancements, AI is set to become a cornerstone of innovative and resilient tourism business models. Therefore, Artificial Intelligence (AI) presents a significant potential for tourism businesses by revolutionizing various aspects of their operations and customer engagement (WTTC, 2024). AI-powered tools can analyse extensive datasets to understand customer behaviour and preferences, enabling businesses to offer personalized recommendations for destinations, accommodations, and activities, thereby enhancing customer satisfaction and loyalty. Furthermore, AI facilitates dynamic pricing strategies, allowing businesses to optimize revenue by adjusting prices based on demand and market trends. Automation of routine tasks such as booking management and customer service through AI-driven chatbots can lead to increased operational efficiency and reduced costs, freeing up human resources for more complex and strategic initiatives. Ultimately, the strategic integration of AI can empower tourism businesses to deliver more tailored, efficient, and innovative services, fostering growth and improving their competitive edge in the industry.

Challenges in AI Adoption in Tourism Businesses

While Artificial Intelligence (AI) offers significant benefits for the tourism industry, its adoption comes with a range of challenges that can hinder its effective implementation. One of the most critical barriers is the high initial cost associated with developing and integrating AI technologies into existing tourism operations, which can be particularly burdensome for small and medium-sized enterprises (SMEs) (Ivanov & Webster, 2019). Many tourism businesses also face a lack of technical expertise and skilled personnel to manage and maintain AI systems, leading to underutilization of these technologies. Resistance to change is another common issue, especially among employees concerned about job displacement due to automation and robotics (Zeng et al., 2020).

Additionally, concerns regarding data privacy and cybersecurity are prominent, as AI systems often rely on collecting and analysing vast amounts of customer data, raising ethical and legal questions. Cultural and language differences can also complicate the implementation of AI-driven tools, particularly in destinations with diverse international tourist populations. Moreover, the absence of clear regulatory frameworks and industry standards for AI usage in tourism contributes to uncertainty and inconsistent practices across the sector. The effectiveness of AI also heavily depends on the quality of data input; poor data can lead to biased or inaccurate outcomes, which can negatively impact customer experiences and decision-making (Wirtz et al., 2018). Therefore, addressing these challenges requires a multifaceted approach involving investment in infrastructure and training, strong data governance policies, ethical guidelines, and collaboration between public and private stakeholders to create an AI-ready tourism ecosystem.

One of the other significant challenges is the cost and complexity of AI technology, making it difficult for smaller businesses to invest and compete (Buhalis & Sinarta, 2019). Moreover, many tourism professionals lack the technical knowledge required to operate and maintain AI systems, leading to underutilization. There's also resistance from employees who fear that AI will replace human jobs, especially in customer-facing roles (Zeng et al., 2020). Ethical concerns arise from the collection and use of personal data, as tourists may be uncomfortable with surveillance or AI decisions made without transparency. In regions with limited infrastructure or poor internet connectivity, implementing advanced AI tools may be impractical. Additionally, without standardized regulations and policies governing AI in tourism, businesses may face uncertainty and compliance issues. To overcome these challenges, tourism stakeholders must invest in digital literacy, promote responsible AI use, and develop partnerships that support innovation while protecting both workers and consumers.

Furthermore, the integration of AI systems with existing legacy infrastructure can be complex and costly, often requiring significant system upgrades or replacements. Another challenge lies in the potential for job displacement as AI automates certain tasks, necessitating workforce retraining and adaptation strategies (Tussyadiah & Griffin, 2018). Addressing these challenges through supportive policies, accessible technology solutions, and a focus on ethical and responsible AI implementation is crucial for the successful integration of AI within tourism businesses (World Economic Forum, 2023). Expanding on the challenges of AI adoption in tourism, the lack of skilled personnel capable of developing, implementing, and maintaining AI systems poses a significant impediment (Jafari, 2019). Many tourism businesses, particularly smaller ones, may struggle to attract and retain talent with the necessary AI expertise.

Furthermore, ensuring the accuracy and reliability of AI algorithms is crucial; biases in training data can lead to discriminatory outcomes and negatively impact customer experiences. Resistance to change within organizations and among employees can further slowdown the adoption process, highlighting the importance of effective change management strategies and demonstrating the tangible benefits of AI implementation. Successfully navigating these multifaceted challenges requires a strategic and holistic approach, encompassing investment in talent development, robust data governance frameworks, ethical guidelines, and proactive change management initiatives. The growing investment in AI by businesses across the industries is making it inevitable to adopt more AI based tools in future. It has multifaceted benefits for the business, but it brings in new challenges for the employees. Because of advancing technology in AI, the employees are under constant pressure to learn these evolving technologies, compete as per the performance of these AI tools and it leads to anxiety, sadness and psychological trauma (Hamarat et al., 2024).

Ethical Perspectives on the Use of AI in Tourism Businesses

In recent years, ethical studies on AI and robotics have witnessed a notable increase in research (Nurik et al., 2021; Arbelaez Ossa et al., 2024). Within tourism and hospitality industry, the ethical considerations of AI and robotics are vital for several concerns, such as AI technologies operating independently on large-scale data, which brings ethical issues of supervision, accountability, equity, and credibility (Ivanov & Umbrello, 2021). AI systems manage private data and confidential information, leads to concerns of users' confidentiality. It can have a disproportionate effect on willingness to pay, which could lead to legal and ethical challenges. Tourism and hospitality often lack control over the design of AI systems. The expanding control of AI by a few providers poses a threat to competition, risks limiting market competition, however, AI systems need to be sensitive to cultural diversity in global tourism.

Although AI—driven enhancements in tourism foster tourism experiences, however, privacy and data usage remain prominent among tourists (Ferhataj & Memaj, 2024). Ivanov & Umbrello (2021) stressed that the balance between technological growth and privacy rights is evident as privacy is being overshadowed by security demands. Governance, ethical concerns, regulatory compliance, and mitigating potential job displacement are a few major key challenges (Mihai Gabriel & Cosmin, 2024). In response to these challenges, ensuring transparency in AI algorithms, bolstering data privacy, and creating detailed privacy engineering standards. It is very crucial to ensure that the technological innovations are carefully aligned with ethical principles in the tourism sector. Moreover, nowadays AI-enabled hiring can optimise hiring processes, but it can reinforce bias based on attributes such as gender, race, and individual personality characteristics (Chen, 2023). Robust AI model evaluation can also be used to decrease the biasness and ensure fairness by integrating algorithm-level interventions and prepare unbiased dataset (Srivastava & Sinha, 2023). Law & Lei (2025) and Srivastava & Sinha (2023) underscore the ethical use of AI to foster resilience, boost AI systems and merge needful collaborations among all stakeholders.

Transparency and explainability among others are the rising concerns across diverse sectors, including tourism and hospitality (Waltl, 2021; Warner & Sloan, 2021). Hassija et al. (2024) highlighted the "black box" models, hindering both roles of regulators and users, calling for essential methods that demystify how AI reaches its conclusions. Huang et al. (2022) designed a seven-dimensional framework to evaluate resilient mindset, context adaptability, stakeholder collaboration and developed different AI technologies with search engines and chatbots that rank high in adoption potential. Some proposed a conceptual model signifying the integration of human-centric values into AI that integrates stakeholder engagement and external environmental factors. The merge of AI in tourism extends to transform tourism across multiple areas, from visitors' experiences, destination management, and enterprise operations (Tuo et al., 2025). Studies suggest that responsible AI needs a balance between innovation and regulation, context-sensitive AI design, and active engagement from all stakeholders.

METHODOLOGY

The present study aims to understand the emerging role of AI in business perspectives, specially in tourism businesses. The present study adopted a review methodology to go through the extensive literature on AI and its application in the businesses or how it is shaping the general and specialized areas of functioning and operations within the tourism businesses. It has covered the literature collected through different sources and of different time span to also understand how the research and academic discourse is taking place in the area of study. The chapter also studies how tourism businesses strategically approach the integration of AI and dive deep into predicting the future of AI for tourism businesses.

STRATEGIC APPROACHES TO AI INTEGRATION IN TOURISM BUSINESSES

Adoption of AI in tourism is motivated by strategic goals to foster their competitiveness and enrich customer experiences. Tourism companies leverage AI and big data analytics to create innovative services, manage vast datasets, and gain a sustainable competitive advantage (Stroumpoulis et al., 2022). Smart tourism destinations (STDS) use AI to leverage innovative business models, better marketing tactics, and enhance operational workflows. Yet, the significant financial and continuous maintenance requirements challenge some destinations (Florido-Benítez & Alcázar Martínez, 2024). Integration of AI-powered personalisation and emotional intelligence in tourism can lead to engaging and satisfying experiences.

Across the business landscape, AI has transitioned from a novel technology to a core element of customer-focused strategies, powering personalised experiences and predictive analytics, fostering an interdependent relationship between AI capabilities and human expertise (Tamanine et al., 2024; Nannelli et al., 2023). Incorporating AI into business development in tourism-based companies strengthens data analysis, supports better decisions, and enriches customer insights. In flourishing business strategies, adaptability, innovation, and customer orientation demonstrate the advancement of AI's capacity. To increase efficiency, productivity, and profitability among tourism service providers, AI and big data strategies can increase demand and foster behaviour pattern analysis of consumer experiences (Doborjeh et al., 2022; Samara et al., 2020). These effective AI technology demands data governance, employing AI experts, and nurturing innovation driven environment (Olutimehin et al., 2024). AI holds the ability to analyse the emotional data of the consumers and can help the businesses to take better and informed decisions about the consumers (Kim et al., 2024).

Through the "Tourism 3.0" model, research highlighted that the usage of AI in recommendation personalisation, smart itinerary planning, and real-time information can enhance travellers' engagement at every stage of the journey. Moreover, AI transformation in the tourism industry needs to compare pre- and post-AI trends, including ethical issues, resilience and consumer preferences. To gauge this AI adoption, Huang et al. (2022) seven-dimensional framework based on diffusion theory to assess AI adoption in tourism through search and booking engines, virtual assistants, and chatbots, showing strong adoption potential.

The Role of AI in Shaping Ethical and Strategic Transformation in the Tourism and Hospitality Industry

AI can streamline business operations by significantly facilitating pricing, sales management, demand prediction, and revenue optimisation. As the future of AI in tourism highlights the transformation of personalised tourist experiences through AI, it will enhance the efficiency and empower smaller businesses to compete more effectively with established firms. (Tuo et al., 2025). AI is innovating, which could foster the growth of smart tourism destinations by refining both marketing efforts and operational processes. The incorporation of AI in destination management can develop adaptive business models to meet changing consumer needs. From niche tool to central element of tourist experience strategies, powered by advances in data analytics, predictive models, and emotional intelligence, AI is progressing, resulting in fostering resilience and consumer-centric enterprises (Sunday Tubokirifuruar Tula et al., 2024). AI expansion heightened AI applications based on data privacy, surveillance, and user manipulation concerns (Putera et al., 2022).

Mitigating algorithmic bias based on race, gender, and other attributes will demand robust strategies, including transparent algorithms and bias-free datasets is critical (Srivastava & Sinha, 2023). Ethical governance and regulatory compliance need to be prioritised in the tourism industry to ensure responsible AI use by developing frameworks that support transparency, accountability, and stakeholder involvement (Huang et al., 2022).

Lastly for AI integration with tourism businesses will demand leadership commitment, employee adaptability, and a strategic emphasis on AI-powered innovation and foster collaboration among stakeholders to optimise AI use and control its risks (Tamanine et al., 2024).

AI and Tourism Businesses: An Outlook

The tourism industry is sure to witness profound implications of AI in all the areas of tourism businesses. As the growth and phenomenal development of AI is continuously evolving, it is expected to play a major central role in shaping the customer experiences, operations, and decision making by rightly predicting the market situations in short term. Two of the most notable advancements poised to influence the tourism industry are Generative AI and Intern of Things (IoT). Still, the future of tourism industry with AI is exciting but uncertain (Kim et al., 2024). The generative AI is often used to create personalized itineraries, real time content for travel advice, in-depth storytelling experiences as per the need and requirements of the different travellers. This allows more engaging and responsive traveller experiences and interaction to further enhance their satisfaction. By leveraging the potential of generative AI, the tourism businesses can enhance the well-being and job enrichment of the employees by providing them more freedom and opportunity to engage with the emerging technology (Dwivedi et al., 2024).

Also, the integration of AI with Internet of Things (IoT) enables to enhance the tourism infrastructure and create smarter environments. The IoT helps in creating highly personalized and augmented experiences for guests in the hotels based on their behaviour and preferences while smart cities equipped with advanced technologies can provide travellers with seamless transportation, navigation, real-time updates and better local recommendations. These technologies promise to create a better and advanced holistic tourism ecosystem for tourism businesses and for other stakeholders as well.

Not only this, but the AI is also expected to disrupt the traditional old style general tourism business models. AI driven platforms will change the way the travel agencies used to function by offering end-to-end services in planning, booking, coordinating, suggesting and resolving customer issues without any human induced intervention. AI will enable the tourism businesses to target niche and personalized services rather than targeting the mass. AI also has the potential to play a pivotal role in sustainable tourism development. It can help in predicting the tourist flow, weather forecasting, measuring the impact on destinations, and monitoring the environment being damaged by over tourism etc. AI can also assist the tourism businesses and tourists in planning for eco-conscious travel choices by properly guiding and suggesting the options. It can not only contribute to business innovation rather can also help in long-term sustainability of the tourism industry. AI in coming years will not only disrupt the already operational tourism businesses but will have a major impact on the way the start-ups emerge and operate. Most of the start-ups in tourism will be driven by technologies such as AR, VR, blockchain, voice recognition, predictive personalization etc. (Ghouse & Chaudhary, 2024). Thus, it can be articulated that AI will have a very positive impact on the businesses by providing opportunities to their staff to enhance their efficiency, improve quality of services through personalization, and to remain competitive with an edge (Indaryanto et al., 2023). AI will enable the businesses to improve resource allocation, revenue streamlining, drive innovation and revolutionaries the overall functioning of tourism industry in long run (Ma, 2024).

RECOMMENDATION

The extensive review of the literature and by analysing the data available through different sources, a sound understanding was developed on the growth of AI and its application in the businesses across the industries. The current scenario revolving around AI, challenges associated with its use and its prospects for businesses in future was thoroughly discussed. The study recommends the growing need for businesses to adopt the AI and its evolving new technological elements to support the business functions which may make it ease to plan, coordinate, analyse and implement the decisions using AI. The businesses are also recommended to analyse the challenges posed by AI and find solutions to overcome them. One of the most significant aspects of using the AI is to consider ethical perspectives and data privacy. It is also important to know in what context and to what limit it should be used in businesses so that it supplements the human intelligence rather than being a slave of artificial intelligence.

CONCLUSION

The chapter explored the diverse and multifaceted role of AI which is transforming the business operations and their approach. AI technologies including predictive analytics, dynamic pricing tools, chatbots etc. are enhancing the decision making of the businesses along with customer engagement and satisfaction. Although AI has huge potential for the businesses in near future, it also faces substantial challenges like financial barriers, skill gaps, data privacy, algorithmic fairness, workforce displacement among others.

The integration of AI in businesses is not only a technological upgrade for the businesses but also a strategic imperative. The businesses need to be proactive in adopting the growing applications of these AI technologies, but also improve approach in staff training and embracing innovation. Policymakers also have to adopt a critical role in advancing the ethical and regulatory aspects of AI adoption in the businesses to ensure better services, competitive environment and improved outcomes for all. They also need to ensure to protect the customer data and promote transparency. The academic institutions have a greater responsibility to prepare the future tourism workforce as per the dynamic nature of the industry with respect to changing needs to adopt the technology and AI into their curricula. As AI is becoming integral part of almost all the industries, the industry must balance innovation and upholding ethics in adopting it and managing its implications.

REFERENCES

Arbelaez Ossa, L., Lorenzini, G., Milford, S. R., Shaw, D., Elger, B. S., & Rost, M. (2024). Integrating ethics in AI development: A qualitative study. *BMC Medical Ethics*, *25*(1), 10. DOI: 10.1186/s12910-023-01000-0 PMID: 38262986

Buhalis, D., & Sinarta, Y. (2019). Real-time co-creation and nowness: Challenges and opportunities for tourism management. *Information and Communication Technologies in Tourism*, *2019*, 317–330.

Chen, Z. (2023). Ethics and discrimination in artificial intelligence-enabled recruitment practices. *Humanities & Social Sciences Communications*, *10*(1), 1–12. DOI: 10.1057/s41599-023-02079-x

Doborjeh, N. Z., Moghavvemi, S., Musa, G., & Tengku Arshad, T. N. B. (2022). Chatbot adoption in tourism: A systematic review of the literature. *Tourism Management Perspectives*, *41*, 100944.

Doborjeh, Z., Hemmington, N., Doborjeh, M., & Kasabov, N. (2022). Artificial intelligence: A systematic review of methods and applications in hospitality and tourism. *International Journal of Contemporary Hospitality Management*, *34*(3), 1154–1176. DOI: 10.1108/IJCHM-06-2021-0767

Dwivedi, Y. K., Pandey, N., Currie, W., & Micu, A. (2024). Leveraging ChatGPT and other generative artificial intelligence (AI)-based applications in the hospitality and tourism industry: Practices, challenges and research agenda. *International Journal of Contemporary Hospitality Management*, *36*(1), 1–12. DOI: 10.1108/IJCHM-05-2023-0686

Ferhataj, A., & Memaj, F. (2024). Challenges and opportunities of ai implementation in tourism: an ethical and technological perspective. *Studijos–verslas–visuomenė: dabartis ir ateities įžvalgos, 1*(IX), 217-231.

Florido-Benítez, L., & del Alcázar Martínez, B. (2024). How artificial intelligence (AI) is powering new tourism marketing and the future agenda for smart tourist destinations. *Electronics (Basel)*, *13*(21), 4151. DOI: 10.3390/electronics13214151

Ghouse, S. M., & Chaudhary, M. (2024). Artificial Intelligence (AI) for tourism Start-Ups. In *Advances in hospitality, tourism and the services industry (AHTSI) book series* (pp. 161–178). DOI: 10.4018/979-8-3693-2019-8.ch010

Gretzel, U., Sigala, M., Xiang, Z., & Koo, C. (2015). Smart tourism: Foundations and developments. *Electronic Markets*, *25*(3), 179–188. DOI: 10.1007/s12525-015-0196-8

Hamarat, H., Sahin, H., Koç Apuhan, A., & İnan, R. (2024). The dark side of artificial intelligence: Threats to tourism workers. *Worldwide Hospitality and Tourism Themes, 16*(2), 127–137. DOI: 10.1108/WHATT-03-2024-0061

Hassija, V., Chamola, V., Mahapatra, A., Singal, A., Goel, D., Huang, K., & Hussain, A. (2024). Interpreting black-box models: A review on explainable artificial intelligence. *Cognitive Computation, 16*(1), 45–74. DOI: 10.1007/s12559-023-10179-8

Huang, A., Chao, Y., de la Mora Velasco, E., Bilgihan, A., & Wei, W. (2022). When artificial intelligence meets the hospitality and tourism industry: An assessment framework to inform theory and management. *Journal of Hospitality and Tourism Insights, 5*(5), 1080–1100. DOI: 10.1108/JHTI-01-2021-0021

Indaryanto, A., Harijadi, B. D., & Sinaga, E. (2023). The growing use and impact of artificial intelligence technologies in the tourism industry. *Sustainable Engineering and Innovation ISSN 2712-0562, 5*(2), 189–204. DOI: 10.37868/sei.v5i2.id238

Ivanov, S., & Umbrello, S. (2021). The ethics of artificial intelligence and robotisation in tourism and hospitality conceptual framework and research agenda. *Journal of Smart Tourism, 1*(4), 9–18. DOI: 10.52255/smarttourism.2021.1.4.3

Ivanov, S., & Webster, C. (2019). *Robots, artificial intelligence, and service automation in travel, tourism and hospitality*. Emerald Publishing Limited. DOI: 10.1108/9781787566873

Jafari, J. (2019). The need for a new tourism ethics. *Journal of Tourism Futures, 5*(3), 195–198.

Kim, H., So, K. K. F., Shin, S., & Li, J. (2024). Artificial intelligence in hospitality and tourism: Insights from industry practices, research literature, and expert opinions. *Journal of Hospitality & Tourism Research (Washington, D.C.), 49*(2), 366–385. DOI: 10.1177/10963480241229235

Law, R., Ye, H., & Lei, S. S. I. (2025). Ethical artificial intelligence (AI): Principles and practices. *International Journal of Contemporary Hospitality Management, 37*(1), 279–295. DOI: 10.1108/IJCHM-04-2024-0482

Ma, S. (2024). Enhancing tourists' satisfaction: Leveraging artificial intelligence in the tourism sector. *Pacific International Journal, 7*(3), 89–98. DOI: 10.55014/pij.v7i3.624

Mariani, M., & Wamba, S. F. (2020). Exploring how consumer information systems produce information and effect consumer behavior: The role of artificial intelligence and big data. *Information & Management, 57*(8), 103370.

Mihai Gabriel, C. R. I. S. T. I. A. N., & Cosmin, T. I. L. E. A. G. Ă. (2024). Challenges and perspectives of AI in sustainable tourism. *Management of Sustainable Development, 16*(2).

Murphy, J., Hofacker, C. F., & Gretzel, U. (2017). Dawning of the age of robots in hospitality and tourism: Challenges for teaching and research. *European Journal of Tourism Research, 15*, 104–111. DOI: 10.54055/ejtr.v15i.265

Nannelli, M., Capone, F., & Lazzeretti, L. (2023). Artificial intelligence in hospitality and tourism. State of the art and future research avenues. *European Planning Studies, 31*(7), 1325–1344. DOI: 10.1080/09654313.2023.2180321

Nurik, C. L., Gauthier, M., Sawchuk, K., DeJong, S., Gómez, S., Canzonetta, J., & Cox, C. M. (2021). *Intersectional automations: robotics, AI, algorithms, and Equity*. Rowman & Littlefield.

Olutimehin, D. O., Ofodile, O. C., Ejibe, I., Odunaiya, O. G., & Soyombo, O. T. (2024). Implementing AI in business models: Strategies for efficiency and innovation. *International Journal of Management & Entrepreneurship Research, 6*(3), 863–877. DOI: 10.51594/ijmer.v6i3.940

Putera, N. S. F. M. S., Saripan, H., Bajury, M. S. M., & Yaâ, S. N. (2022). Artificial Intelligence in the Tourism Industry: A privacy impasse. *Environment-Behaviour Proceedings Journal, 7*(SI7), 433–440. DOI: 10.21834/ebpj.v7iSI7.3812

Samara, D., Magnisalis, I., & Peristeras, V. (2020). Artificial intelligence and big data in tourism: A systematic literature review. *Journal of Hospitality and Tourism Technology, 11*(2), 343–367. DOI: 10.1108/JHTT-12-2018-0118

Srivastava, S., & Sinha, K. (2023). From bias to fairness: A review of ethical considerations and mitigation strategies in artificial intelligence. *International Journal for Research in Applied Science and Engineering Technology, 11*(3), 2247–2251. DOI: 10.22214/ijraset.2023.49990

Stroumpoulis, A., Kopanaki, E., & Varelas, S. (2022). Role of artificial intelligence and big data analytics in smart tourism: A resource-based view approach. *WIT Transactions on Ecology and the Environment, 256*(2022), 99-108. method. *Mathematics, 10*(13), 2190.

Tamanine, R., Fahim, I., Houssas, M. B., Elkandoussi, F., & Baazizi, Y. (2024). Assessing the Determinants of AI Integration in Tourism SMEs. International Journal of Business and Management Research Tuo, Y., Wu, J., Zhao, J., & Si, X. (2025). Artificial intelligence in tourism: Insights and future research agenda. *Tourism Review, 80*(4), 793–812.

Tula, S. T., Kess-Momoh, A. J., Omotoye, G. B., Bello, B. G., & Daraojimba, A. I. (2024). AI-enabled customer experience enhancement in business. Computer Science & IT Research Journal, 5(2), 365-389.

Tussyadiah, I. P. (2020). A review of research into automation in tourism: Launching the Annals of Tourism Research Curated Collection on Artificial Intelligence and Robotics in Tourism. *Annals of Tourism Research, 81*, 102883. DOI: 10.1016/j.annals.2020.102883

Tussyadiah, I. P., & Griffin, T. (2018). The implications of AI for tourism: A discussion. In *Information and communication technologies in tourism 2018* (pp. 637–641). Springer.

UNWTO. (2022). Tourism and COVID-19: Unprecedented economic impacts. World Tourism Organization. https://www.unwto.org/covid-19-and-tourism-2022

Waltl, B. (2021). Explainability and transparency of machine learning in ADM systems. In *Research handbook on big data law* (pp. 341–356). Edward Elgar Publishing. DOI: 10.4337/9781788972826.00023

Warner, R., & Sloan, R. H. (2021). Making artificial intelligence transparent: Fairness and the problem of proxy variables. *Criminal Justice Ethics*, *40*(1), 23–39. DOI: 10.1080/0731129X.2021.1893932

Wirtz, J., Patterson, P. G., Kunz, W. H., Gruber, T., Lu, V. N., Paluch, S., & Martins, A. (2018). Brave new world: Service robots in the frontline. *Journal of Service Management*, *29*(5), 907–931. DOI: 10.1108/JOSM-04-2018-0119

World Economic Forum. (2023). *Global Cybersecurity Outlook 2023*. Retrieved from https://www.weforum.org/reports/global-cybersecurity-outlook-2023/

WTTC. (2024). *AI Set to Shape the Future of Travel & Tourism, Says WTTC*. Retrieved from https://wttc.org/news/ai-set-to-shape-the-future-of-travel-and-tourism-says-wttc

Zeng, Z., Chen, P. J., Lew, A. A., & Lin, H. (2020). From high-touch to high-tech: COVID-19 drives robotics adoption. *Tourism Geographies*, *22*(3), 724–734. DOI: 10.1080/14616688.2020.1762118

KEY TERMS AND DEFINITIONS

AI: Artificial Intelligence, which is an automated system fed commands and instructions helpful in getting the system generated responses to the inputs.

DMO: Destination Management Organization. The organizations like these are responsible for the overall planning, development and marketing of any tourist attraction.

IoT: Internet of Things. There are devices which are interconnected with each other and function as per the instructions fed in the system or devices enabling its working.

OTA: Online Travel Agencies. These agencies operated through online mode without catering to the needs of customers in physical, one-on-one basis.

SME: Small Medium Enterprise

STD: Smart Tourism Destination. These are the tourist destinations which feature and implement the latest technologies to monitor, guide, measure the impact of visitors to the destination.

Chapter 14
Reframing Digital Health Sovereignty in the Era of Ethical AI:
Challenges, Risks, and Transformation Pathways

Ikram Boudallaa
https://orcid.org/0000-0001-7770-1698
Ibn Tofail University, Morocco

Rachid Dekiki
Mohammed V University, Morocco

ABSTRACT

The digital transformation of healthcare has accelerated in recent years, reshaping access to care and medical data management. This shift, amplified by the spread of telemedicine, has raised many challenges, particularly around equitable access between urban and rural areas. The COVID-19 crisis further underscored Morocco's urgent need to strengthen digital sovereignty to secure health data and enhance system resilience. Grounded in a systematic review, this study proposes a strategic model of digital sovereignty adapted to the Moroccan context. Findings indicate that sovereignty is a crucial lever for ensuring the ethical and secure management of health data, while also promoting interoperability with international standards. Special attention is given to ethical artificial intelligence, both as a driver and a challenge for responsible transformation, emphasizing the need to mitigate bias and safeguard equity. By mobilizing data governance, cybersecurity, innovation, ethical regulation, and regional cooperation, Morocco can advance a resilient and sovereign digital health system.

DOI: 10.4018/979-8-3373-3104-1.ch014

INTRODUCTION

The digital transition in healthcare has sped up in the last few years. This has changed how people look for treatment and how medical data is kept. In the past, people may have assumed that only a few people may benefit from remote medical consultations. However, these days, they are common and provide ongoing therapy with less danger to patients' health. There will be a lot of ethical, practical, and technical problems because of this change. Mahdaoui et al. (2023) and Jallal et al. (2023) say that policymakers are now worried about protecting private information, improving digital services, and controlling technology. As health and digital technologies have grown more common, it has become clearer that healthcare systems rely largely on infrastructure controlled by foreign firms. Because of this dependency, you are open to big strategic problems like cyber assaults, health data breaches, and even involvement from other countries. Some countries are changing their rules to allow their citizens greater freedom online. They are still keeping an eye on multinational digital enterprises, but they are focusing on improving local infrastructure. Health data has become a valuable commodity in the geopolitical setting, leading to arguments about who owns it, how it is utilized, and what its prospective applications are (Jallal et al., 2023). Morocco has a significant challenge in an increasingly linked world: safeguarding patient information and averting system instability caused by technology dependency, while maintaining the long-term digital autonomy of the healthcare sector. Mahdaoui et al. (2023) propose a multifaceted strategy to tackle this issue: first, we must fortify digital infrastructures to avert advanced assaults on confidential medical data. We also need to build up our own research and innovation skills so that we don't have to rely on foreign technology as much. Thirdly, we must build rigorous regulatory frameworks that strike a compromise between individuals' rights and the modernization of the health system. Finally, we need to promote alliances amongst countries in the MENA area so they may share resources and become stronger together. For digital technologies to work well in healthcare, it is important to fix the differences in access to healthcare, especially between cities and rural regions. Telemedicine has a lot of promise, but it has to be regulated to make sure that all patients receive the care they need. Studies indicate that those with low income and those residing in rural areas have more difficulties in accessing necessary telemedicine services, exacerbating existing disparities (Ko et al., 2023; Sheets et al., 2020). This study performs an extensive literature evaluation to examine the mechanisms for an effective and ethical digital transformation in Morocco's healthcare industry. It seeks to find a way for Morocco to have digital health sovereignty that balances technical innovation, digital sovereignty, and data protection. It also wants Morocco to be a significant role in the digital world, both in the region and around the world.

METHODS

This study employs a systematic review approach, inspired by the methodology of Petticrew and Roberts (2006), to examine the contribution of digital sovereignty models to the transformation of Morocco's healthcare system. In parallel, we propose a model tailored to this specific context.

Inclusion and Exclusion Criteria

Since 2000, articles and institutional reports that deal with the subject of this study have been added. Conversely, publications predating this date, interviews, scientific communications, personal accounts, and any content derived from subjective or unconfirmed opinions were omitted. The articles that were chosen were in either French or English.

Data Sources and Literature Search

A comprehensive search was conducted in several databases, including PubMed, Scopus, and Google Scholar, to identify relevant studies. The following keywords were combined using the Boolean operators "AND", "OR", and "NOT" to generate various search equations: "Digital sovereignty", "Health data", "Interoperability", "Cybersecurity in health", "Data protection", "Digital independence", "Health data governance", and "Digital health infrastructure".

The retrieved articles were cross-checked to ensure that none appeared in predatory journals (verified using the list available at https://predatoryjournals.org/the-list).

Quality Assessment of Studies

Each included study underwent a rigorous assessment based on the following criteria: (1) clear description of the study protocol; (2) adherence to the established protocol; (3) precise formulation of the research question; (4) detailed explanation of the analytical method used; (5) use of objective and unbiased measurement criteria; (6) transparency regarding conflicts of interest and consideration of potential biases.

Study Selection and Data Collection Process

The analysis of results adhered to the technique established by Petticrew and Roberts (2006), encompassing a systematic classification of research, a comprehensive examination of each, and a narrative synthesis of the outcomes. We wrote down each stage of the procedure in an Excel file, along with the reasoning for including or leaving out articles.

We utilized Zotero, a bibliographic management program, to go through references and find duplication. The PRISMA flow diagram was used to find articles that met the criteria. The author independently evaluated the titles and abstracts obtained from the search, omitting those deemed irrelevant. Full texts of papers that might be eligible were collected and evaluated according to the inclusion criteria.

The literature search found a total of 813 studies. After getting rid of duplicates and works that weren't related, 56 studies were kept for closer study. Of these, 33 studies were considered suitable for inclusion in the review.

Figure 1. Flow diagram of the study selection process

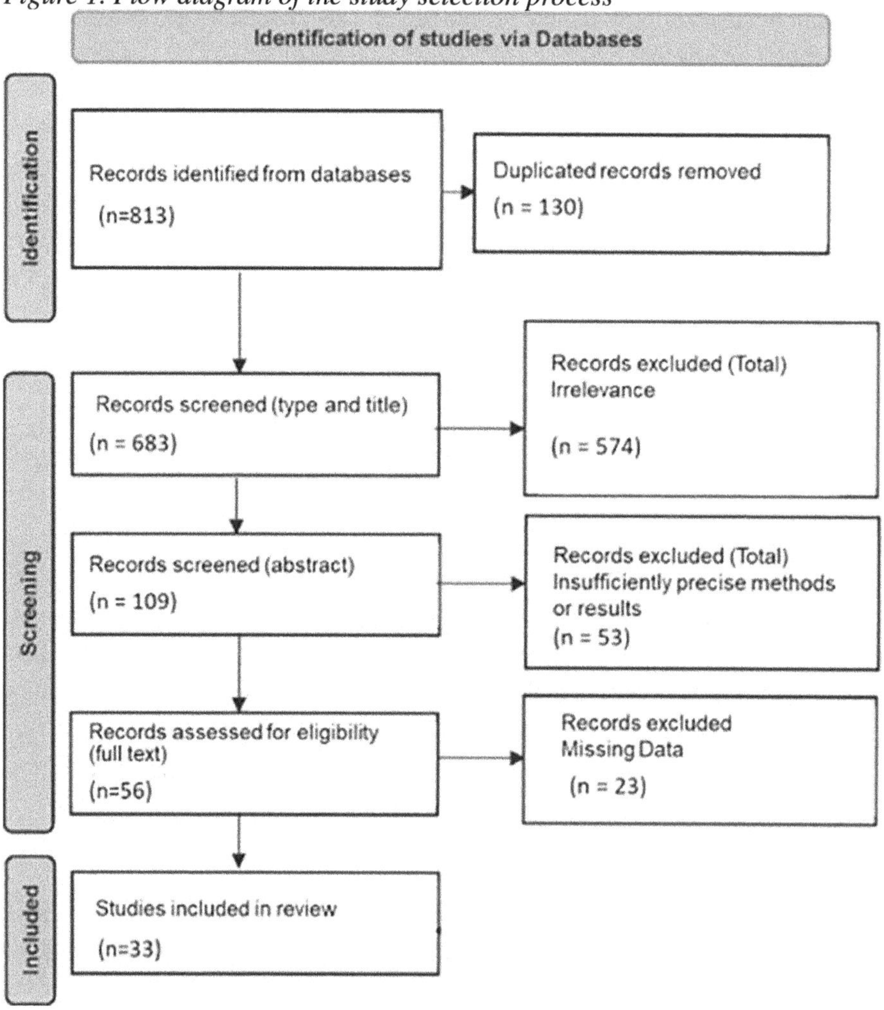

Source: Authors' Original Creation

Data Analysis

The data extracted from the selected studies were synthesized and analyzed qualitatively. Recurrent themes and best practices related to digital sovereignty in health were identified to construct a conceptual framework and formulate recommendations tailored to the Moroccan context.

RESULTS AND DISCUSSION

I. Digital Transformation of Healthcare in Morocco

The digital transformation of Morocco's healthcare system is a major step toward a model that is more efficient, integrated, and easy to get to. But this change comes with technological, legal, and organizational problems that need a planned and coordinated approach to get the most out of it. In this sense, Morocco's digital sovereignty in the health sector is shown by efforts to expand access to care and strengthen technical independence.

The COVID-19 epidemic showed how important it is to have a strong healthcare system. El Otmani et al. (2021) and Economic, Social and Environmental Council (CESE) (2021) say that this made Morocco start making its own medical devices and putting money into digital solutions. The "Wiqaytna" app, which keeps track of COVID-19 contacts, was the first step in giving people more control over their health. Following the findings of Boudallaa et al. (2025), the Mohammed V Foundation for Solidarity initiated projects like Connected Mobile Medical Units and Proximity Medical Centers, which enabled the management of medical data through the integration of hospital information systems. The main objective of this ethical and inclusive digital transformation is to make healthcare better while also keeping people's privacy safe (MSPS, 2020; El Otmani et al., 2021). Since then, a number of efforts have been made to improve Morocco's healthcare system. One example of this kind of system is Shared Medical Records (SMR). It is an important part of continuity of care because it puts a patient's medical records in one place and makes them easier for healthcare professionals to access (Er-Rays & M'dioud, 2024). The automated claim form makes it easier to deal with the CNSS. This makes it faster and easier to process administrative data without mistakes or delays. The gradual removal of Hospital Information Systems (HIS) will make it easier for public and private organizations to share and manage health data. In the end, this will lead to a National Information System (MSPS, 2023; El Otmani et al., 2021). (Yamani & El Merouani, 2020). Moreover, the rapid growth of telemedicine has made it feasible to eliminate medical deserts, which might help close the gap in access to healthcare, particularly in rural regions. A strong digital infrastructure and better interoperability across information systems are important but not enough for healthcare to be more democratic (Abdellatif et al., 2020).

Digitalization does indeed offer one of its most important benefits: making treatment more accessible to everyone. The widespread use of mandatory health insurance (AMO) and family allowances, along with useful digital tools, helps to close the gap between rich and poor and between different parts of the country. Digitalization improves care operations by making systems more responsive and cutting down on paperwork. Automating tasks, keeping records in one place, and using systems that can work together all help improve operational efficiency and make data safer (Mahdaoui et al., 2023; OECD, 2018).

But there are still problems, especially when it comes to digital governance and coordinating digitization activities in the public sector. Morocco is still strengthening its digital sovereignty in healthcare by merging new technologies with improvements to medical facilities (Omrana et al., 2012).

Health Data: A Strategic Resource Between Opportunity and Vulnerability

Health data are now regarded as a true digital goldmine, owing to their strategic value for medical innovation, scientific research, hospital management, and public health policies (Boudallaa, 2021, 2022, 2023, 2025). They feed predictive models, facilitate personalized medicine, and enable faster and more effective decision-making (Kaushik et al., 2025).

But this wealth comes with enormous hazards, like privacy violations, unethical exploitation, targeted cyberattacks, and loss of control because we depend on services or infrastructure from other countries. These hazards not only put personal information at risk of being exposed, but they also put national security at risk in a very critical area. To safeguard its digital sovereignty, Morocco has to create technology independence by employing safe national platforms, hosting that is independent of other countries, and rigorous local regulations. It is also crucial for health information systems to be able to work together so that stakeholders, institutions, and places may share data safely and easily (Jayathissa & Hewapathirana, 2023). This entails applying the same laws, making sure the data formats are the same, and obeying the rules of other countries. Finally, we need to think about protecting health data from all sides, such as technical infrastructure, institutional governance, legal frameworks, and making both professionals and the general public more aware of the issue. So, an integrated health data strategy is a fantastic way to develop trust, get new ideas, and make Morocco's healthcare system work better and be more stable (Scheibner et al., 2020).

The Digital Future of Morocco's Healthcare Sector

Digital technology will change how healthcare works in Morocco in the future. This is part of a significant transformation that is happening because of emerging technologies like AI, blockchain, advanced data analytics, and telemedicine. These innovative ideas make it possible to improve the quality, accessibility, and efficiency of care (El Otmani et al., 2021; Abdellatif et al., 2020). For them to work successfully together, we need to keep spending money on research and development, constantly altering the rules, and make our infrastructure and people better (El-Gazzar et al., 2023). The issue is both social and geographic: Morocco needs to take an inclusive approach that closes the gap in access to technology between urban and rural areas and raises public awareness of the benefits and uses of digital health.

The COVID-19 health crisis accelerated the adoption of digital health services while also revealing structural weaknesses, particularly in terms of technical, logistical, legal, and ethical aspects. Morocco thus faces technological dependence on foreign solutions, raising concerns regarding sovereignty, health data confidentiality, and service continuity (El Otmani et al., 2021). In response, many countries, including Morocco, are reorienting their strategy toward strengthening digital autonomy through the establishment of local platforms, the development of sovereign data centers, and the implementation of stricter regulatory policies (El-Gazzar et al., 2023).

In this case, digital sovereignty becomes a strategic priority for both the world and the country.

Model of Digital Sovereignty in Morocco's Healthcare Sector

Digital sovereignty has become a major issue for countries when it comes to protecting the independence, integrity, and efficient administration of health information. Morocco has to take a comprehensive strategy to make sure that its digital health ecosystem is well-managed. There is a rising need for technology from other countries, cyber threats are becoming more common, and AI is everywhere. Our team's model shows the most important parts of digital sovereignty in Morocco's healthcare industry. This makes it easier to find important leverage points and see how technology, politics, the economy, and legislation are all interrelated. The purpose of this thorough review is to discover how useful the model really is and how complete it is.

Figure 2. Model of Digital Sovereignty in Morocco's Healthcare Sector

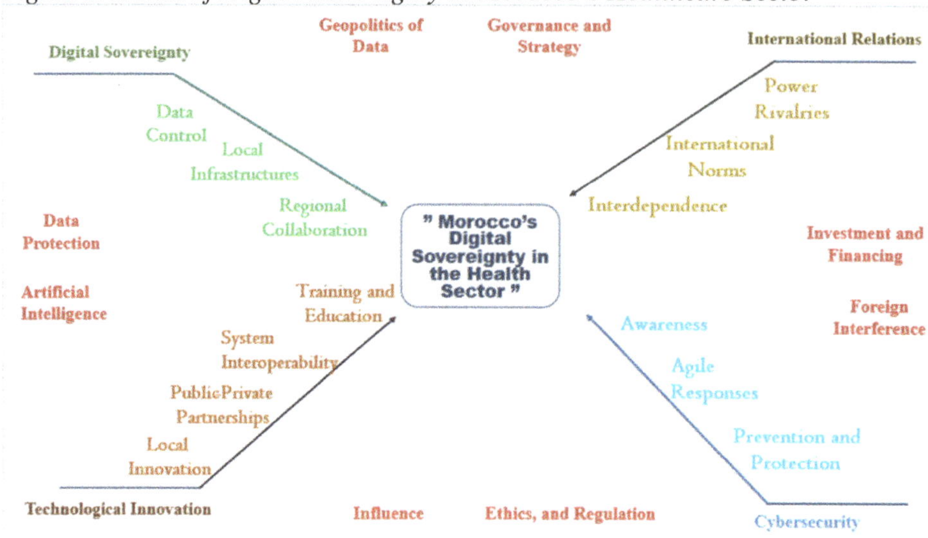

Source: Authors' Original Creation

This model's systematic analytical framework may help us understand all the many parts of Morocco's healthcare system that have to do with digital sovereignty. It is based on six main ideas: digital sovereignty, cybersecurity, international relations, technical innovation, ethics and regulation, and strategic governance. The model demonstrates the interdependence of geostrategic, political, and technological challenges. To preserve digital sovereignty in the long run, people think that data management, building up local infrastructure, and working with nearby regions are all important. The advise also stresses the need of having strong security and preventative measures, as well as reacting quickly and in an organized way to the rise in cyber threats. Morocco's health system has to become better by working with other systems, teaching the public, and coming up with innovative ideas.

II. Pillars of Digital Sovereignty in Morocco's Healthcare Sector

As the healthcare system becomes more digital, health data is becoming a significant resource for countries and tech corporations. The owner of the data is also in charge of medical research, predicting pandemics, and making decisions for the government (Marelli et al., 2021). Morocco has to be more open about its aspirations to take more control over digital sovereignty. For this to happen, there has to be a clear national strategy, public policies that include all sectors, uniform standards, and fair representation in international cooperation mechanisms, especially in African and multilateral organizations (Scheibner et al., 2020). Protecting private information is an important part of our freedom. ISO 27799 and the General Data Protection Regulation (GDPR) are two global standards that concentrate on protecting health information. Following these rules is very important for creating a reliable, safe, and ethical framework (De Freitas et al., 2021). To get digital sovereignty in healthcare, we will need a lot more money. Morocco needs money from both the public and commercial sectors to buy new technologies, build its own infrastructure (such data centers and national health platforms), and improve cybersecurity. To make sure that the digital transformation of the sector goes smoothly and lasts, institutions need strong leadership, strategic partnerships, and support for digital health businesses (Laukka et al., 2022). Morocco's digital sovereignty in the health sector is closely linked to the fact that international ties are always evolving in our interconnected world. Health information systems need to work together to share epidemiological data and keep treatment continuing. The World Health Organization (WHO) strongly recommends open interoperability protocols like HL7 FHIR to make it easier for health systems throughout the world to talk to each other. This is why a well-coordinated worldwide strategy is needed. Major Chinese and American IT corporations, such the Chinese IT behemoths and GAFAM, are fighting for the Moroccan market. This is why it's important to find a balance in the policy of technological alliances so that innovation may grow throughout the world without giving too much influence to any one country (WHO, 2023).

Research indicates that the implementation of data security measures, privacy regulations, and communication protocols significantly influences the interoperability of patient health information (Shrivastava et al., 2021). This proposal would not only make Morocco more digitally independent, but it would also make the country a leader in building African digital sovereignty based on the ideals of solidarity, interoperability, and common security. The IMPaCT-Data program in Europe believes we need to discover a safe way to transmit electronic health records. This could be a model for building comparable systems in Morocco (Rodríguez-Mejías et al., 2024).

Technological innovation is an important part of the Moroccan health system's ability to bounce back and grow

Morocco's health system may get stronger and keep developing if it uses new technologies as a strategic instrument. Making individualized digital solutions makes it easier for patients to get the care they need and makes healthcare delivery work better. Artificial intelligence, or AI, is a brilliant new notion that might change medicine in a big way by helping with diagnosis, running hospitals more efficiently, and using predictive medicine. But it needs strict rules because of its moral, legal, and liability issues (Tafenzi et al., 2025).

For a digital ecosystem to work well, systems must be able to communicate with one another. Standardizing data formats and making sure that health platforms can work together are both important for making it easier to transfer information, making sure that care continues, and promoting telemedicine (Mandl et al., 2024). In this circumstance, it is very important for the public and private sectors to work together. The state, Moroccan business owners, schools, and big IT businesses all work together to accelerate up the digital transition and make sure that the area gets value. In the end, it seems that encouraging invention among native people is an important part of independence. Promoting the creation of digital solutions tailored to Morocco's unique health system diminishes technological dependence, rejuvenates the digital economy, and enhances national expertise in e-health (Shrivastava et al., 2021 ; Rodríguez-Mejías et al., 2024).

Cybersecurity is becoming a crucial feature of digital sovereignty since it protects key infrastructure and makes people feel comfortable. Cyberattacks are especially detrimental for the healthcare industry since it handles private information and is crucial for the economy (Gilbert et al., 2024). . Morocco needs to take a proactive and all-encompassing approach that encompasses prevention, training, and the ability to respond to problems.

For a culture of digital awareness to grow, health care practitioners and patients need to know more about the risks of cybersecurity. According to a study (Carello et al., 2023), health staff need to be well-informed and pay attention in order to create a culture of digital awareness. Health institutions need government aid to enhance their responsiveness, in addition to digital resilience solutions, emergency protocols, and monitoring units. Studies show that hospitals and other healthcare facilities may improve their digital resilience by setting up monitoring units and backup plans (Sardi et al., 2021). To defend their systems, platforms, and networks, healthcare businesses need to spend a lot of money on cybersecurity. For the medical field in particular, this means following international cybersecurity regulations, making national cyber governance stronger, and working together in a smart way. One study found that following international standards is very important for the safety of the health system (Newaz et al., 2020).

Digital sovereignty in healthcare must not infringe upon citizens' fundamental rights. Instead, it should be built on strict laws and regulations that find a balance between protecting people's rights and making the most of new technologies. The increased use of digital technology in healthcare has raised concerns about data governance, informed consent, and privacy (Scheibner et al., 2020). Morocco has to change its laws to safeguard health data in a way that is both in accordance with international standards and works for Morocco (Rumbold & Pierscionek, 2017).

Such a legislation should explicitly spell out the duties of stakeholders, the regulations for accessing, using, and storing data, and the penalties for not following them (Marelli et al., 2023). Also, to create democratic governance of health data, it is important to acknowledge patients' rights to know how their data is being used, to accept or disagree with certain digital processes, and to see their medical records (Holly et al., 2023). These ethical principles should be included in digital public policy to build public confidence and confirm the changes that are happening in the healthcare business. Digital sovereignty that is ethical, open, and long-lasting needs both new ideas and rules (Scheibner et al., 2020).

III. The Value Chain of Digital Sovereignty in Morocco's Health Sector

Digital sovereignty means that a country may safely, independently, and in line with its goals govern its own data and infrastructure online. As health information systems become more important as strategic assets, this method has become increasingly important in the age of digital health and large medical data.Digital sovereignty means that Morocco can produce and use its own software, infrastructure, data, and skills without having a lot of aid from other countries. It also makes sure that the health information of Moroccan citizens stays private. It is important for making the health system stronger, encouraging local innovation, and making sure that everyone can get treatment in the digital era.

Like many other countries, Morocco is becoming more aware of these problems. This is especially true because the COVID-19 pandemic sped up the health sector's digital transition.

There are seven parts to the value chain of digital sovereignty in health: physical infrastructure, software and digital services, governance, human capital, health data and artificial intelligence (AI), uses and applications, resilience, and innovation. Digital sovereignty cannot be fully realized without each of these components, which make up a basic structure.

Physical Infrastructure

The strength of physical infrastructure is the most important part of any digital health sovereignty strategy. This comprises telecommunications networks (like high-speed internet and 4G/5G mobile networks), data centers (like national data centers), IT technology in healthcare institutions, and a steady supply of electricity to keep these systems running.

Morocco has a fairly robust base for the growth of information and communication technologies (ICT). As of 2017, the country had the fourth best internet connection (including fiber optics, 3G/4G, and the current implementation of 5G) and mobile network coverage in Africa. In addition, the electrification rate is over 99%, including in distant rural areas. This gives the country a big edge over many African countries who have problems with electricity shortages. The Digital Development Agency (ADD) was set up to speed up the digital transformation of public services and coordinate e-government projects (El Otmani et al., 2021). This has helped these infrastructure achievements.

Even with these strengths, it is still hard to get full control over digital health infrastructures. Health data hosting and cloud services still rely significantly on foreign suppliers, which raises important considerations regarding where health data centers should be located. To have true sovereignty, a country must be able to store and process sensitive data locally, preferably in secure national data centers. This is necessary to maintain legal and operational control over the data. At recent digital events in Morocco, new projects were unveiled, like public-private collaborations to build secure local health clouds.

But rural connectivity has become better, even if it is still inconsistent and doesn't have enough bandwidth. To make sure that all areas get the most out of e-health solutions, it is important to close the digital gap across regions. Many African countries' rural regions don't have good enough internet access for telehealth services to grow. Rwanda and Kenya are only two examples of nations that have proven how building national fiber-optic backbones and expanding 4G coverage in rural areas may speed up the process of getting everyone online for healthcare (Riazul Islam et al., 2015). Europe has some of the best infrastructure in the world, with broad access to high-speed internet and continuous deployment of national or European cloud infrastructures via projects like Gaia-X, which intends to create a federated cloud ecosystem in Europe. The European Union has also stressed the need of a sovereign cloud infrastructure for storing sensitive data, such medical records, as part of the European Health Data Space (EHDS). The value chain goes on with an emphasis on digital services, apps, and software that are relevant to health. There are several examples, including as online appointment scheduling tools, electronic prescription systems, telemedicine platforms, hospital information systems, mHealth apps, and electronic medical records. Moroccan healthcare workers and inhabitants should be able to create or manage the digital tools they use so that they don't have to rely on foreign solutions. This would give them more control over this area. In recent years, several projects, many of which were done on their own, have tried to digitize Morocco's healthcare system. Hospitals and clinics have only partly embraced computerization. Even while many healthcare institutions employ electronic health records, they are still in their early stages on a national level. The initiatives are separate and do not work together (Parks et al., 2019).

This "siloed" way of doing things has made it harder for a unified digital health system to work (Parks et al., 2019). In response, suggestions have been made to create a national integrated plan for EMRs, which would include interoperability standards and a single framework for technical governance (Parks et al., 2019; Bouhaddou et al., 2013). As early as the 2010s, Moroccan pioneers in medical informatics stressed the necessity for national standards and an interoperable framework backed by a national coordinating office (Bouhaddou et al., 2013). This suggestion is still very useful today since it helps keep the system from breaking up any more.

At the same time, new digital services have come out. The mental health e-health portal is a good example. It has an Electronic Appointment System (EAS) for those with mental health issues. Patients have not widely adopted this service; many have only used it once before going back to phone or in-person communication (Khalil et al., 2024). This shows that just having software isn't enough; users need to trust it and find it easy to use, which will be talked about further later. Teleconsultation platforms are another type of digital service. During the COVID-19 pandemic, the Moroccan Ministry of Health started Tbib 24, a free tele-advisory platform with more than 100 volunteer doctors. This allowed those who were stuck at home to get remote consultations. Many startups and private clinics also came out with video teleconsultation options at this important time, using the interim regulatory freedom to their advantage as they waited for a new legal framework. These changes show how Morocco can come up with new ideas and quickly roll out digital services in times of crisis. European countries, on the other hand, are ahead of the US in using health software. For instance, almost all general practitioners in France, Germany, and the UK use electronic health records. Telehealth has also grown a lot since COVID, with millions of teleconsultations happening every year. Still, Europe has had problems with software sovereignty, especially when it comes to storing health data in the cloud. The French Health Data Hub, which was first hosted by Microsoft Azure, caused a lot of debate and led to a move toward open-source and independent European or national alternatives. A number of African countries have chosen open-source software for their medical information systems. Tunisia and Senegal, for example, have used DHIS2 to manage public health data in order to obtain more independence (Alegana et al., 2023). Morocco could gain from these experiences by adopting more indigenous or open-source solutions that are more suited to its needs (Bouhaddou et al., 2013).

Management and governance

Governance includes all the rules, regulations, strategies, and institutions that have been put in place to guide the digital transformation of the health sector. It is a key part of the value chain because it sets the direction, lays out the rules for how people can work together, and makes sure that everyone is on the same page. Morocco is a leader in digital health governance, especially when it comes to rules and regulations.

This shows that the government is sensitive to the needs of digital health implementation, data security, and privacy. Governance goes beyond telemedicine to include the larger e-health plan. Digital health is now a part of the Ministry of Health's strategic plans, such as Health Plan 2025, as a way to increase access, quality, and efficiency of care across the board. The Ministry has set up a separate Directorate for Health Digitalization to keep an eye on health information system (HIS) projects. Coordination between sectors is also very important. For example, the Digital Development Agency (ADD), which is part of the Ministry of Digital Transition, works with the Ministry of Health on health programs that use e-government.

Morocco's personal data protection law (Law No. 09-08 of 2009) also extends to health data. The country was one of the first in Africa to set up a special agency (CNDP) to make sure that privacy rules are followed online. This early knowledge of challenges related to data sovereignty and cybersecurity is a big plus for government. Many African countries are still working on or adopting national e-health policies at the regional level. A recent study found that more than 80% of African countries had at least one strategic document around digital health or HIS. The WHO or the African Union often helped draft these documents. Nonetheless, execution and governance continue to be inadequate (Alegana et al., 2023). Some of the shortcomings that have been found are the lack of clear rules for health data governance, system interoperability, and the safe use of new technology. For example, not many countries have passed rules that control the use of AI in healthcare or mobile health apps.

Morocco is a remarkable exception because it has a strong telemedicine statute and an active data protection authority. The governance of digital health in Europe is changing quickly thanks to big projects like the proposed European Health Data Space (EHDS) Regulation. This regulation aims to give people more control over their health data and make it easier for people to share data safely for care and research across the EU (Marelli et al., 2023). Critics, on the other hand, say that the current form of the EHDS could make things harder for healthcare workers and take away patients' real control over their data, which shows that the rules need to be improved (Marelli et al., 2023). The EU has made it plain that it wants to improve regional governance of health data as part of its larger digital sovereignty goal (Alegana et al., 2023).

Human Capital

The fourth component, human capital, is perhaps the most cross-cutting: it refers to the people involved, such as healthcare professionals, engineers, technicians, and decision-makers, whose skills, training, and involvement are necessary for digital sovereignty. A health system can only become digitally sovereign if its workers know how to use digital tools, understand the problems that need to be solved, and are large enough to implement and maintain digital solutions.

Morocco has a general lack of healthcare workers, no matter how digitalization affects things. The World Health Organization (WHO) says that there should be 23 doctors for every 10,000 people in a country. However, there are only 7.8 doctors for every 10,000 people in this country (Jallal et al., 2023). This scarcity is made worse by the fact that most of Morocco's doctors work along the Rabat-Casablanca corridor, which means that many rural communities are not getting the care they need (Jallal et al., 2023). There are two main effects of this situation: Digital health technologies (like telemedicine) are becoming essential to make up for the lack of doctors in the field. At the same time, the lack of staff makes it harder to use digital tools because busy professionals have less time to learn and use new tools.

So, it's really important to hire more healthcare staff and teach them how to use technology in their jobs.

In this regard, Morocco has made early progress. In 2008, it established a medical specialization in health informatics as part of the restructuring of the medical curriculum. In 2010, the Moroccan Association for Medical Informatics was established (Bouhaddou et al., 2013). In 2012, there was a national convention on health informatics. This marked the start of a network of professionals and scholars who work in this field (Bouhaddou et al., 2013). These efforts have created a local group of digital health experts, which has made it possible for IT experts and doctors to work together and start experimental projects.

By 2023, the results of these efforts are becoming more clear. For example, several Moroccan scholars are publishing in this field (many of the works cited in this chapter are by Moroccan academics), and there are departments of medical informatics in hospitals (like the University Hospital of Casablanca) and health schools. Nonetheless, these competencies require enhancement and broader dissemination (Elkachradi et al., 2023). A suggested action is to incorporate fundamental training in health informatics and information systems security into all healthcare curricula (medicine, nursing, pharmacy) (Innovation Center E-health, 2022). This would make sure that all caregivers know how to use common digital tools and are aware of recommended practices, such as cyber hygiene and keeping data private. This advice is in line with what has been found in Africa: e-health education is still rare in medical and nursing institutions in most African nations, and it has to be created quickly to educate a technologically competent future health workforce (Alegana et al., 2023).

In addition to technical talents, the cultural and management aspects of human capital must also be taken into account. Healthcare organizations need to change the way they think and how they work in order to adopt digital technology. This necessitates ongoing professional development, designed to mitigate resistance to change and illustrate the enhanced utility of digital technologies in routine practice. Khalil et al. (2024) found that psychiatric patients need to trust technology and have technical help in order to use e-health services. A recent poll of young Moroccan doctors found that they are largely amenable to using artificial intelligence (AI) in healthcare and know about its possible benefits. Specifically, 58% think AI can speed up clinical tasks, and 51% think it can improve clinical performance (et al., 2023). Almost half of them already use AI-based apps in their daily life, such as translation tools and diagnostic aids (Berrami et al., 2023). This suggests a new generation of professionals who are good with computers. Nevertheless, 63% articulated apprehensions regarding AI, especially the potential diminishment of human engagement with patients (Berrami et al., 2023). It is crucial to engage professionals in the co-design of digital solutions to ensure they comprehend their functionality and limitations, and to underscore that technology is intended to support, not supplant, them.

Health Data, Artificial Intelligence and Ethical using

Patient records, clinical data, medical imaging, genomic information, and data from linked devices are all examples of health data that has become the new "black gold" of the medical field. Digital sovereignty is all about being able to fully use these data assets. This means having safe places to store national data, being able to manage who can access it and how it may be used, and being able to make the most of data, especially through AI, to improve treatment and research. This pillar is closely related to the others, such as storage infrastructure, data management software, access rights governance, data analysis capabilities, and more.

Morocco recognized the strategic relevance of health data early on, especially with the passage of Law 09-08 on personal data protection (2009), which specifically protects sensitive health data. The CNDP, or National Commission for the Protection of Personal Data, makes ensuring that e-health programs follow privacy rules. The 2021 update to the telemedicine order made it clear that CNDP approval is needed for almost any new technology that collects patient data, such as a telemedicine app. This law makes sure that Moroccan laws still apply to data even when the technology provider is from another country. But technical sovereignty is still a problem. Today, a lot of Morocco's health data may go through servers housed in other countries, such as general-purpose cloud services. There is still no national Health Data Hub or sovereign health cloud that can bring together and analyze vast amounts of data for research or public health objectives. Establishing a publicly managed data warehouse would allow Moroccan researchers to utilize this data resource while complying with ethical and security norms, a strategy currently under consideration in Europe, notwithstanding its difficulties (e.g., France's Health Data Hub).

AI, or artificial intelligence, is one of the most promising ways to get useful information from health data. It can help with diagnostics (such interpreting medical images and tests), predictive medicine, epidemiological surveillance, and making healthcare systems work better. Digital sovereignty means not just using AI solutions from other countries, but also building algorithms that are trained on local data to make sure that models fit Morocco's unique genetic and epidemiological profiles. This area is showing signs of early progress. Also, the number of surgical robots used (for example, Da Vinci systems for minimally invasive surgery) went grown by 30% from 2020 to 2024 (Idaomar et al., 2024). These impressive results show that Morocco is using more and more complex technologies. The same study says that these kinds of new ideas are still very centralized in big cities, which shows that there is still a big difference in territory (Idaomar et al., 2024). Morocco's progress in health-related AI is seen as "acceptable" compared to that of other rising nations, but there are still problems to be solved, especially when it comes to making sure that these technologies are used fairly (Idaomar et al., 2024).

Another important problem is the quality and interoperability of data. Algorithms need structured, trustworthy data from connected systems in order to work well. Morocco has recognized this necessity; the national e-health plan underscores the significance of establishing standards and reference frameworks (e.g., unique patient identification, disease coding systems). As previously noted, interoperability is still limited since software ecosystems are not very well connected. Many nations have the same problems. For example, almost all of Africa's national digital health strategies say that systems (such patient records, labs, and pharmacies) don't work together, which makes it harder to combine data (Alegana et al., 2023). The European Union is working on this issue with the European Health Data Space (EHDS). The EHDS encourages countries and systems to utilize standard data exchange formats so that care may continue and data can be reused on a broad scale for research (Marelli et al., 2023). The secondary use problem is very sensitive: the EHDS wants to change how health data is handled in Europe, but it has led to discussions about how to preserve patient rights while still being open with data (Marelli et al., 2023). Morocco must also think about how to encourage data-driven innovation (such partial open data and data science projects) while yet protecting its sovereignty and moral standards.

This pillar is closely related to data security and cybersecurity. Cyberattacks like ransomware and data breaches are most likely to happen in healthcare facilities. A major event involving data held or handled by foreign companies could put national sovereignty at risk. Morocco's cybersecurity system has gotten stronger recently, thanks to the 2021 cybersecurity law and the establishment of the DGSSI, a national watchdog. In the health industry, it is important to make sure that security regulations follow international standards, such ISO 27001, and that technology vendors have to follow these rules. The Digital Development Agency is also aiming to promote sovereign cloud solutions that meet national security and compliance criteria. This is another way to reach this goal. To trust data and AI systems, they need to be cyber resilient, which we will talk about in more depth later.

Technology is only useful when it is used in a practical way. The sixth link in the value chain, use and adoption, is about how patients, the general public, and healthcare professionals actually use digital health services in their daily lives. Khalil et al. (2024) conducted a study on mental health patients that exemplifies this point: even though the University Psychiatric Hospital in Casablanca introduced an Electronic Appointment System (EAS), relatively few people used it. A lot of people only utilized the site once before going back to more traditional ways of making appointments, like phone calls and in-person visits. Patients recognized the prospective advantages of the online method (e.g., time efficiency, convenience). But many said they didn't get enough technical and social support, like help learning how to use the site, feeling alone, and technological problems that made them feel bad. The primary determinants of enhanced uptake were perceived utility (the conviction that the tool enhances the service), trust in technology, and the existence of enabling conditions (availability of assistance and accessibility) (Khalil et al., 2024). On the other hand, perceived complexity and social influence did not have a big effect on adoption in this case. These findings indicate that to enhance the utilization of digital services among patients, it is essential to convey specific benefits, foster confidence (regarding security and dependability), and offer user support (including digital mediation and aid for less technologically proficient users).

The same thing is happening with healthcare practitioners. Before 2020, people didn't utilize telemedicine apps much because they weren't sure how to use them or were afraid of breaking the law. The COVID-19 pandemic sped things up. During the harsh lockdown in 2020, doctors had to start doing remote consultations. The success of platforms like Tbib24 and other private efforts showed that there was substantial demand from both patients (especially those with chronic diseases who kept up with follow-up remotely) and doctors (who wanted to keep the therapeutic contact). During this time, many platforms appeared on their own, showing how quickly people adapted to digital tools when they needed to. To allow these activities to continue outside of the original legal framework, the Ministry of Health had to temporarily ease several rules.

Telemedicine is slowly becoming a normal feature of healthcare after the pandemic. The National Medical Council (CNOM) has given the green light for telemedicine, and more and more dermatologists and psychiatrists are using it for initial assessments or follow-up care. This puts Morocco in line with a trend around the world. For example, the number of remote consultations in France went up fifty times between 2019 and 2020, and by 2023, more than 10% of all consultations were still done online.

However, in order for these new ways of doing things to last, some structural problems need to be fixed. One of them is financial coverage: in Morocco, health insurance doesn't yet clearly cover telemedicine services, which could make both patients and providers less likely to use them (El Otmani Dehbi et al., 2021). People are working to make telemedicine a part of the official names and payment systems, which would make it much more appealing to use. Another big problem is the digital divide, which makes it hard for some groups, like older people, people with low literacy, and people who live in remote areas, to use internet resources. There is a big gap in digital literacy in the country, so targeted education and digital mediation activities are needed. For instance, the "Digital Mediators" program might be used in the health care field to help patients use online services like making appointments and telemedicine.

Morocco has a relatively young population and a high smartphone penetration compared to other African countries. Both of these things are good for the adoption of e-health. SMS-based m-health services have shown strong engagement rates (e.g., vaccine reminders, health advice) in numerous West African nations, even among people with low literacy, because SMS messaging is so easy to use. Morocco might learn from these examples and build usage models that include everyone, such voice-based platforms in local dialects for health advice.

Patient portals, which let people in Europe see their medical records and send encrypted messages to their doctors, have become popular. For instance, more over half of people in Estonia and Denmark regularly check their health records online. This culture of giving patients more power is still new in Morocco. The Shared Medical Record (Dossier Médical Partagé, DMP), which is like a digital version of a personal health booklet, is still in its early stages. Its success will depend on getting more people to know how valuable it is (for example, for keeping track of medical history and avoiding duplicate testing).

One of the biggest problems for professionals is having too much work and not enough time. Healthcare personnel may see poorly designed software (slow, non-intuitive) as a burden instead of a help, which could lead to cognitive overload. So, it's clear how important it is to include people in the design process using user-centered design methods. For instance, in the US, the quick adoption of EMRs caused some doctors to burn out because they had to spend so much time entering data. Morocco should aim to avoid this by learning from these experiences.

Innovation

The last part of the value chain, resilience and innovation, looks at how well the digital health system can handle crises and keep changing thanks to local inventiveness. Resilience is the ability of a system to take shocks, whether they are health-related, technical, or geopolitical, and still work well and even come out stronger. Innovation, on the other hand, means trying out and using new (technology or organizational) ideas to make health care better all the time. These two things are very important for sovereignty: a digitally independent country needs to be able to use its resources to come up with new ideas. It needs to develop a system that can handle problems (Cordes et al., 2024).

The COVID-19 pandemic was a huge test of how well countries around the world could handle digital health issues. Morocco was able to quickly use digital technologies to deal with the issue, which showed some strength. As mentioned before, the national response was greatly helped by the use of telemedicine, digital contact tracking (using the Wiqaytna app), and information technologies to handle the logistics of testing and vaccination. The use of a digital system to run the vaccination campaign, which included online appointments, stock tracking, and a mobile health pass app (Yakada liqah / Jawaz Assaha), was praised as a key to success because it made it possible to give millions of vaccine doses quickly and easily (El Otmani Dehbi et al., 2021). This experience made decision-makers and experts more aware of how useful digital solutions can be in times of crisis, which will make them more likely to use them in the future (El Otmani Dehbi et al., 2021). Morocco learned by doing, which sped up its digital maturity at the same time. The crisis also showed how important it is to have flexible rules. The fact that the telemedicine decree could be changed rapidly in 2020–2021 showed that institutions were strong enough to avoid legal gaps that could stifle innovation.

But resilience goes beyond pandemics. It also includes being ready for cyberattacks. Cyberattacks on hospitals are becoming more common around the world. For example, in 2021, ransomware attacked Dax Hospital in France and made it impossible for people to get care. A strong digital sovereignty needs a full digital business continuity strategy. This should include regular backups of health data on secure local servers, backup mechanisms (like paper-based workflows), and a staff to respond to cyber incidents. Morocco can improve the cybersecurity of its health system by learning from what other countries have done. It's a good indicator that a national cybersecurity policy has been put in place and that some hospitals have been named "critical infrastructure." The domestic scientific and intellectual output of a country is a good sign of how well it can innovate on its own. The increasing quantity of Moroccan scientific papers in e-health is an encouraging development, as domestic research generates contextualized knowledge (for instance, studies on patient acceptance and cost-effectiveness evaluations of digital health technologies within the Moroccan framework). Policymakers will be able to decide which new ideas to expand and which to adjust based on this kind of evidence-based information.

In the end, resilience and invention come together in the idea of adaptation. A sovereign system must be capable of rapid evolution in response to emerging requirements or technologies. This necessitates modular, interoperable system architectures that facilitate the integration of new features without necessitating the reconstruction of entire systems. For instance, generative AI (like ChatGPT) could change how patients learn about their conditions or how they talk to doctors over the phone. If Morocco produced a medical language dataset in Moroccan Arabic dialects, it could train its own AI models to help patients instead of using foreign AI solutions that might not work well with the culture or language. This is a new area for sovereign innovation: creating "Made in Morocco" health AI solutions that are trained on national datasets and meet the demands of the local population (for example, a diabetes screening algorithm made for Moroccans).

Evidence-Informed Recommendations for Advancing Moroccan Digital Health Sovereignty

This systematic study presents evidence-based recommendations for enhancing Moroccan digital health sovereignty across four dimensions: data sovereignty, artificial intelligence and governance, digital transformation, and risk management. Lessons from the European Health Data Space show both the potential and the limits of this approach (Raab et al., 2023; European Commission, 2025; Marelli et al., 2023). Therefore, Morocco should strengthen national control over infrastructure and make sure that data is kept private. There are many well-known problems with fairness and transparency in AI and governance (Alderman et al., 2025; Chin et al., 2023; Sikstrom et al., 2022). At the same time, AI sovereignty is becoming a topic of discussion in European policy (Ryan, 2024; Mügge, 2024). Therefore, Morocco should promote sovereign, ethical AI and protect basic rights. For digital transformation to work, we need strong models for maturity (Duncan et al., 2022; Kim et al., 2025), systemic (Ruotsalainen et al., 2025), adoption (Greenhalgh et al., 2017), and evaluation (Rouleau et al., 2024). This means building strong institutional frameworks and encouraging local innovation. Finally, the risks of relying on the cloud (Webster, 2019), using commercial datasets (Alberto et al., 2023), and dealing with global governance problems (Kickbusch et al., 2021) show that Morocco needs to strengthen its strategic independence, improve its infrastructure, and make its cybersecurity stronger.

Study Limitations: This is a wonderful place to start because it is the first full examination of digital sovereignty in the Moroccan health business. But the analysis isn't perfect. The suggested method mostly focuses on health data and digital infrastructure, which could mean that it misses other essential aspects, like education and public awareness.

CONCLUSION

For a long-term process to attain digital health sovereignty, there must be a defined aim, sustained investment, and people working together. Every link in the value chain must be made stronger together because a weak link might break the whole system. Morocco has come a long way and has a lot of good things in place, such a strong political will, new skills from its own people, and exciting pilot initiatives. The next ten years will be highly significant for going from testing to full-scale implementation and from being somewhat dependent to being strategically independent. A fully realized digital health sovereignty will ultimately lead to a more effective, equitable, and resilient health system—one that can provide Moroccan citizens with the benefits of cutting-edge technologies while maintaining the core values of ethics, solidarity, and privacy that are essential to a human-centered approach to healthcare.

This model, which was built after a rigorous and detailed analysis, aims to describe digital health sovereignty in Morocco. Morocco can construct a strong, open, and independent digital health system by using crucial building blocks including data control, cybersecurity, local innovation, ethical regulation, and working together with other countries. But for this transition to work, everyone needs to be involved in a coordinated and cross-cutting fashion. This includes people from civic society, researchers, public agencies, healthcare institutions, startups, and huge IT companies. This communal dynamic must be based on the principles of openness, involvement, and collaboration. The present study aims to progress by analyzing the mechanisms of participatory governance that would effectively engage all stakeholders in Morocco's ambitious and essential digital transformation of the health sector.

CONFLICTS OF INTEREST

The writers claim that there is no conflict of interest.

REFERENCES

Abdellatif, A. A., Al-Marridi, A. Z., Mohamed, A., Erbad, A., Chiasserini, C. F., & Refaey, A. (2020). SSHealth: Toward secure, blockchain-enabled healthcare systems. *arXiv preprint arXiv:2006.10843*. https://arxiv.org/abs/2006.10843

Alberto, I. R. I., Ghosh, A. K., Jain, B., Jayakumar, S., Martinez-Martin, N., McCague, N., Moukheiber, D., Moukheiber, L., Moukheiber, M., Moukheiber, S., Yaghy, A., Zhang, A., & Celi, L. A. (2023). The impact of commercial health datasets on medical research and health-care algorithms. *The Lancet. Digital Health*, *5*(5), e288–e294. DOI: 10.1016/S2589-7500(23)00025-0 PMID: 37100543

Alderman, J. E., Kapadia, D., Charitakis, K., Speed, K., Mathur, R., Banerjee, A., Denaxas, S., & Hemingway, H. (2025). Tackling algorithmic bias and promoting transparency in health datasets: The STANDING Together consensus recommendations. *The Lancet. Digital Health*, *7*(1), e16–e25. DOI: 10.1016/S2589-7500(24)00224-3 PMID: 39701919

Alegana, V. A., Frank, L. J., Makubalo, L. E., Okeibunor, J. C., Abimbola, S., Abdalla, S. M., & Kasilo, O. M. J. (2023). Digital health in the African region should be integral to health system strengthening. *Mayo Clinic Proceedings. Digital Health*, *1*(3), 425–434. DOI: 10.1016/j.mcpdig.2023.06.003 PMID: 37744952

Berrami, H., Serhier, Z., Jallal, M., & Bennani Othmani, M. (2023). Understanding and use of artificial intelligence among doctors in a university hospital in Morocco. *Studies in Health Technology and Informatics*, *309*, 215–219. DOI: 10.3233/SHTI230781 PMID: 37869845

Boudallaa, I. (2025). What model of hospital attractiveness for the Moroccan hospital? *Moroccan Journal of Public Health*, *7*(1). https://revues.imist.ma/index.php/MJPH/article/view/53076/29618

Boudallaa, I., Elkachradi, R., Assarag, B., Hassan Chrifi, H., & Kadouri, A. (2022). The challenges of hospital reform in Morocco. *International Journal of Advanced Research*, *10*(4), 733–742. DOI: 10.21474/IJAR01/14608

Boudallaa, I., Elkachradi, R., & Kadouri, A. (2021). Le management des connaissances comme levier stratégique des organisations: État de l'art. *International Journal of Innovation in Management and Applied Sciences*, *4*(1). https://www.researchgate.net/publication/355439860

Boudallaa, I., Elkachradi, R., & Kadouri, A. (2022). Organizational learning: A necessity in the hospital and environmental responsibility. In *E3S Web of Conferences* (Vol. 351, p. 01094). EDP Sciences., https://www.e3s-conferences.org/articles/e3sconf/abs/2022/18/e3sconf_icies2022_01094/e3sconf_icies2022_01094.html DOI: 10.1051/e3sconf/202235101094

Boudallaa, I., Elkachradi, R., & Kadouri, A. (2023). Change management in hospitals: A way to learn from climate change. In *E3S Web of Conferences* (Vol. 412, p. 01058). EDP Sciences., https://www.e3s-conferences.org/articles/e3sconf/abs/2023/49/e3sconf_icies2023_01058/e3sconf_icies2023_01058.html DOI: 10.1051/e3sconf/202341201058

Boudallaa, I., Oughannou, Z., & Bourekkadi, S. (2025). Knowledge management and digital technologies in Moroccan hospitals. *Proceedings on Engineering*, *7*(1), 79–84. https://doaj.org/article/731f8da2d17345049a3965150eb5eaed. DOI: 10.24874/PES07.01.010

Bouhaddou, O., Bennani Othmani, M., & Diouny, S. (2013). Medical informatics in Morocco. *Yearbook of Medical Informatics*, *8*, 190–196. https://pubmed.ncbi.nlm.nih.gov/23974571/ PMID: 23974571

Carello, M. P., Marchetti Spaccamela, A., Querzoni, L., & Angelini, M. (2023). A systematization of cybersecurity regulations, standards and guidelines for the healthcare sector. *arXiv preprint arXiv:2304.14955*. https://arxiv.org/abs/2304.14955

Chin, M. H., Afsar-Manesh, N., Bierman, A. S., Chang, C., Colón-Rodríguez, C. J., Dullabh, P., & Ohno-Machado, L. (2023). Guiding principles to address the impact of algorithm bias on racial and ethnic disparities in health and health care. *JAMA Network Open*, *6*(12), e2345050. DOI: 10.1001/jamanetworkopen.2023.45050 PMID: 38100101

Cordes, A., Bak, M., Gospodarevskaya, E., Glozier, N., Thorpe, A., McMillan, F., & Westphaln, K. K. (2024). Competing interests: Digital health and Indigenous data sovereignty. *npj. Digital Medicine*, *7*(1), 166. Advance online publication. DOI: 10.1038/s41746-024-01171-z PMID: 38909083

De Freitas, C., Dos Reis, V., Silva, S., Videira, P. A., Morava, E., & Jaeken, J. (2021). Public and patient involvement in health data governance (DATAGov): Protocol of a people-centred, mixed-methods study on data use and sharing for rare diseases care and research. *BMJ Open, 11*(3), e044289. DOI: 10.1136/bmjopen-2020-044289 PMID: 33722870

Duncan, R., Eden, R., Woods, L., Wong, I., & Sullivan, C. (2022). Synthesizing dimensions of digital maturity in hospitals: Systematic review. *Journal of Medical Internet Research, 24*(3), e32994. DOI: 10.2196/32994 PMID: 35353050

Economic, Social and Environmental Council. (2021). *Towards a responsible and inclusive digital transformation* [Original title: *Vers une transformation digitale responsable et inclusive*]. https://www.cese.ma/docs/vers-une-transformation-digitale-responsable-et-inclusive/

El-Gazzar, R., Watat, J. K., & Agbozo, E. (2023). Blockchain technology in healthcare within the Global South: Mapping the area and developing a research scenario. *ResearchGate.* https://www.researchgate.net/publication/367482578

El Otmani Dehbi, Z., Sedrati, H., Chaqsare, S., Idrissi Azami, A., Merzouki, M., Raji, M., Rhalem, W., Al Idrissi, N., Nejjari, C., Amzazi, S., & Ghazal, H. (2021). Moroccan digital health response to the COVID-19 crisis. *Frontiers in Public Health, 9*, 690462. DOI: 10.3389/fpubh.2021.690462 PMID: 34485221

Elkachradi, R., Boudallaa, I., Hillali, M., & Kadouri, A. (2023). Lifelong learning among healthcare professionals in public hospitals: Historical analysis and multiple case studies in Morocco. In *E3S Web of Conferences* (Vol. 412, p. 01078). EDP Sciences., https://www.e3s-conferences.org/articles/e3sconf/abs/2023/49/e3sconf_icies2023_01078/e3sconf_icies2023_01078.html DOI: 10.1051/e3sconf/202341201078

Er-Rays, Y., & M'dioud, M. (2024). Assessment of technical efficiency in the Moroccan public hospital network: Using the DEA method. *arXiv preprint arXiv:2402.14940.* https://arxiv.org/abs/2402.14940

European Commission. (2025). European Health Data Space (EHDS) – Regulation overview page. *Public Health.*

Gilbert, S., Baca-Motes, K., Quer, G., Topol, E. J., & Steinhubl, S. R. (2024). Citizen data sovereignty is key to wearables and wellness data reuse for the common good. *npj. Digital Medicine, 7*(1), 13. Advance online publication. DOI: 10.1038/s41746-024-01004-z PMID: 38225423

Greenhalgh, T., Wherton, J., Papoutsi, C., Lynch, J., Hughes, G., A'Court, C., Hinder, S., Fahy, N., Procter, R., & Shaw, S. (2017). Beyond adoption: A new framework for theorizing and evaluating nonadoption, abandonment, and challenges to the scale-up, spread, and sustainability of health and care technologies. *Journal of Medical Internet Research, 19*(11), e367. DOI: 10.2196/jmir.8775 PMID: 29092808

Holly, L., Thom, S., Elzemety, M., Murage, B., Mathieson, K., & Petralanda, M. I. I. (2023). Strengthening health data governance: New equity and rights-based principles. *International Journal of Health Governance, 28*(3), 287–301. DOI: 10.1108/IJHG-11-2022-0104

Idaomar, C., Idaomar, D., Hannaoui, M., & Chafik, K. (2024). Applications of artificial intelligence in Morocco's healthcare sector: A springboard to medical excellence. *Journal of Computer and Communications, 12*(9), 63–77. DOI: 10.4236/jcc.2024.129004

Innovation Center E-health. Université Mohammed V de Rabat. (2022). *White paper on e-health in Morocco: Realities, challenges and development levers* [White paper]. https://smcmaroc.org/wp-content/uploads/2023/01/White-paper-on-e-health-in-Morocco.pdf

Jallal, M., Serhier, Z., Berrami, H., & Bennani Othmani, M. (2023). Current state and prospects of telemedicine in Morocco: Analysis of challenges, initiatives, and regulatory framework. *Cureus, 15*(12), e50963. DOI: 10.7759/cureus.50963 PMID: 38259391

Jallal, M., Serhier, Z., Berrami, H., & Bennani Othmani, M. (2023). Telemedicine: The situation in Morocco. *Cureus, 15*(12), e50963. DOI: 10.7759/cureus.50963 PMID: 38259391

Jayathissa, P., & Hewapathirana, R. (2023). Enhancing interoperability among health information systems in low- and middle-income countries: A review of challenges and strategies. *arXiv preprint arXiv:2309.12326*. https://arxiv.org/abs/2309.12326

Kaushik, A., Barcellona, C., Mandyam, N. K., Tan, S. Y., & Tromp, J. (2025). Challenges and opportunities for data sharing related to artificial intelligence tools in health care in low- and middle-income countries: Systematic review and case study from Thailand. *Journal of Medical Internet Research, 27*, e58338. DOI: 10.2196/58338 PMID: 39903508

Khalil, L., Serhier, Z., & Bennani Othmani, M. (2024). Exploring patients' acceptance of mental health e-services in Morocco: A quantitative approach. *Cureus, 16*(12), e76143. DOI: 10.7759/cureus.76143 PMID: 39835084

Kickbusch, I., Piselli, D., Agrawal, A., Balicer, R., Banner, O., Adelhardt, M., Capobianco, E., Fabian, C., Gill, A. S., Lupton, D., Medhora, R. P., Ndili, N., Ryś, A., Sambuli, N., Settle, D., Swaminathan, S., Vega Morales, J., Wolpert, M., Wyckoff, A. W., & Xue, L. (2021). The Lancet and Financial Times Commission on governing health futures 2030: Growing up in a digital world. *The Lancet, 398*(10312), 1727–1776. DOI: 10.1016/S0140-6736(21)01824-9

Kim, G. J., Namkoong, K., Kreps, G. L., Huh, J., Park, E., & Rimal, R. (2025). Developing the digital health communication maturity model. *Journal of Medical Internet Research, 27*, e68344. DOI: 10.2196/68344 PMID: 40228239

Ko, J. S., El-Toukhy, S., Quintero, S. M., Wilkerson, M. J., Nápoles, A. M., Stewart, A. L., & Strassle, P. D. (2023). Disparities in telehealth access, not willingness to use services, likely explain rural telehealth disparities. *The Journal of Rural Health, 39*(3), 617–624. DOI: 10.1111/jrh.12759 PMID: 37042413

Laukka, E., Pölkki, T., & Kanste, O. (2022). Leadership in the context of digital health services: A concept analysis. *Journal of Nursing Management, 30*(7), 2763–2780. DOI: 10.1111/jonm.13763 PMID: 35942802

Mahdaoui, A., Seffar, M., El Kettani, A., Belayachi, J., & Abouqal, R. (2023). Morocco's healthcare system: Achievements, challenges, and perspectives. *Cureus, 15*(7), e42706. DOI: 10.7759/cureus.41143 PMID: 37519608

Mandl, K. D., Gottlieb, D., & Mandel, J. C. (2024). Integration of AI in healthcare requires an interoperable digital data ecosystem. *Nature Medicine, 30*(3), 631–634. DOI: 10.1038/s41591-023-02783-w PMID: 38291298

Marelli, L., Stevens, M., Sharon, T., Van Hoyweghen, I., Boeckhout, M., Colussi, I., & Shabani, M. (2023). The European Health Data Space: Too big to succeed? *Health Policy (Amsterdam), 135*, 104861. DOI: 10.1016/j.healthpol.2023.104861 PMID: 37399677

Ministry of Health and Social Protection of the Kingdom of Morocco (MSPS). (2020). Wiqaytna mobile application for the fight against. *COVID, 19*, •••. https://www.sante.gov.ma

Ministry of Health and Social Protection of the Kingdom of Morocco (MSPS). (2023). Santé en chiffres. https://www.sante.gov.ma

Mügge, D. (2024). EU AI sovereignty: For whom, to what end, and with what effect? *Journal of European Public Policy, 31*(8), 2200–2225. Advance online publication. DOI: 10.1080/13501763.2024.2318475

Newaz, A. I., Sikder, A. K., Rahman, M. A., & Uluagac, A. S. (2020). A survey on security and privacy issues in modern healthcare systems: Attacks and defenses. arXiv preprint arXiv:2005.07359. https://arxiv.org/abs/2005.07359

Omrana, H., Nassiri, S., Belouadha, F.-Z., & Roudiés, O. (2012). Design for distributed Moroccan hospital pharmacy information environment with service-oriented architecture. arXiv preprint arXiv:1207.3277. https://arxiv.org/abs/1207.3277

Organisation for Economic Co-operation and Development. (2018). *Digital government review of Morocco*. https://www.oecd.org/content/dam/oecd/fr/publications/reports/2018/07/digital-government-review-of-morocco_g1g8d20f/9789264299917-fr.pdf

Parks, R., Wigand, R. T., Bennani Othmani, M., Serhier, Z., & Bouhaddou, O. (2019). Electronic health records implementation in Morocco: Challenges of silo efforts and recommendations for improvements. *International Journal of Medical Informatics*, *129*, 430–437. DOI: 10.1016/j.ijmedinf.2019.05.026 PMID: 31445288

Petticrew, M., & Roberts, H. (2006). *Systematic reviews in the social sciences: A practical guide*. Blackwell Publishing., DOI: 10.1002/9780470754887

Raab, R., Küderle, A., Zakreuskaya, A., Stern, A. D., Klucken, J., Kaissis, G., Rueckert, D., Boll, S., Eils, R., Wagener, H., & Eskofier, B. M. (2023). Federated electronic health records for the European Health Data Space. *The Lancet. Digital Health*, *5*(11), e840–e847. DOI: 10.1016/S2589-7500(23)00156-5 PMID: 37741765

Riazul Islam, S. M., Kwak, D., Kabir, M. H., Hossain, M., & Kwak, K. S. (2015). The Internet of Things for health care: A comprehensive survey. *IEEE Access : Practical Innovations, Open Solutions*, *3*, 678–708. DOI: 10.1109/ACCESS.2015.2437951

Rodríguez-Mejías, S., Degli-Esposti, S., González-García, S., & Parra-Calderón, C. L. (2024). Toward the European Health Data Space: The IMPaCT-Data secure infrastructure for EHR-based precision medicine research. *Journal of Biomedical Informatics*, *156*, 104670. DOI: 10.1016/j.jbi.2024.104670 PMID: 38880235

Rouleau, G., Wu, K., Ramamoorthi, K., Boxall, C., Liu, R. H., Maloney, S., Zelmer, J., Scott, T., Larsen, D., Wijeysundera, H. C., Ziegler, D., Bhatia, S., Kishimoto, V., Steele Gray, C., & Desveaux, L. (2024). Mapping theories, models, and frameworks to evaluate digital health interventions: Scoping review. *Journal of Medical Internet Research*, *26*, e51098. DOI: 10.2196/51098 PMID: 38315515

Rumbold, J. M. M., & Pierscionek, B. K. (2017). A critique of the regulation of data science in healthcare research in the European Union. *BMC Medical Ethics*, *18*(1), 27. Advance online publication. DOI: 10.1186/s12910-017-0184-y PMID: 28388916

Ruotsalainen, P., & Blobel, B. (2025). A system model and requirements for transformation to human-centric digital health. *Journal of Medical Internet Research, 27*, e68661. DOI: 10.2196/68661 PMID: 40294402

Ryan, M. (2024). Will the real data sovereign please stand up? An EU policy analysis. *International Journal of Law and Information Technology*. Advance online publication. DOI: 10.1093/ijlit/eaae006

Sardi, A., Rizzi, A., Sorano, E., & Guerrieri, A. (2021). Cyber risk in health facilities: A systematic literature review. arXiv preprint arXiv:2102.04093. https://arxiv.org/abs/2102.04093

Scheibner, J., Raisaro, J. L., Troncoso-Pastoriza, J. R., Ienca, M., Fellay, J., Vayena, E., & Hubaux, J. P. (2020). Revolutionizing medical data sharing using advanced privacy enhancing technologies: Technical, legal and ethical synthesis. arXiv preprint arXiv:2010.14445. https://arxiv.org/abs/2010.14445 DOI: 10.2196/preprints.25120

Sheets, L. R., Wallach, E., Khairat, S., Mutrux, R., Edison, K., & Becevic, M. (2020). Similarities and differences between rural and urban telemedicine utilization. *Perspectives in Health Information Management, 18*(Winter), 1e. https://www.ncbi.nlm.nih.gov/pmc/articles/PMC7883358/ PMID: 33633515

Shrivastava, U., Song, J., Han, B. T., & Dietzman, D. (2021). Do data security measures, privacy regulations, and communication standards impact the interoperability of patient health information? *International Journal of Medical Informatics, 148*, 104401. DOI: 10.1016/j.ijmedinf.2021.104401 PMID: 33571743

Sikstrom, L., Maslej, M. M., Hui, K., Findlay, Z., Buchman, D. Z., & Hill, S. L. (2022). Conceptualising fairness: Three pillars for medical algorithms and health equity. *BMJ Health & Care Informatics, 29*(1), e100459. DOI: 10.1136/bmjhci-2021-100459 PMID: 35012941

Tafenzi, H. A., Essaadi, I., & Belbaraka, R. (2025). Digital oncology in Morocco: Embracing artificial intelligence in a new era. *JCO Global Oncology, 11*(11), e2400583. DOI: 10.1200/GO-24-00583 PMID: 39787448

Webster, P. (2019). Patient data in the cloud. *The Lancet. Digital Health, 1*(9), e385. DOI: 10.1016/S2589-7500(19)30202-X

World Health Organization. (2023). 2023 emerging technologies and scientific innovations: A global public health perspective. https://www.who.int/publications/i/item/9789240073876

Yamani, F. Z., & El Merouani, M. (2020). Review the enterprise resource planning in Moroccan healthcare organizations. arXiv preprint arXiv:2010.06989. https://arxiv.org/abs/2010.06989

Conclusion

The collective insights presented in *Navigating Responsible Business Practices Through Ethical AI* underscore a shared conviction: the integration of Artificial Intelligence into business and societal processes must be guided by principles of responsibility, inclusivity, and sustainability. Across the nineteen chapters, this book has brought together diverse perspectives, case studies, and research findings that explore how ethical considerations can be embedded in AI applications spanning multiple sectors and contexts. The contributions demonstrate that AI's transformative power is evident in finance, supply chains, human resources, marketing, scientific research, healthcare, tourism, taxation, and beyond. Yet, this power carries with it an equally significant responsibility to safeguard human values, protect individual rights, and ensure equitable access to technological benefits. The research presented here confirms that ethical AI is not merely a regulatory requirement; it is a strategic imperative that fosters trust, strengthens stakeholder relationships, and promotes long-term resilience.

Several chapters have highlighted the operational dimensions of ethical AI from optimizing financial operations and enterprise resource planning, to ensuring fairness in automated decision-making, mitigating biases, and reinforcing sustainability within supply chains. Others have addressed sector-specific challenges, such as in tourism, green marketing, intellectual property management, and scientific research, illustrating that each domain faces unique ethical and practical considerations, yet all share the need for clear governance and human oversight. A recurring theme throughout the volume is the tension between innovation and ethics, and the ways in which organizations can navigate this balance. Whether in cross-border bankruptcy contexts, public sector management control, or the pursuit of digital health sovereignty, the chapters illustrate actionable strategies for aligning technological adoption with societal goals. These include designing transparent algorithms, engaging diverse stakeholders, promoting inclusivity in research, and leveraging AI for sustainable socio-economic development.

Importantly, the book has emphasized that responsible AI is not the sole domain of technologists. It requires multidisciplinary collaboration bringing together expertise from law, economics, sociology, psychology, environmental studies, and beyond. The diversity of our contributors, both in geographic representation and disciplinary background, reflects the global and cross-sectoral nature of this challenge. Looking forward, the lessons distilled here have broad implications for academia, industry, and policy. For scholars, the findings point to new research avenues on the measurement of AI's ethical impact, cross-cultural frameworks for responsible AI governance, and the intersection between emerging technologies and corporate social responsibility. For practitioners, the chapters provide concrete tools for embedding ethics into AI strategies, from risk assessment methodologies to bias mitigation protocols. For policymakers, the work reinforces the urgency of developing adaptive, context-aware regulatory frameworks that balance innovation with protection of public interest.

Navigating Responsible Business Practices Through Ethical AI contributes meaningfully to the global dialogue on shaping AI for the common good. It reaffirms that ethical AI is not a constraint on progress but a pathway to more robust, resilient, and human-centered innovation. The integration of AI into the fabric of business and governance is inevitable; ensuring that this integration advances equity, transparency, and sustainability is a collective responsibility and perhaps the defining challenge of our time.

Compilation of References

Aamri, E., & Bouaziz, S. M. (2025). L'impact des technologies émergentes sur les systèmes de contrôle de gestion: un modèle conceptuel pour une intégration équilibrée. Journal of Economics, Finance and Management (JEFM), 4(3), 637–652. https://journal-efm.fr/index.php/JEFM/article/view/234/159

Abd Hakim, T., Karia, A. A., David, J., Ginsad, R., Lokman, N., & Zolkafli, S. (2022). Impact of direct and indirect taxes on economic development: A comparison between developed and developing countries. *Cogent Economics & Finance*, *10*(1), 2141423. Advance online publication. DOI: 10.1080/23322039.2022.2141423

Abdelaal Mahmoud, A. (2010). FDI, local financial markets, employment and poverty alleviation. *African Economic Review, 22*(3), 77–94.

Abdelkader, M. A. A. R. O. U. F. I., & Hanane, B. R. A. B. I. J. E. (2025). L'Impact de l'Intelligence Artificielle sur le Contrôle Interne et la Gestion des Risques. Revue Internationale de la Recherche Scientifique (Revue-IRS), 3(3), 2718-2727.

Abdellatif, A. A., Al-Marridi, A. Z., Mohamed, A., Erbad, A., Chiasserini, C. F., & Refaey, A. (2020). SSHealth: Toward secure, blockchain-enabled healthcare systems. *arXiv preprint arXiv:2006.10843*. https://arxiv.org/abs/2006.10843

Abelnica, K., Petrovskis, G., Vindecs, A., & Romanovs, A. (2024). IT governance in the artificial intelligence age: Trends and practices. In J. Grabis, A. Romanovs, & M. Narigina (Eds.), *2024 IEEE 65th International Scientific Conference on Information Technology and Management Science of Riga Technical University, ITMS 2024* (pp. 92–97). IEEE. DOI: 10.1109/ITMS64072.2024.10741935

Abhivardhan. (2020, May). The Ethos of Artificial Intelligence as a Legal Personality in a Globalized Space: Examining the Overhaul of the Post-liberal Technological Order. In IFIP International Conference on Artificial Intelligence Applications and Innovations (pp. 3-14). Cham: Springer International Publishing.

Abhulimen, A. O., & Ejike, O. G.Angela Omozele AbhulimenOnyinye Gift Ejike. (2024). Ethical considerations in AI use for SMEs and supply chains: Current challenges and future directions. *International Journal of Applied Research in Social Sciences*, *6*(08), 1653–1679. DOI: 10.51594/ijarss.v6i8.1391

Abossolo, A. C., & Koumetio Kenfack, M. (2024). Le rôle du contrôle de gestion dans le pilotage des activités opérationnelles des formations sanitaires au Cameroun. *Revue Française d'Économie et de Gestion*, *5*(7), 1–20.

Abouamama, M. (2007). L'implantation des systèmes ERP dans le secteur public marocain: Enjeux et limites. *Revue Marocaine de Comptabilité*, *3*(1), 41–55.

Acemoglu, D., Autor, D. H., Hazell, J., & Restrepo, P. (2022). Artificial Intelligence and Jobs: Evidence from Online Vacancies. *Journal of Labor Economics*, *40*(S1), S293–S340. DOI: 10.1086/718327

Afroogh, S., Mostafavi, A., Akbari, A., Pouresmaeil, Y., Goudarzi, S., Hajhosseini, F., & Rasoulkhani, K. (2024). Embedded ethics for responsible artificial intelligence systems (EE-RAIS) in disaster management: A conceptual model and its deployment. *AI and Ethics*, *4*(4), 1117–1141. DOI: 10.1007/s43681-023-00309-1

Aguirre, S., & Rodriguez, A. (2017, August). Automation of a business process using robotic process automation (RPA): A case study. In *Workshop on engineering applications* (pp. 65-71). Cham: Springer International Publishing. DOI: 10.1007/978-3-319-66963-2_7

Ahamjik, I., Agbani, A., Abik, M., Khayli, M., Galzim, N., Berrada, J., & Bouslikhane, M. (2024). Contribution of artificial intelligence for understanding animal rabies epidemiology in Morocco: What are the perspectives of an innovative and predictive approaches? *One Health*, *19*, 100874. Advance online publication. DOI: 10.1016/j.onehlt.2024.100874 PMID: 39247759

Akinsola, K. (2025). Regulatory Compliance and the Role of Corporate Governance in Preventing Financial Misstatements. Available at *SSRN* 5126627.

Akter, S., Dwivedi, Y. K., Sajib, S., Biswas, K., Bandara, R. J., & Michael, K. (2022). Algorithmic bias in machine learning-based marketing models. *Journal of Business Research*, *144*, 201–216. DOI: 10.1016/j.jbusres.2022.01.083

Akter, S., Sultana, S., Mariani, M., Fosso Wamba, S., Spanaki, K., & Dwivedi, Y. K. (2023). Advancing algorithmic bias management capabilities in AI-driven marketing analytics research. *Industrial Marketing Management*, *114*, 243–261. DOI: 10.1016/j.indmarman.2023.08.013

Al Khoury, G., Thrassou, A., Papasolomou, I., & Vrontis, D. (2023). The requisite role of emotional intelligence in customer service in the retail banking sector. *The International Journal of Organizational Analysis*, *31*(7), 3315–3347. DOI: 10.1108/IJOA-04-2022-3229

Albaroudi, E., Mansouri, T., & Alameer, A. (2024). A comprehensive review of AI techniques for addressing algorithmic bias in job hiring. *AI*, *5*(1), 383–404. DOI: 10.3390/ai5010019

Alberto, I. R. I., Ghosh, A. K., Jain, B., Jayakumar, S., Martinez-Martin, N., McCague, N., Moukheiber, D., Moukheiber, L., Moukheiber, M., Moukheiber, S., Yaghy, A., Zhang, A., & Celi, L. A. (2023). The impact of commercial health datasets on medical research and health-care algorithms. *The Lancet. Digital Health*, *5*(5), e288–e294. DOI: 10.1016/S2589-7500(23)00025-0 PMID: 37100543

Alderman, J. E., Kapadia, D., Charitakis, K., Speed, K., Mathur, R., Banerjee, A., Denaxas, S., & Hemingway, H. (2025). Tackling algorithmic bias and promoting transparency in health datasets: The STANDING Together consensus recommendations. *The Lancet. Digital Health*, *7*(1), e16–e25. DOI: 10.1016/S2589-7500(24)00224-3 PMID: 39701919

Alegana, V. A., Frank, L. J., Makubalo, L. E., Okeibunor, J. C., Abimbola, S., Abdalla, S. M., & Kasilo, O. M. J. (2023). Digital health in the African region should be integral to health system strengthening. *Mayo Clinic Proceedings. Digital Health*, *1*(3), 425–434. DOI: 10.1016/j.mcpdig.2023.06.003 PMID: 37744952

Alioto, M. (2022). From Less Batteries to Battery-Less: Enabling A Greener World through Ultra-Wide Power-Performance Adaptation down to pWs. ESSCIRC 2022- IEEE 48th European Solid State Circuits Conference (ESSCIRC), 33–40. DOI: 10.1109/ESSCIRC55480.2022.9911465

ALL, E. (2019). *Trust in science and changing landscapes of communication.*

AllahRakha, N.AllahRakha. (2024). UNESCO's AI Ethics Principles: Challenges and Opportunities. *International Journal of Law and Policy*, *2*(9), 24–36. DOI: 10.59022/ijlp.225

Allam, K., & Lalaoui, S. (2025). Determinants of artificial intelligent adoption in emerging economies: Evidence from the information technology industry in Morocco. In Zimmermann, R., Rodrigues, J. C., Simoes, A., & Dalmarco, G. (Eds.), *Human-Centred Technology Management For A Sustainable Future, VOL 2* (pp. 43–50). IAMOT., DOI: 10.1007/978-3-031-72494-7_5

Alogogianni, E., & Virvou, M. (2023). Interpretable Machine Learning for Undeclared Work Prediction. 2023 14th International Conference on Information, Intelligence, Systems & Applications (IISA), 1–8. DOI: 10.1109/IISA59645.2023.10345943

Alvarez, J. M., Colmenarejo, A. B., Elobaid, A., Fabbrizzi, S., Fahimi, M., Ferrara, A., Ghodsi, S., Mougan, C., Papageorgiou, I., Reyero, P., Russo, M., Scott, K. M., State, L., Zhao, X., & Ruggieri, S. (2024). Policy advice and best practices on bias and fairness in AI. *Ethics and Information Technology*, *26*(2), 31. DOI: 10.1007/s10676-024-09746-w

Amisse, M., Faur, M., Gonard, L., & Orcesi, A. (2024). Promouvoir des modèles d'intelligence artificielle frugale pour et par les politiques publiques.

Anderson, T. W. (1971). The statistical analysis of time series. *Wiley Classics Library Edition Published.*https://books.google.co.ma/books?hl=fr&lr=&id=rCOzXIC8ZLkC&oi=fnd&pg=PR11&dq=Anderson,+T.+W.+(1971).+The+statistical+analysis+of+time+series.+Wiley.,+p.1-728.&ots=Iv36rgUEOs&sig=mvB8bQ2016i_ERAvCKWdllUNxs8&redir_esc=y#v=onepage&q&f=false

André, E. (2021). Socially Interactive Artificial Intelligence: Past, Present and Future. *International Conference on Multimodal Interfaces*, 4. DOI: 10.1145/3462244.3480862

Andresen, S. L. (2002). John McCarthy: Father of AI. *IEEE Intelligent Systems*, *17*(5), 84–85. DOI: 10.1109/MIS.2002.1039837

Angwin, J., Larson, J., Mattu, S., & Kirchner, L. (2016). *Machine bias*. ProPublica.

Anthony, R. N. (1965). Planning and control systems: A framework for analysis. Division of Research, Graduate School of Business Administration, Harvard University. https://archive.org/details/planningcontrols0000anth

Aouragh, M., & SETTA, B. (2025). Les Secrets de l'ia dans l'octroi de prêts: Décryptage des critères non financiers. *Dossiers de Recherches en Économie et Gestion*, *13*(1), 11–28.

Araujo, M. (2024). L'impact de l'intelligence artificielle sur le métier de comptable et les sociétés fiduciaires.

Arbelaez Ossa, L., Lorenzini, G., Milford, S. R., Shaw, D., Elger, B. S., & Rost, M. (2024). Integrating ethics in AI development: A qualitative study. *BMC Medical Ethics*, *25*(1), 10. DOI: 10.1186/s12910-023-01000-0 PMID: 38262986

Arharbi, N., & El Aissaoui, Y. (2024). Vers un contrôle de gestion agile: L'intelligence Artificielle et le Big Data pour la performance en environnement incertain. IJDAM. *International Journal of Digitalization and Applied Management, 1*(2), 195–210. DOI: 10.23882/ijdam.24141

Arner, D. W., Barberis, J., & Buckley, R. P. (2015). The evolution of Fintech: A new post-crisis paradigm. *SSRN, 47*, 1271. DOI: 10.2139/ssrn.2676553

Aro, O. E.Opeyemi E Aro. (2024). Data analytics as a driver of digital transformation in financial institutions. *World J. Adv. Res. Rev., 24*(1), 1054–1072. DOI: 10.30574/wjarr.2024.24.1.3124

Arrieta, A. B., Díaz-Rodríguez, N., Del Ser, J., Bennetot, A., Tabik, S., Barbado, A., & Herrera, F. (2020). Explainable Artificial Intelligence (XAI): Concepts, taxonomies, opportunities and challenges toward responsible AI. *Information Fusion, 58*, 82–115. DOI: 10.1016/j.inffus.2019.12.012

Asif, M., Searcy, C., & Castka, P. (2022). Exploring the role of industry 4.0 in enhancing supplier audit authenticity, efficacy, and cost effectiveness. *Journal of Cleaner Production, 331*, 129939. DOI: 10.1016/j.jclepro.2021.129939

Asmundson, I., Dorsey, T. W., Khachatryan, A., Niculcea, I., & Saito, M. (2011). Trade and trade finance in the 2008-09 financial crisis.

Aspray, W. (1997). The Intel 4004 microprocessor: What constituted invention? *IEEE Annals of the History of Computing, 19*(3), 4–15. DOI: 10.1109/85.601727

Augias, D. (2023). La nouvelle agence nationale de la cohésion des territoires en France, levier de politiques publiques contractualisées entre État et collectivités locales? Politiques & Management Public, 40(2), 145–163. https://www.jle.com/fr/revues/pmp/article/6442

Bader, M. A. (2006). *Intellectual property management in R&D collaborations: the case of the service industry sector*. Physica-Verlag HD.

Bahoo, S., Cucculelli, M., & Qamar, D. (2023). Artificial intelligence and corporate innovation: A review and research agenda. *Technological Forecasting and Social Change, 188*, 122264. DOI: 10.1016/j.techfore.2022.122264

Baker, R. S., & Hawn, A. (2021). Algorithmic bias in education. *International Journal of Artificial Intelligence in Education*, •••, 1–41.

Bal, M., & Jaouhari, L. (2024). Le contrôle de gestion au service de la gouvernance des collectivités territoriales. *Revue Internationale des Sciences de Gestion, 7*(3), 33–47.

Bal, M., Stok, M., Bombaerts, G., Huijts, N., Schneider, P., Spahn, A., & Buskens, V. (2023). A fairway to fairness: Toward a richer conceptualization of fairness perceptions for just energy transitions. *Energy Research & Social Science, 103*, 103213. DOI: 10.1016/j.erss.2023.103213

Bandara, R. J., Biswas, K., Akter, S., Shafique, S., & Rahman, M. (2025). Addressing algorithmic bias in AI-driven HRM systems: Implications for strategic HRM effectiveness. *Human Resource Management Journal*, 1748-8583.12609. Advance online publication. DOI: 10.1111/1748-8583.12609

Barocas, S., Hardt, M., & Narayanan, A. (2019). Fairness and Machine Learning: Limitations and Opportunities. http://fairmlbook.org

Barocas, S., Hardt, M., & Narayanan, A. (2020). *Fairness and machine learning: Limitations and opportunities*.

Baum, S. D., & Owe, A. (2023). Artificial Intelligence Needs Environmental Ethics. *Ethics Policy\& Environment, 26*(1), 139–143. DOI: 10.1080/21550085.2022.2076538

Behera, J., & Sharma, R. (2024). Understanding superstar firms' phenomenon. *Competitiveness Review*. Advance online publication. DOI: 10.1108/CR-09-2023-0231

Belenguer, L. (2022). AI bias: Exploring discriminatory algorithmic decision-making models and the application of possible machine-centric solutions adapted from the pharmaceutical industry. *AI and Ethics*, *2*(4), 771–787. DOI: 10.1007/s43681-022-00138-8 PMID: 35194591

Belkadi, Y., SOUAF, M., & BELKADI, S. (2025). L'intelligence artificielle et la performance de l'entreprise: Une revue de la littérature sur la redéfinition des pratiques et des interactions sociales. *Revue Française d'Economie et de Gestion*, *6*(1).

Bellamy, R. K. E., Dey, K., Hind, M., Hoffman, S. C., Houde, S., Kannan, K., ... & Zhang, Y. (2019). AI Fairness 360: An extensible toolkit for detecting, understanding, and mitigating unwanted algorithmic bias. *IBM Journal of Research and Development*, 63(4/5), 4:1–4:15.

BENDAOUD, H., & EL OUAZZANI, S. T. (2025). L'intelligence artificielle au sein de l'administration fiscale: Revue de la littérature. *International Journal of Economics and Management Sciences*, *4*(1), 173–185.

Benhmama, A., & Bennani, Y. B. (2024). Factors driving the adoption of artificial intelligence technology in the recruitment process in Morocco. *Access-Access To Science Business Innovation In The Digital Economy*, *5*(3), 387–406. DOI: 10.46656/access.2024.5.3(1)

Benjamins, R. (2020). Towards organizational guidelines for the responsible use of AI. *arXiv preprint arXiv:2001.09758*. DOI: 10.3233/FAIA200433

Bennour, M. (2024). Le design comme outil pour humaniser les politiques publiques et favoriser l'intelligence collective. ModACT, 12(1), 77–95. https://popups.uliege.be/3041-4687/index.php?id=177&lang=fr

Bensalah, M. (2021). Toward an ethical code of AI and human rights in Morocco. *Arribat-International Journal of Human Rights Published by CNDH Morocco, 1*(2), 187–203.

Berrami, H., Serhier, Z., Jallal, M., & Bennani Othmani, M. (2023). Understanding and use of artificial intelligence among doctors in a university hospital in Morocco. *Studies in Health Technology and Informatics, 309*, 215–219. DOI: 10.3233/SHTI230781 PMID: 37869845

Besse, P. (2022). Conformité européenne des systèmes d'ia: outils statistiques élémentaires. Statistique et société.

Beta, W. (2019). *Lithium-ion batteries: a machine-generated summary of current research.* Springer.

Beulen, E., Plugge, A., & van Hillegersberg, J. (2022). Formal and relational governance of artificial intelligence outsourcing. *Information Systems And E-Business Management, 20*(4, SI), 719–748. DOI: 10.1007/s10257-022-00562-7

Bhimani, A. (2021). Accounting disrupted: How digitalization is changing finance. Wiley. https://www.wiley.com/enus/Accounting+Disrupted%3A+How+Digitalization+Is+Changing+Finance-p-9781119720065

Bibri, S. E., & Allam, Z. (2022). The Metaverse as a virtual form of data-driven smart cities: The ethics of the hyper-connectivity, datafication, algorithmization, and platformization of urban society. *Computational Urban Science, 2*(1), 22. DOI: 10.1007/s43762-022-00050-1 PMID: 35915731

Bietti, E. (2020). From ethics washing to ethics bashing: a view on tech ethics from within moral philosophy. *Proceedings of the 2020 Conference on Fairness, Accountability, and Transparency*, 210–219. DOI: 10.1145/3351095.3372860

Binns, R. (2018, January). Fairness in machine learning: Lessons from political philosophy. In *Conference on fairness, accountability and transparency* (pp. 149-159). PMLR.

Binns, R. (2020). On the apparent conflict between individual and group fairness. *ACM Conference on Fairness, Accountability, and Transparency*, 514–524. DOI: 10.1145/3351095.3372864

Binns, R., Veale, M., Van Kleek, M., & Shadbolt, N. (2021). 'It's reducing a human being to a percentage': Perceptions of justice in algorithmic decisions. *Social Science Computer Review*, *39*(2), 278–296.

Bird, S., Dudik, M., Edgar, R., Horn, B., Luttin, J., Milan, V., & Walker, K. (2020). *Fairlearn: A toolkit for assessing and improving fairness in AI.* Microsoft Research.

Biroğul, S., & Şahin, Ö. Exploring the Impact of ISO/IEC 42001. (2023). AI Management Standard on Organizational Practices. *Advances in Artificial Intelligence Research*, *5*(1), 14–22.

Bogucki, A. (2025). Ethical, Legal, and Socioeconomic Aspects of Implementing Artificial Intelligence in Tax Administration. *Acta Universitatis Lodziensis. Folia Iuridica*, *110*, 19–36. DOI: 10.18778/0208-6069.110.12

Böhme, R., Christin, N., Edelman, B., & Moore, T. (2015). Bitcoin: Economics, technology, and governance. *The Journal of Economic Perspectives*, *29*(2), 213–238. DOI: 10.1257/jep.29.2.213

Boon, C., Eckardt, R., Lepak, D. P., & Boselie, P. (2017). Integrating strategic human capital and strategic human resource management. *International Journal of Human Resource Management*, *29*(1), 34–67. DOI: 10.1080/09585192.2017.1380063

Borger, J. G., Ng, A. P., Anderton, H., Ashdown, G. W., Auld, M., Blewitt, M. E., Brown V, D., Call, M. J., Collins, P., Freytag, S., Harrison, L. C., Hesping, E., Hoysted, J., Johnston, A., Mcinneny, A., Tang, P., Whitehead, L., Jex, A., & Naik, S. H. (2023). Artificial intelligence takes center stage: exploring the capabilities and implications of ChatGPT and other AI-assisted technologies in scientific research and education. *Immunology And Cell Biology, 101*(10, SI), 923–935. DOI: 10.1111/imcb.12689

BOUBOUH, I., & GHANIM, M. (2024). L'Intelligence Artificielle au Service du Contrôle Interne: Vers une Performance Organisationnelle Optimisée. *Revue Internationale des Sciences de Gestion*, *7*(4).

Bouchard, A. (2025). # workinprogress: IA générative et outils de recherche de littérature académique. Urfistinfo.

Bouckaert, G. (2022). From NPM to NWS in Europe. Transylvanian Review of Administrative Sciences, 18(SI), 5–21. https://doi.org/DOI: 10.24193/tras.SI2022.2

Boudallaa, I. (2025). What model of hospital attractiveness for the Moroccan hospital? *Moroccan Journal of Public Health, 7*(1). https://revues.imist.ma/index.php/MJPH/article/view/53076/29618

Boudallaa, I., Elkachradi, R., Assarag, B., Hassan Chrifi, H., & Kadouri, A. (2022). The challenges of hospital reform in Morocco. *International Journal of Advanced Research, 10*(4), 733–742. DOI: 10.21474/IJAR01/14608

Boudallaa, I., Elkachradi, R., & Kadouri, A. (2021). Le management des connaissances comme levier stratégique des organisations: État de l'art. *International Journal of Innovation in Management and Applied Sciences, 4*(1). https://www.researchgate.net/publication/355439860

Boudallaa, I., Elkachradi, R., & Kadouri, A. (2022). Organizational learning: A necessity in the hospital and environmental responsibility. In *E3S Web of Conferences* (Vol. 351, p. 01094). EDP Sciences., https://www.e3s-conferences.org/articles/e3sconf/abs/2022/18/e3sconf_icies2022_01094/e3sconf_icies2022_01094.html DOI: 10.1051/e3sconf/202235101094

Boudallaa, I., Oughannou, Z., & Bourekkadi, S. (2025). Knowledge management and digital technologies in Moroccan hospitals. *Proceedings on Engineering, 7*(1), 79–84. https://doaj.org/article/731f8da2d17345049a3965150eb5eaed. DOI: 10.24874/PES07.01.010

Boufakir, A. I., & Akrich, S. (2023). La contribution du système de contrôle de gestion à l'évaluation du capital immatériel des entreprises innovantes: Modèle conceptuel théorique. [RMd]. *Revista Multidisciplinar, 5*(3), 291–308. DOI: 10.23882/rmd.23173

Bouhaddou, O., Bennani Othmani, M., & Diouny, S. (2013). Medical informatics in Morocco. *Yearbook of Medical Informatics, 8*, 190–196. https://pubmed.ncbi.nlm.nih.gov/23974571/ PMID: 23974571

Boyd, D., & Ellison, N. B. (2007). Social network sites: Definition, history, and scholarship. *IEEE Engineering Management Review, 38*(3), 16–31. DOI: 10.1109/EMR.2010.5559139

Bratic, H. (2006). Tax Expenditures: A Theoretical Review. *Financial Theory and Practice, 30*(2), 113–127.

Braun, V., & Clarke, V. (2006). Using thematic analysis in psychology. *Qualitative Research in Psychology, 3*(2), 77–101. DOI: 10.1191/1478088706qp063oa

Brin, S., & Page, L. (1998). *The anatomy of a large-scale hypertextual Web search engine. 30*, 107–117. DOI: 10.1016/S0169-7552(98)00110-X

Brintrup, A., Kosasih, E., Schaffer, P., Zheng, G., Demirel, G., & MacCarthy, B. L. (2024). Digital supply chain surveillance using artificial intelligence: Definitions, opportunities and risks. *International Journal of Production Research, 62*(13), 4674–4695. DOI: 10.1080/00207543.2023.2270719

Brogaard, J. (2010). High frequency trading and its impact on market quality. *Northwestern University Kellogg School of Management Working Paper, 66*(10).

Brynjolfsson, E. (2022). The Turing Trap: The Promise & Peril of Human-Like Artificial Intelligence. *Daedalus, 151*(2), 272–287. DOI: 10.1162/daed_a_01915

Brynjolfsson, E., & Mcafee, A. N. D. R. E. W. (2017). Artificial intelligence, for real. *Harvard Business Review, 1*(1), 1–31.

Buchak, G., Matvos, G., Piskorski, T., & Seru, A. (2018). Fintech, regulatory arbitrage, and the rise of shadow banks. *Journal of Financial Economics, 130*(3), 453–483. DOI: 10.1016/j.jfineco.2018.03.011

Budhwar, P., Malik, A., De Silva, M. T. T., & Thevisuthan, P. (2022). Artificial intelligence – challenges and opportunities for international HRM: A review and research agenda. *International Journal of Human Resource Management, 33*(6), 1065–1097. DOI: 10.1080/09585192.2022.2035161

Buhalis, D., & Sinarta, Y. (2019). Real-time co-creation and nowness: Challenges and opportunities for tourism management. *Information and Communication Technologies in Tourism, 2019*, 317–330.

Buolamwini, J., & Gebru, T. (2018). Gender shades: Intersectional accuracy disparities in commercial gender classification. *Proceedings of the 1st Conference on Fairness, Accountability and Transparency*, 77–91.

Burrell, J. (2016). How the machine 'thinks': Understanding opacity in machine learning algorithms. *Big Data \& Society, 3*(1), 2053951715622512.

Cachat-Rosset, G., & Klarsfeld, A. (2023). *Diversity,* Equity, and Inclusion in Artificial Intelligence: An Evaluation of Guidelines. *Applied Artificial Intelligence, 37*(1), 2176618. Advance online publication. DOI: 10.1080/08839514.2023.2176618

Caiumi, A. (2011). The evaluation of the effectiveness of tax expenditures - A novel approach: An application to the regional tax incentives for business investments in Italy. *OECD Taxation Working Papers, 5. OECD Publishing.* https://ideas.repec.org/p/oec/ctpaaa/5-en.html

Calacci, D., & Pentland, A. (2022). Bargaining with the Black-Box. Proceedings of the ACM on Human-Computer Interaction, 6(CSCW2), 1–24. DOI: 10.1145/3570601

Cameron, R., Herrmann, H., & Nankervis, A. (2024). Mapping the evolution of algorithmic HRM (AHRM): A multidisciplinary synthesis. *Humanities & Social Sciences Communications, 11*(303), 303. Advance online publication. DOI: 10.1057/s41599-024-02786-z

Camilleri, A. (2017). Corporate sustainability, social responsibility and environmental management. *Corporate Governance Review, 11*(4), 214–229.

Cantens, T., & Tourpe, H. (2025). How AI Can Help Both Tax Collectors and Taxpayers.*IMF blog.* https://www.imf.org/en/Blogs/Articles/2025/02/25/how-ai-can-help-both-the-taxman-and-the-taxpayer

Capelli, G., Verdi, D., Frigerio, I., Rashidian, N., Ficorilli, A., Grasso, S. V., Majidi, D., Gumbs, A. A., Spolverato, G., & Ethics, A. I. S. E. B. S. G. (2023). White paper: Ethics and trustworthiness of artificial intelligence in clinical surgery. *Artificial Intelligence Surgery, 3*(2), 111–122. DOI: 10.20517/ais.2023.04

Capocci, C. (2024). L'impact de la crise financière de 2007 sur les réglementations des hedge funds. *Revue Internationale d'Économie et de Finance, 15*(1), 45–62. https://matheo.uliege.be/handle/2268.2/19635

Carello, M. P., Marchetti Spaccamela, A., Querzoni, L., & Angelini, M. (2023). A systematization of cybersecurity regulations, standards and guidelines for the healthcare sector. *arXiv preprint arXiv:2304.14955.* https://arxiv.org/abs/2304.14955

Catalini, C., & Gans, J. S. (2020). Some simple economics of the blockchain. *Communications of the ACM, 63*(7), 80–90. DOI: 10.1145/3359552

Centobelli, P., Cerchione, R., Del Vecchio, P., Oropallo, E., & Secundo, G. (2022). Blockchain technology for bridging trust, traceability and transparency in circular supply chain. Information \& Management, 59(7), 103508.

Chaboud, A. P., Chiquoine, B., Hjalmarsson, E., & Vega, C. (2014). Rise of the machines: Algorithmic trading in the foreign exchange market. *The Journal of Finance, 69*(5), 2045–2084. DOI: 10.1111/jofi.12186

Chamboko-Mpotaringa, M., & Manditereza, B. (2024). Artificial Intelligence in higher education: A decade's bibliometric snapshot, emerging themes and future research. *International Journal of Research In Business and Social Science, 13*(8), 192–202. DOI: 10.20525/ijrbs.v13i8.3708

Chan, K. T. (2022). Emergence of the 'Digitalized Self' in the Age of Digitalization. *Computers in Human Behavior Reports, 6*, 100191. DOI: 10.1016/j.chbr.2022.100191

Chao, C.-H. (2019). Ethics Issues in Artificial Intelligence. *2019 International Conference On Technologies And Applications Of Artificial Intelligence (TAAI).* DOI: 10.1109/TAAI48200.2019.8959925

Chegri, M., & Torra, M. (2024). Le contrôle de gestion au service du pilotage de la performance de l'entreprise. *Revue Française d'Économie et de Gestion, 5*(4), 102–117.

Chen, P., Wu, L., & Wang, L. (2023). AI fairness in data management and analytics: A review on challenges, methodologies, and applications. *Applied Sciences (Basel, Switzerland), 13*(18), 10258. DOI: 10.3390/app131810258

Chen, S., Fan, L., Meng, G., Su, T., Xue, M., Xue, Y., & Xu, L. (2020, June). An empirical assessment of security risks of global android banking apps. In *Proceedings of the ACM/IEEE 42nd International Conference on Software Engineering* (pp. 1310-1322). DOI: 10.1145/3377811.3380417

Chen, Y., & Bellavitis, C. (2020). Blockchain disruption and decentralized finance: The rise of decentralized business models. *Journal of Business Venturing Insights, 13*, e00151. DOI: 10.1016/j.jbvi.2019.e00151

Chen, Z. (2023). Collaboration among recruiters and artificial intelligence: Removing human prejudices in employment. *Cognition Technology and Work, 25*(1), 135–149. DOI: 10.1007/s10111-022-00716-0 PMID: 36187287

Chen, Z. (2023). Ethics and discrimination in artificial intelligence-enabled recruitment practices. *Humanities & Social Sciences Communications, 10*(567), 567. DOI: 10.1057/s41599-023-02079-x

Chen, Z., Chen, C., Yang, G., He, X., Chi, X., Zeng, Z., & Chen, X. (2024). Research integrity in the era of artificial intelligence: Challenges and responses. *Medicine, 103*(27), e38811. Advance online publication. DOI: 10.1097/MD.0000000000038811 PMID: 38968491

Cheong, B. C. (2024). Transparency and accountability in AI systems: Safeguarding wellbeing in the age of algorithmic decision-making. *Frontiers in human dynamics*, 6.

Chhillar, D., & Aguilera, V. R. (2022). An Eye for Artificial Intelligence: Insights Into the Governance of Artificial Intelligence and Vision for Future Research. *Business & Society, 61*(5), 1197–1241. DOI: 10.1177/00076503221080959

Chiappa, S., & Gillam, T. P. (2021). Path-specific counterfactual fairness. *Proceedings of the AAAI Conference on Artificial Intelligence, 35*(9), 7936–7944.

CHIKHI, D. E. B., & RAMDANI, B. (2023). L'influence de l'intelligence artificielle sur l'efficacité de l'audit financier: Tendances, défis et opportunités. *Revue Française d'Economie et de Gestion*, *4*(12).

Chilwalo, M. (2016). Multinational Corporations: Corporate Social Responsibility versus Environmental Problems. *European Scientific Journal*, *12*(17), 241. DOI: 10.19044/esj.2016.v12n17p241

Chin, M. H., Afsar-Manesh, N., Bierman, A. S., Chang, C., Colón-Rodríguez, C. J., Dullabh, P., & Ohno-Machado, L. (2023). Guiding principles to address the impact of algorithm bias on racial and ethnic disparities in health and health care. *JAMA Network Open*, *6*(12), e2345050. DOI: 10.1001/jamanetworkopen.2023.45050 PMID: 38100101

Chisita, C. T., Durodolu, O. O., & Rusero, A. M. (2025). Data capitalism in the milieu of the surveillance economy: What can libraries do? *IFLA Journal*, *51*(2), 339–349. Advance online publication. DOI: 10.1177/03400352241286170

Chiwalo, M. (2016). Multinational corporations: Corporate social responsibility versus environmental problems. *Global Business and Sustainability Review, 8*(3), 56–71.

Choi, T. M., Wallace, S. W., & Wang, Y. (2018). Big data analytics in operations management. *Production and Operations Management*, *27*(10), 1868–1883. DOI: 10.1111/poms.12838

Chomanski, B. (2021). The missing ingredient in the case for regulating big tech. *Minds and Machines*, *31*(2), 257–275. DOI: 10.1007/s11023-021-09562-x

Choudhary, E., Narayanan, S., & Khan, F. (2024). Legal and Regulatory Landscape. *Advances in Healthcare Information Systems and Administration Book Series*, 222–239. DOI: 10.4018/979-8-3693-7452-8.ch014

Choung, H., Seberger, J. S., & David, P. (2023). When AI is Perceived to Be Fairer than a Human: Understanding Perceptions of Algorithmic Decisions in a Job Application Context. *International Journal of Human-Computer Interaction*, •••, 1–18. DOI: 10.1080/10447318.2023.2266244

Chow, J. C. (2018). Analysis of financial credit risk using machine learning. arXiv preprint arXiv:1802.05326.

Cho, W., Choi, S., & Choi, H. (2023). Human resources analytics for public personnel management: Concepts, cases, and caveats. *Administrative Sciences*, *13*(2), 1–22. DOI: 10.3390/admsci13020041

Christiaens, J., Vanhee, C., Manes Rossi, F., Aversano, N., & Van Cauwenberge, P. (2015). The effect of IPSAS on reforming governmental financial reporting: An international comparison. *International Review of Administrative Sciences, 81*(1), 158–177. DOI: 10.1177/0020852314546580

CIAT. (2021). Use of Big Data in Tax Administrations. https://www.ciat.org/use-of-big-data-in-tax-administrations/?lang=en

Cihon, P., Maas, M. M., & Kemp, L. (2020). Should artificial intelligence governance be centralised? Design Lessons from History. *Proceedings Of The 3rd Aaai/Acm Conference On AI, Ethics, and Society AIES 2020*, 228–234. DOI: 10.2139/ssrn.3761636

Cobbe, J., Veale, M., & Singh, J. (2023, June). Understanding accountability in algorithmic supply chains. In *Proceedings of the 2023 ACM Conference on Fairness, Accountability, and Transparency* (pp. 1186-1197). DOI: 10.1145/3593013.3594073

Collins, D. (2012). Sustainable international investment law after the Pax Americana: The BOOT on the other foot. *Journal of International Economic Law, 15*(3), 433–456.

Corbett-Davies, S., & Goel, S. (2018). The measure and mismeasure of fairness: A critical review of fair machine learning. *arXiv preprint arXiv:1808.00023.*

Cordes, A., Bak, M., Gospodarevskaya, E., Glozier, N., Thorpe, A., McMillan, F., & Westphaln, K. K. (2024). Competing interests: Digital health and Indigenous data sovereignty. *npj. Digital Medicine*, *7*(1), 166. Advance online publication. DOI: 10.1038/s41746-024-01171-z PMID: 38909083

Cottingham, J. (1992). *The Cambridge Companion to Descartes*. Cambridge University Press., https://www.amazon.com/Cambridge-Companion-Descartes-Companions-Philosophy/dp/0521366968 DOI: 10.1017/CCOL0521366232

Couzineau-Zegwaard, E., & Tarnaud, L. (2025). ['IA ou la transition par la donnée: Interview de Sandrine Tarnaud,«Head of data and AI» chez Microsoft. Le temps des transitions.]. *Chapitre, 10*, L.

Cowls, J., & Floridi, L. (2018). Prolegomena to a white paper on an ethical framework for a good AI society. *Minds and Machines, 28*(4), 689–707. DOI: 10.1007/s11023-018-9482-5 PMID: 30930541

Crane, A., & Matten, D. (2021). *Business ethics: Managing corporate citizenship and sustainability in the age of globalization* (6th ed.). Oxford University Press.

Crawford, K. (2021). *Atlas of AI: Power, politics, and the planetary costs of artificial intelligence*. Yale University Press.

Crawford, K. (2021). *The atlas of AI: Power, politics, and the planetary costs of artificial intelligence*. Yale University Press.

Cui, X., Xu, B., & Razzaq, A. (2022). Can Application of Artificial Intelligence in Enterprises Promote the Corporate Governance? *Frontiers in Environmental Science*, *10*, 944467. Advance online publication. DOI: 10.3389/fenvs.2022.944467

Culot, G., Podrecca, M., & Nassimbeni, G. (2024). Artificial intelligence in supply chain management: A systematic literature review of empirical studies and research directions. *Computers in Industry*, *162*, 104132. DOI: 10.1016/j.compind.2024.104132

Dalal, R. S., Howard, D. J., Bennett, R. J., Posey, C., Zaccaro, S. J., & Brummel, B. J. (2021). Organizational science and cybersecurity: Abundant opportunities for research at the interface. *Journal of Business and Psychology*, *37*(1), 1–29. DOI: 10.1007/s10869-021-09732-9 PMID: 33564206

Dapra, J., Wright, R., & Wu, C. (2019). FDI 101: An overview of foreign direct investment. *International Review of Economics and Business, 27*(1), 13–29.

Dastin, J. (2018). *Amazon scrapped "sexist AI" recruiting tool*. Reuters.

Dastin, J. (2020). *Apple, Goldman revise credit practices after bias complaints*. Reuters.

Davenport, J. (2025). Enhancing legal document analysis with large language models: A structured approach to accuracy, context preservation, and risk mitigation. *Open Journal of Modern Linguistics*, *15*(2), 232–280. DOI: 10.4236/ojml.2025.152016

Davenport, T. H., & Mittal, N. (2022). *Working with AI: Real stories of human-machine collaboration*. MIT Press. DOI: 10.7551/mitpress/14453.001.0001

de Almeida, P. G. R., dos Santos, C. D., & Farias, J. S. (2021). Artificial Intelligence Regulation: A framework for governance. *Ethics and Information Technology*, *23*(3), 505–525. DOI: 10.1007/s10676-021-09593-z

De Freitas, C., Dos Reis, V., Silva, S., Videira, P. A., Morava, E., & Jaeken, J. (2021). Public and patient involvement in health data governance (DATAGov): Protocol of a people-centred, mixed-methods study on data use and sharing for rare diseases care and research. *BMJ Open*, *11*(3), e044289. DOI: 10.1136/bmjopen-2020-044289 PMID: 33722870

De Spiegeleer, J., Madan, D. B., Reyners, S., & Schoutens, W. (2018). Machine learning for quantitative finance: Fast derivative pricing, hedging and fitting. *Quantitative Finance, 18*(10), 1635–1643. DOI: 10.1080/14697688.2018.1495335

Dekri, M. (2024). Repenser le système de contrôle de gestion dans le secteur public à l'ère du Post New Public Management. Dossiers de Recherches en Économie et Gestion, 12(2), 11–36. https://doi.org/DOI: 10.34874/IMIST.PRSM/doreg-v12i2.49385

Delabre, I., Alexander, A., & Rodrigues, C. (2020). Strategies for tropical forest protection and sustainable supply chains: Challenges and opportunities for alignment with the UN sustainable development goals. *Sustainability Science, 15*(6), 1637–1651. DOI: 10.1007/s11625-019-00747-z

Dempere, J., Udjo, E., & Mattos, P. (2024). The Entrepreneurial Impact of the European Directive on Corporate Sustainability Due Diligence. *Administrative Sciences, 14*(10), 266. DOI: 10.3390/admsci14100266

Denzin, N. K., & Lincoln, Y. S. (2011). *The Sage handbook of qualitative research.* sage.

Desjardins, G. & Gould, A. (2024). Fausses promesses et stupidité organisationnelle: pourquoi l'ia n'est pas synonyme de performance. Ad machina.

Diakon Debebe, A. (2018). The role of foreign direct investment in technology transfer for local employees: Evidence from SNP, Hawassa Town, Ethiopia. *Ethiopian Journal of Economics, 26*(2), 101–120.

Díaz-Rodríguez, N., Del Ser, J., Coeckelbergh, M., López de Prado, M., Herrera-Viedma, E., & Herrera, F. (2023). Connecting the dots in trustworthy artificial intelligence: From AI principles, ethics, and key requirements to responsible AI systems and regulation. *Information Fusion, 99*, 101896. DOI: 10.1016/j.inffus.2023.101896

Dignam, A. (2020). Artificial intelligence, tech corporate governance and the public interest regulatory response. *Cambridge Journal Of Regions Economy and Society, 13*(1, SI), 37–54. DOI: 10.1093/cjres/rsaa002

Dignum, V. (2022). Responsible artificial intelligence---from principles to practice: A keynote at theWebConf 2022. In *ACM SIGIR Forum* (Vol. 56, No. 1, pp. 1-6). New York, NY, USA: ACM.

DiMaggio, P. J., & Powell, W. W. (1983). The iron cage revisited: Institutional isomorphism and collective rationality in organizational fields. *American Sociological Review, 48*(2), 147–160. DOI: 10.2307/2095101

Ding, M. (2012). Business diplomacy management as a key role in the sustainable development and stakeholder management in the multinational corporations: DaimlerChrysler case. *International Journal of Corporate Governance, 5*(3), 33–51.

Dispas, C., Kayanakis, G., Servel, N., & Striukova, L. (2021). *Innovation and financial markets*. John Wiley & Sons. DOI: 10.1002/9781119819912

Djellaba, A., Seifeddine,A., & Osman, A.A.M. (2024). Artificial intelligence in tax administration: benefits and challenges. Faculty of Economics, Business and Management Sciences, Chadli Bendjedid University.

Djerassi, C., Smith, D. H., Crandell, C. W., Gray, N. A. B., Nourse, J. G., & Lindley, M. R. (1982). The DENDRAL project: Computational aids to natural products structure elucidation. *Pure and Applied Chemistry*, *54*(12), 2425–2442. DOI: 10.1351/pac198254122425

Doborjeh, N. Z., Moghavvemi, S., Musa, G., & Tengku Arshad, T. N. B. (2022). Chatbot adoption in tourism: A systematic review of the literature. *Tourism Management Perspectives*, *41*, 100944.

Doborjeh, Z., Hemmington, N., Doborjeh, M., & Kasabov, N. (2022). Artificial intelligence: A systematic review of methods and applications in hospitality and tourism. *International Journal of Contemporary Hospitality Management*, *34*(3), 1154–1176. DOI: 10.1108/IJCHM-06-2021-0767

dos Santos Sequeira, A. C. (2024). A Step Towards Corporate Sustainability—The New Directive on Due Diligence (pp. 17–33). DOI: 10.1007/978-3-031-77531-4_2

Doshi-Velez, F., & Kim, B. (2017). Towards a Rigorous Science of Interpretable Machine Learning. ArXiv preprint. https://arxiv.org/abs/1702.08608

Doshi-Velez, F., & Kim, B. (2017). Towards a rigorous science of interpretable machine learning. *arXiv*.

Du Québec, G. (2025). Consultation sur la transformation des milieux de travail par le numérique.

Duncan, R., Eden, R., Woods, L., Wong, I., & Sullivan, C. (2022). Synthesizing dimensions of digital maturity in hospitals: Systematic review. *Journal of Medical Internet Research*, *24*(3), e32994. DOI: 10.2196/32994 PMID: 35353050

Dwivedi, D. N., Mahanty, G., & Dora, T. (2024). Hyperautomation in Financial Services: Revolutionizing Banking and Investment Processes. In *Hyperautomation in Business and Society* (pp. 184-199). IGI Global.

Dwivedi, Y. K., Hughes, D. L., Ismagilova, E., Aarts, G., Coombs, C., Crick, T., Duan, Y., Dwivedi, R., Edwards, J., Eirug, A., Galanos, V., Ilavarasan, P. V., Janssen, M., Jones, P., Kar, A. K., Kizgin, H., Kronemann, B., Lal, B., Lucini, B., & Williams, M. D. (2021). Artificial intelligence (AI): Multidisciplinary perspectives on emerging challenges, opportunities, and agenda for research, practice and policy. *International Journal of Information Management*, *57*, 101994. DOI: 10.1016/j.ijinfomgt.2019.08.002

Dwivedi, Y. K., Kshetri, N., Hughes, L., Slade, E. L., Jeyaraj, A., Kar, A. K., Baabdullah, A. M., Koohang, A., Raghavan, V., Ahuja, M., Albanna, H., Albashrawi, M. A., Al-Busaidi, A. S., Balakrishnan, J., Barlette, Y., Basu, S., Bose, I., Brooks, L., Buhalis, D., & Wright, R. (2023). Opinion Paper:"So what if ChatGPT wrote it?" Multidisciplinary perspectives on opportunities, challenges and implications of generative conversational AI for research, practice and policy. *International Journal of Information Management*, *71*, 102642. DOI: 10.1016/j.ijinfomgt.2023.102642

Dwivedi, Y. K., Pandey, N., Currie, W., & Micu, A. (2024). Leveraging ChatGPT and other generative artificial intelligence (AI)-based applications in the hospitality and tourism industry: Practices, challenges and research agenda. *International Journal of Contemporary Hospitality Management*, *36*(1), 1–12. DOI: 10.1108/IJCHM-05-2023-0686

Dwork, C., Hardt, M., Pitassi, T., Reingold, O., & Zemel, R. (2012). Fairness through awareness. *Proceedings of the 3rd Innovations in Theoretical Computer Science Conference*, 214–226. DOI: 10.1145/2090236.2090255

Economic, Social and Environmental Council. (2021). *Towards a responsible and inclusive digital transformation* [Original title: *Vers une transformation digitale responsable et inclusive*]. https://www.cese.ma/docs/vers-une-transformation-digitale-responsable-et-inclusive/

Ehrgott, M. (2012). Vilfredo Pareto and multi-objective optimization. *Doc. math*, *8*, 447-453.

Ejjami, R. (2024). Adopting artificial intelligence and big data tools across industry sectors in Morocco: An integrative literature review. *International Journal of Environment. Workplace and Employment*, *8*(2), 171–198. DOI: 10.1504/IJEWE.2024.141574

Ekuma, K. (2024). Artificial Intelligence and Automation in Human Resource Development: A Systematic Review. *Human Resource Development Review*, *23*(2), 199–229. DOI: 10.1177/15344843231224009

El Abed, A., & Bellaaj, M. (2025). *Perception de l'utilisation de l'intelligence artificielle par les dirigeants tunisiens: motivations, freins et avantages perçus* [CG1.1]. Revue Française d'Economie et de Gestion.

El Kezazy, H. (2022). Contrôle de gestion et la bonne gouvernance dans les collectivités territoriales: Une revue de littérature. *Revue Internationale des Sciences de Gestion*, *5*(2), 77–92.

El Kezazy, H., & Hilmi, Y. (2024). Le contrôle de gestion territorial: Levier de la bonne gouvernance. *Alternatives Managériales Économiques*, *6*(1), 55–72.

El Korchi, C., & Akhlaffou, M. (2024). La transformation de la gestion publique à la lumière des innovations managériales. *Revue Française d'Économie et de Gestion*, *5*(5), 133–149.

El Mehdi, B., Hind, C., Slimane, S., & Abdelali, L. (2025). Vers une modélisation conceptuelle de la performance globale des entreprises de services à travers le contrôle de gestion, les systèmes d'information et l'intelligence artificielle. *Revue Internationale de la Recherche Scientifique*, *3*(3), 2599–2620.

El Otmani Dehbi, Z., Sedrati, H., Chaqsare, S., Idrissi Azami, A., Merzouki, M., Raji, M., Rhalem, W., Al Idrissi, N., Nejjari, C., Amzazi, S., & Ghazal, H. (2021). Moroccan digital health response to the COVID-19 crisis. *Frontiers in Public Health*, *9*, 690462. DOI: 10.3389/fpubh.2021.690462 PMID: 34485221

El-Gazzar, R., Watat, J. K., & Agbozo, E. (2023). Blockchain technology in healthcare within the Global South: Mapping the area and developing a research scenario. *ResearchGate.* https://www.researchgate.net/publication/367482578

Elkington, J., & Rowlands, I. H. (1999). Cannibals with forks: The triple bottom line of 21st century business. *Alternatives Journal*, *25*(4), 42.

Emon, M. M. H. (2025b). Leveraging AI on Social Media & Digital Platforms to Enhance English Language Teaching and Learning. In AI-Powered English Teaching (pp. 211–238). IGI Global Scientific Publishing. DOI: 10.4018/979-8-3373-1952-0.ch007

Emon, M. M. H. (2025c). Overcoming Hurdles: Navigating Challenges in the Adoption of Smart Logistics. In Emerging Trends in Smart Logistics Technologies (A. Hlali, pp. 197–228). IGI Global Scientific Publishing. DOI: 10.4018/979-8-3373-2434-0.ch007

Emon, M. M. H., & Chowdhury, M. S. A. (2025). AI and IoT-Powered Smart Logistics: Transforming Supply Chains for Efficiency and Sustainability. In Emerging Trends in Smart Logistics Technologies (A. Hlali, pp. 45–72). IGI Global Scientific Publishing. DOI: 10.4018/979-8-3373-2434-0.ch002

Emon, M. M. H., & Khan, T. (2025c). The Rise of Circular Supply Chains: Recycling and Closed-Loop Systems Could Mitigate Inflationary Pressures. In Supply Chain Disruptions and Impact on Global Inflation (pp. 195–226). IGI Global. DOI: 10.4018/979-8-3373-1982-7.ch008

Emon, M. M. H. (2025a). The Mediating Role of Supply Chain Responsiveness in the Relationship Between Key Supply Chain Drivers and Performance: Evidence from the FMCG Industry. *Brazilian Journal of Operations & Production Management*, *22*(1), 2580. DOI: 10.14488/BJOPM.2580.2025

Emon, M. M. H., & Khan, T. (2025a). The mediating role of attitude towards the technology in shaping artificial intelligence usage among professionals. *Telematics and Informatics Reports*, *17*, 100188. DOI: 10.1016/j.teler.2025.100188

Emon, M. M. H., & Khan, T. (2025b). The transformative role of Industry 4.0 in supply chains: Exploring digital integration and innovation in the manufacturing enterprises. *Journal of Open Innovation*, *11*(2), 100516. DOI: 10.1016/j.joitmc.2025.100516

Emon, M. M. H., & Khan, T. (2025d). Corporate Social Responsibility for Sustainable Development: A Systematic Review of Business Contributions to Address Global Challenges. *Review of Business and Economics Studies*, *13*(2), 6–39. DOI: 10.26794/2308-944X-2025-13-2-6-39

Erman, E., & Furendal, M. (2024). Artificial intelligence and the political legitimacy of global governance. *Political Studies*, *72*(2), 421–441. DOI: 10.1177/00323217221126665

Er-Rays, Y., & M'dioud, M. (2024). Assessment of technical efficiency in the Moroccan public hospital network: Using the DEA method. *arXiv preprint arXiv:2402.14940*. https://arxiv.org/abs/2402.14940

Esmaeilzadeh, P. (2024). Challenges and strategies for wide-scale artificial intelligence (AI) deployment in healthcare practices: A perspective for healthcare organizations. *Artificial Intelligence in Medicine*, *151*, 102861. DOI: 10.1016/j.artmed.2024.102861 PMID: 38555850

Esplugues Mota, C. (2018). A more targeted approach to foreign direct investment: The establishment of screening systems on national security grounds. *Journal of International Trade Policy, 9*(4), 221–239.

Eubanks, V. (2018). *Automating inequality: How high-tech tools profile, police, and punish the poor*. St. Martin's Press.

European Commission. (2019). *Ethics guidelines for trustworthy AI*. High-Level Expert Group on Artificial Intelligence (AI HLEG), European Commission. https://ec.europa.eu/futurium/en/ai-alliance-consultation

European Commission. (2021). *Horizon Europe – The EU research and innovation programme (2021–2027): General annexes*. European Commission. https://ec.europa.eu/programmes/horizon2020/

European Commission. (2021). Proposal for a regulation laying down harmonised rules on artificial intelligence (Artificial Intelligence Act). Official Journal of the European Union. COM(2021) 206 final. https://eur-lex.europa.eu/legal-content/EN/TXT/?uri=CELEX%3A52021PC0206

European Commission. (2021). *Proposal for a regulation laying down harmonised rules on artificial intelligence*. Artificial Intelligence Act.

European Commission. (2021). *Proposal for a regulation of the European Parliament and of the Council laying down harmonised rules on artificial intelligence (Artificial Intelligence Act) and amending certain Union legislative acts (COM(2021) 206 final)*. European Commission. https://eur-lex.europa.eu/legal content/EN/TX-T/?uri=CELEX%3A52021PC0206

European Commission. (2025). European Health Data Space (EHDS) – Regulation overview page. *Public Health*.

European Commission. S., & others. (2019). Ethics guidelines for trustworthy AI. *Publications Office*.

Fadoua, A.Z.I.O.U.A.L., & Samir, B. A. (2025). Fintechs et IA: nouveaux paradigmes bancaires face à la nécessité d'une régulation juridique. Revue Internationale de la Recherche Scientifique (Revue-IRS), 3(3), 2573-2590.

Fähndrich, J., & Pedell, B. (2025). Evaluating the influencing factors and effects of the digitalization of management control. *Journal of Accounting and Organizational Change*, *21*(2), 278–311. https://www.emerald.com/insight/content/doi/10.1108/JAOC-06-2023-0116/full/html. DOI: 10.1108/JAOC-07-2023-0125

Farooq, K., & Vatyan, A. (2025). AI to modernize tax administration: the story behind Armenia's success. *World Bank*. https://blogs.worldbank.org/en/governance/ai-to-modernize-tax-administration--the-story-behind-armenia-s-s

Farooque, M., Zhang, A., Thürer, M., Qu, T., & Huisingh, D. (2019). Circular supply chain management: A definition and structured literature review. *Journal of Cleaner Production, 228*, 882–900. DOI: 10.1016/j.jclepro.2019.04.303

Feizabadi, J. (2022). Machine learning demand forecasting and supply chain performance. *International Journal of Logistics, 25*(2), 119–142. DOI: 10.1080/13675567.2020.1803246

Feki, M., & Dudézert, A. (2024, May). Exploration du potentiel de chatgpt dans la Supply Chain: revue de littérature et pistes de recherche. In AIM 2024: 29e Conférence de l'Association Information et Management.

Felício, T., Samagaio, A., & Rodrigues, R. (2021). Adoption of management control systems and performance in public sector organizations. *Journal of Business Research, 124*, 593–602. DOI: 10.1016/j.jbusres.2020.10.069

Fenwick, A., Molnar, G., & Frangos, P. (2024). The critical role of HRM in AI-driven digital transformation: A paradigm shift to enable firms to move from AI implementation to human-centric adoption. *Discover Artificial Intelligence, 4*(34), 34. Advance online publication. DOI: 10.1007/s44163-024-00125-4

Ferhataj, A., & Memaj, F. (2024). Challenges and opportunities of ai implementation in tourism: an ethical and technological perspective. *Studijos–verslas–visuomenė: dabartis ir ateities įžvalgos, 1*(IX), 217-231.

Ferrara, E. (2024). Fairness and bias in artificial intelligence: A brief survey of sources, impacts, and mitigation strategies. *Sci, 6*(1), 3. DOI: 10.3390/sci6010003

Ferrara, E. (2024). The butterfly effect in artificial intelligence systems: Implications for AI bias and fairness. *Machine Learning with Applications, 15*, 100525. DOI: 10.1016/j.mlwa.2024.100525

Filgueiras, F. (2024). Artificial intelligence and education governance. *Education, Citizenship and Social Justice, 19*(3), 349–361. DOI: 10.1177/17461979231160674

Fjeld, J., Achten, N., Hilligoss, H., Nagy, A., & Srikumar, M. (2020). Principled artificial intelligence: Mapping consensus in ethical and rights-based approaches to principles for AI. *Berkman Klein Center Research Publication*, (2020-1).

Floridi, L. (2019). Translating principles into practices of digital ethics: Five risks of being unethical. *Philosophy \& Technology, 32*(2), 185–193.

Floridi, L., & Cowls, J. (2019). A Unified Framework of Five Principles for AI in Society. *Harvard Data Science Review, 1*(1). Advance online publication. DOI: 10.1162/99608f92.8cd550d1

Floridi, L., & Cowls, J. (2019, July). A unified framework of five principles for AI in society. *Harvard Data Science Review*, *1*(1).

Florido-Benítez, L., & del Alcázar Martínez, B. (2024). How artificial intelligence (AI) is powering new tourism marketing and the future agenda for smart tourist destinations. *Electronics (Basel)*, *13*(21), 4151. DOI: 10.3390/electronics13214151

Fosch-Villaronga, E., & Poulsen, A. (2022). Diversity and inclusion in artificial intelligence. *Law and artificial intelligence: Regulating AI and applying AI in legal practice*, 109-134.

Freeman, R. E. (2010). *Strategic management: A stakeholder approach*. Cambridge University Press. DOI: 10.1017/CBO9781139192675

Frost, J. (2020). The economic forces driving fintech adoption across countries. *The technological revolution in financial services: how banks, fintechs, and customers win together, 838*(II), 70-89.

Fuster, A., Goldsmith-Pinkham, P., Ramadorai, T., & Walther, A. (2022). Predictably unequal? The effects of machine learning on credit markets. *The Journal of Finance*, *77*(1), 5–47. DOI: 10.1111/jofi.13090

Galiana, I., Gudino, L. C., & Gonzalez, P. M. (2024). Ethics and artificial intelligence. *Revista Clínica Española*, *224*(3), 178–186. DOI: 10.1016/j.rce.2024.01.007 PMID: 38355097

Garawit, S., & El Gnaoui, L. (2024). Principes éthiques internationaux de l'intelligence artificielle (IA): Un essai d'analyse dans l'entreprise au Maroc. International Journal of Strategic Management and Economic Studies (IJSMES).

Garg, R., Kiwelekar, A. W., Netak, L. D., & Bhate, S. S. (2021). *Potential Use-Cases of Natural Language Processing for a Logistics Organization*., DOI: 10.1007/978-3-030-68291-0_13

Gauckler, C., & Werner, M. H. (2024). Artificial intelligence: A challenge to scientific communication. *Klinische Monatsblätter für Augenheilkunde*, *241*(12), 1309–1321. DOI: 10.1055/a-2418-5238 PMID: 39637910

Gebru, T., Morgenstern, J., Vecchione, B., Vaughan, J. W., Wallach, H., Daumé, H.III, & Crawford, K. (2021). Datasheets for datasets. *Communications of the ACM*, *64*(12), 86–92. DOI: 10.1145/3458723

Georgiadis, G., & Poels, G. (2022). Towards a privacy impact assessment methodology to support the requirements of the general data protection regulation in a big data analytics context: A systematic literature review. *Computer Law & Security Review, 44*, 105640. DOI: 10.1016/j.clsr.2021.105640

Ge, S. S., Dongjie, Z., Li, D., Mao, X., & Nemati, A. (2020). Historical and futuristic perspectives of robotics. *Artificial Life and Robotics, 25*(3), 393–399. DOI: 10.1007/s10015-020-00613-7

Ghoubach, S., & El Amine, B. (2024). L'adoption de l'ia et du smart logistique à l'optimisation des coûts d'approvisionnement. Revue du contrôle, de la comptabilité et de l'audit, 8(4).

Ghouse, S. M., & Chaudhary, M. (2024). Artificial Intelligence (AI) for tourism Start-Ups. In *Advances in hospitality, tourism and the services industry (AHTSI) book series* (pp. 161–178). DOI: 10.4018/979-8-3693-2019-8.ch010

Giantini, G. (2023). The sophistry of the neutral tool. Weaponizing artificial intelligence and big data into threats toward social exclusion. *AI and Ethics, 3*(4), 1049–1061. DOI: 10.1007/s43681-023-00311-7

Gilbert, S., Baca-Motes, K., Quer, G., Topol, E. J., & Steinhubl, S. R. (2024). Citizen data sovereignty is key to wearables and wellness data reuse for the common good. *npj. Digital Medicine, 7*(1), 13. Advance online publication. DOI: 10.1038/s41746-024-01004-z PMID: 38225423

Giray, L., Jacob, J., & Gumalin, D. L. (2024). Strengths, weaknesses, opportunities, and threats of using ChatGPT in scientificrResearch. *International Journal Of Technology In Education, 7*(1, SI), 40–58. DOI: 10.46328/ijte.618

Girmay, A. (2018). FDI and gender equality: TNC responsibilities to address gender vulnerabilities in developing countries. *Gender and Development Review, 14*(2), 65–80.

Gniaka, K. J., & Khassim, D. (2024). *Afrique, la gestion des situations complexes au niveau individuel favorise-t-elle un processus d'apprentissage chez les acteurs.* Revue Congolaise de Gestion.

Gokhale, N., Gajjari, A., Kaye, R., & Kuder, D. (2019). AI leaders in financial services: common traits of frontrunners in the artificial intelligence race.

Golchha, C., & Nagariya, R. J. (2025). Exploring Consumer Choices and Shopping Patterns. In Cultural Sensitivity, Cross-Border Logistics, and E-Commerce in Global Marketing (pp. 181–214). IGI Global. DOI: 10.4018/979-8-3373-0528-8.ch009

Goldstein, I., Jiang, W., & Karolyi, G. A. (2019). To FinTech and beyond. *Review of Financial Studies*, *32*(5), 1647–1661. DOI: 10.1093/rfs/hhz025

Gomber, P., Kauffman, R. J., Parker, C., & Weber, B. W. (2018). On the fintech revolution: Interpreting the forces of innovation, disruption, and transformation in financial services. *Journal of Management Information Systems*, *35*(1), 220–265. DOI: 10.1080/07421222.2018.1440766

Gonçalves, S., Dias, P., & Correia, A.-P. (2020). *Nomophobia and lifestyle: Smartphone use and its relationship to psychopathologies. 2*, 100025. DOI: 10.1016/j.chbr.2020.100025

González-Esteban, E. (2023). The ETHNA system and support tools. In *Ethics and Responsible Research and Innovation in Practice: The ETHNA System Project* (pp. 19–37). Springer. DOI: 10.1007/978-3-031-33177-0_2

Gonzalez-Esteban, E., & Calvo, P. (2022). Ethically governing artificial intelligence in the field of scientific research and innovation. *Heliyon*, *8*(2), e08946. Advance online publication. DOI: 10.1016/j.heliyon.2022.e08946 PMID: 35243068

Gopalakrishnan, K. V. (2009). James Watt: Father of steam power. *Resonance*, *14*(6), 522–529. DOI: 10.1007/s12045-009-0062-0

Gorichanaz, T. (2023). Accused: How students respond to allegations of using ChatGPT on assessments. *Learning*, *9*(2), 183–196. DOI: 10.1080/23735082.2023.2254787

Greenhalgh, T., Wherton, J., Papoutsi, C., Lynch, J., Hughes, G., A'Court, C., Hinder, S., Fahy, N., Procter, R., & Shaw, S. (2017). Beyond adoption: A new framework for theorizing and evaluating nonadoption, abandonment, and challenges to the scale-up, spread, and sustainability of health and care technologies. *Journal of Medical Internet Research*, *19*(11), e367. DOI: 10.2196/jmir.8775 PMID: 29092808

Gretzel, U., Sigala, M., Xiang, Z., & Koo, C. (2015). Smart tourism: Foundations and developments. *Electronic Markets*, *25*(3), 179–188. DOI: 10.1007/s12525-015-0196-8

Grevisse, Y. R. & Monique-Stéphanie, K. M. (). … De La Performance Et Gestion Des Risques Dans Les Entreprises Congolaises: Prise De Décision Data-Driven, Analyse Prédictive, Outils Technologiques

Grodzinsky, F. S., Miller, K. W., & Wolf, M. J. (2008). The ethics of designing artificial agents. *Ethics and Information Technology*, *10*(2), 115–121. DOI: 10.1007/s10676-008-9163-9

Guidotti, R., Monreale, A., Ruggieri, S., Turini, F., Giannotti, F., & Pedreschi, D. (2018). A survey of methods for explaining black box models. *ACM Computing Surveys, 51*(5), 1–42. DOI: 10.1145/3236009

Hacker, P. (2023). The European AI liability directives–Critique of a half-hearted approach and lessons for the future. *Computer Law & Security Review, 51*, 105871.

Haddad, C., & Hornuf, L. (2019). The emergence of the global fintech market: Economic and technological determinants. *Small Business Economics, 53*(1), 81–105. DOI: 10.1007/s11187-018-9991-x

Haddar, M. (2010). Methodology of doctoral research in economics (pp. 1–103). AUFRANCOPHONIE.

Hadzovic, S., Becirspahic, L., & Mrdovic, S. (2024). It is time for artificial intelligence governance. *Internet of Things : Engineering Cyber Physical Human Systems, 27*, 101292. Advance online publication. DOI: 10.1016/j.iot.2024.101292

Hagemann, R. P. (2012). Fiscal Consolidation: Part 6. What are the best instruments for fiscal consolidation? *OECD Economics Department Working Papers* No. 937 https://www.oecd.org/content/dam/oecd/en/publications/reports/2012/01/fiscal-consolidation-part-6-what-are-the-best-policy-instruments-for-fiscal-consolidation_g17a20c0/5k9h28kd17xn-en.pdf

Hamarat, H., Sahin, H., Koç Apuhan, A., & İnan, R. (2024). The dark side of artificial intelligence: Threats to tourism workers. *Worldwide Hospitality and Tourism Themes, 16*(2), 127–137. DOI: 10.1108/WHATT-03-2024-0061

HAMZAOUI, S., MEDOUARI, M., Setta, B., & Lahrach, R. (2025). Crowdfunding social et IA: Un duo gagnant pour l'entrepreneuriat social au Maroc. *Revue Française d'Economie et de Gestion, 6*(1).

Hanel, P. (2006). Intellectual property rights business management practices: A survey of the literature. *Technovation, 26*(8), 895–931. DOI: 10.1016/j.technovation.2005.12.001

Hanisch, M., Goldsby, C. M., Fabian, N. E., & Oehmichen, J. (2023). Digital governance: A conceptual framework and research agenda. *Journal of Business Research, 162*, 113777. Advance online publication. DOI: 10.1016/j.jbusres.2023.113777

Hanna, M. G., Pantanowitz, L., Jackson, B., Palmer, O., Visweswaran, S., Pantanowitz, J., Deebajah, M., & Rashidi, H. H. (2025). Ethical and Bias Considerations in Artificial Intelligence/Machine Learning. *Modern Pathology, 38*(3), 100686. DOI: 10.1016/j.modpat.2024.100686 PMID: 39694331

Hardt, M., Price, E., & Srebro, N. (2016). Equality of opportunity in supervised learning. *Advances in Neural Information Processing Systems*, *29*, 3315–3323.

Hassan, A. E., Lin, D., Rajbahadur, G. K., Gallaba, K., Cogo, F. R., Chen, B., Zhang, H., Thangarajah, K., Oliva, G. A., Lin, J., Abdullah, W. M., & Jiang, Z. M. (2024). *Rethinking Software Engineering in the Era of Foundation Models: A Curated Catalogue of Challenges in the Development of Trustworthy FMware*. DOI: 10.1145/3663529.3663849

Hassani, H., Huang, X., & Silva, E. (2021). Digital transformation and big data analytics: A review of the current landscape and research challenges. *Industrial Marketing Management*, *97*, 123–135.

Hassija, V., Chamola, V., Mahapatra, A., Singal, A., Goel, D., Huang, K., & Hussain, A. (2024). Interpreting black-box models: A review on explainable artificial intelligence. *Cognitive Computation*, *16*(1), 45–74. DOI: 10.1007/s12559-023-10179-8

Heinrichs, B. (2022). Discrimination in the age of artificial intelligence. *AI & Society*, *37*(1), 143–154. DOI: 10.1007/s00146-021-01192-2

Hendershott, T., Jones, C. M., & Menkveld, A. J. (2011). Does algorithmic trading improve liquidity? *The Journal of Finance*, *66*(1), 1–33. DOI: 10.1111/j.1540-6261.2010.01624.x

Hendriks, G. (2017). The sustainable development effects of investment by emerging-market multinationals: Shaping beneficial outcomes for home and host country. *Emerging Markets Journal, 9*(1), 51–67.

Henkel, C. (2021). The impact of artificial intelligence on insolvency law and practice. In *Research Handbook on Corporate Restructuring* (pp. 377–399). Edward Elgar Publishing.

Herrera, F. (2024). Responsible artificial intelligence Systems: From trustworthiness to governance. *2024 Design, Automation & Test*. In *Europe Conference & Exhibition*. Date., DOI: 10.23919/DATE58400.2024.10546553

Hilb, M. (2020). Toward artificial governance? The role of artificial intelligence in shaping the future of corporate governance. *The Journal of Management and Governance*, *24*(4), 851–870. DOI: 10.1007/s10997-020-09519-9

Hilliard, A., Guenole, N., & Leutner, F. (2022). Robots are judging me: Perceived fairness of algorithmic recruitment tools. *Frontiers in Psychology*, *13*, 940456. DOI: 10.3389/fpsyg.2022.940456 PMID: 35959005

Hofmann, J., Katzenbach, C., & Gollatz, K. (2022). AI ethics made in Europe? Critical discourse analysis of the AI HLEG's guidelines and the European Commission's white paper. *AI & Society, 37*, 611–623.

Holly, L., Thom, S., Elzemety, M., Murage, B., Mathieson, K., & Petralanda, M. I. I. (2023). Strengthening health data governance: New equity and rights-based principles. *International Journal of Health Governance, 28*(3), 287–301. DOI: 10.1108/IJHG-11-2022-0104

Holstein, K., Wortman Vaughan, J., Daumé, H.III, Dudik, M., & Wallach, H. (2020). Improving fairness in machine learning systems: What do industry practitioners need? *Proceedings of the 2020 CHI Conference on Human Factors in Computing Systems*, 1–16.

Hood, C. (1991). A public management for all seasons? *Public Administration, 69*(1), 3–19. https://eclass-b.uoa.gr/modules/document/file.php/PSPA108/4NMP%20all%20seasonsfulltext.pdf. DOI: 10.1111/j.1467-9299.1991.tb00779.x

Hooker, S. (2021). Moving beyond "algorithmic bias is a data problem". *Patterns (New York, N.Y.), 2*(4), 100241. DOI: 10.1016/j.patter.2021.100241 PMID: 33982031

Horne, J. (2023). Towards a Solution: The Responses (pp. 323–395). DOI: 10.1007/978-3-031-25444-4_12

Hosseini, H., Cakmak, M., & Ghasemzadeh, H. (2021). On the importance of user education in addressing algorithmic fairness. [TiiS]. *ACM Transactions on Interactive Intelligent Systems, 11*(4), 1–29.

Huang, A., Chao, Y., de la Mora Velasco, E., Bilgihan, A., & Wei, W. (2022). When artificial intelligence meets the hospitality and tourism industry: An assessment framework to inform theory and management. *Journal of Hospitality and Tourism Insights, 5*(5), 1080–1100. DOI: 10.1108/JHTI-01-2021-0021

Huemer, M. (2013). The Traditional Social Contract Theory. In *The Problem of Political Authority*. Palgrave Macmillan., DOI: 10.1057/9781137281661_2

Humerick, M. (2017). Taking AI personally: How the EU must learn to balance the interests of personal data privacy & artificial intelligence. *Santa Clara High-Technology Law Journal, 34*, 393.

Hurley, M., & Adebayo, J. (2017). Credit scoring in the era of big data. *Yale Journal of Law and Technology, 18*(1), 148–216.

Hutter, M. (2012). Can Intelligence Explode. *Journal of Consciousness Studies, 19*(1-2), 143–166. http://hutter1.net/publ/singularity.pdf

Idaomar, C., Idaomar, D., Hannaoui, M., & Chafik, K. (2024). Applications of artificial intelligence in Morocco's healthcare sector: A springboard to medical excellence. *Journal of Computer and Communications, 12*(9), 63–77. DOI: 10.4236/jcc.2024.129004

IEEE. (2020). Ethically Aligned Design, *First Edition: A Vision for Prioritizing Human Well-being with Autonomous and Intelligent Systems.* IEEE Standards Association.

Ilieva, K. Artificial Intelligence Usage in EU Tax Administrations (2025), *IBFD.*1-3. https://www.ibfd.org/sites/default/files/2025-01/artificial-intelligence-usage-in-eu-tax-administrationsv7.pdf

Imane, B. & Elk, Y. (2024). E-gouvernance et transformation numérique au Maroc: défis et perspectives pour des institutions et des services publics innovants. International Journal of Economic Studies.

Imane, R. A. B. H. I., & Samira, Z. I. A. N. I., & BUDAS, I. (2025). Exploiter l'intelligence artificielle pour renforcer la gestion de la qualité hospitalière au Maroc: Vers un modèle stratégique intégré. *Revue Dossiers De Recherches en Économie Et Management Des Organisations, 10*(1), 1–18.

IMF. (2019). Tax Expenditure Reporting and Its Use in Fiscal Management: A Guide for Developing Economies .2019(002).1-19. .DOI: 10.5089/9781498303217.061

Indaryanto, A., Harijadi, B. D., & Sinaga, E. (2023). The growing use and impact of artificial intelligence technologies in the tourism industry. *Sustainable Engineering and Innovation ISSN 2712-0562, 5*(2), 189–204. DOI: 10.37868/sei.v5i2.id238

Innovation Center E-health. Université Mohammed V de Rabat. (2022). *White paper on e-health in Morocco: Realities, challenges and development levers* [White paper]. https://smcmaroc.org/wp-content/uploads/2023/01/White-paper-on-e-health-in-Morocco.pdf

International Public Sector Accounting Standards Board (IPSASB). (2021). Handbook of international public sector accounting pronouncements. International Federation of Accountants. https://www.ipsasb.org/publications/2021-handbook-international-public-sector-accounting-pronouncements

Islam, S. R., Eberle, W., Bundy, S., & Ghafoor, S. K. (2019). Infusing domain knowledge in ai-based" black box" models for better explainability with application in bankruptcy prediction. arXiv preprint arXiv:1905.11474.

Ivanov, D., Dolgui, A., & Sokolov, B. (2019). The impact of digital technology and Industry 4.0 on the ripple effect and supply chain risk analytics. *International Journal of Production Research, 57*(3), 829–846. DOI: 10.1080/00207543.2018.1488086

Ivanov, D., Dolgui, A., Sokolov, B., Werner, F., & Ivanova, M. (2021). A digital supply chain twin for managing the disruption risks and resilience in the era of Industry 4.0. *Production Planning and Control, 32*(9), 761–777. DOI: 10.1080/09537287.2020.1768450

Ivanov, S., & Umbrello, S. (2021). The ethics of artificial intelligence and robotisation in tourism and hospitality conceptual framework and research agenda. *Journal of Smart Tourism, 1*(4), 9–18. DOI: 10.52255/smarttourism.2021.1.4.3

Ivanov, S., & Webster, C. (2019). *Robots, artificial intelligence, and service automation in travel, tourism and hospitality*. Emerald Publishing Limited. DOI: 10.1108/9781787566873

Jabeur, B., & Nammouri, H. (2025). *Comment l'ia, les Big data et l'informatique quantique redessinent les marchés de l'énergie*. Management & Datascience.

Jabir, H., Lagtati, K., & Pohe-Tokpa, D. (2024). Ethical and legal regulation of using artificial intelligence in Morocco. *Journal of Digital Technologies and Law, 2*(2), 450–472. DOI: 10.21202/jdtl.2024.23

Jacob, S. & Brousseau, S. (2025). Quand l'algorithme décide: L'État, l'ia et nous.

Jadan-Guerrero, J., Acosta-Vargas, P., & Gutierrez-De Gracia, N. E. (2024). Enhancing scientific research and paper writing processes by integrating artificial intelligence tools. In Stephanidis, C., Antona, M., Ntoa, S., & Salvendy, G. (Eds.), *HCI International 2024 Posters, PT VII, HCII 2024* (Vol. 2120, pp. 64–74)., DOI: 10.1007/978-3-031-62110-9_7

Jafari, J. (2019). The need for a new tourism ethics. *Journal of Tourism Futures, 5*(3), 195–198.

Jallal, M., Serhier, Z., Berrami, H., & Bennani Othmani, M. (2023). Current state and prospects of telemedicine in Morocco: Analysis of challenges, initiatives, and regulatory framework. *Cureus, 15*(12), e50963. DOI: 10.7759/cureus.50963 PMID: 38259391

Janssen, M., Brous, P., Estevez, E., Barbosa, L. S., & Janowski, T. (2020). Data governance: Organizing data for trustworthy artificial intelligence. *Government Information Quarterly, 37*(3), 101493. Advance online publication. DOI: 10.1016/j.giq.2020.101493

Jayathissa, P., & Hewapathirana, R. (2023). Enhancing interoperability among health information systems in low- and middle-income countries: A review of challenges and strategies. *arXiv preprint arXiv:2309.12326*. https://arxiv.org/abs/2309.12326

Jensen, M. C., & Meckling, W. H. (1976). Theory of the firm: Managerial behavior, agency costs and ownership structure. *Journal of Financial Economics, 3*(4), 305–360. DOI: 10.1016/0304-405X(76)90026-X

JIANG, K., & LI, Q. (2020). On Piaget's epistemological methodology and its contemporary significance. *Acta Psychologica Sinica, 52*(8), 1017–1030.

Jobin, A., Ienca, M., & Vayena, E. (2019). The global landscape of AI ethics guidelines. *Nature Machine Intelligence, 1*(9), 389–399. DOI: 10.1038/s42256-019-0088-2

Johansen, S. (1988). Statistical analysis of cointegration vectors. *Journal of Economic Dynamics & Control, 12*(2-3), 231–254. DOI: 10.1016/0165-1889(88)90041-3

Jørgensen, N., Gunasekaran, S. S., & Ma, Z. G. (2025). Impact of eu laws on ai adoption in smart grids: A review of regulatory barriers, technological challenges, and stakeholder benefits. *Energies, 18*(12), 3002. DOI: 10.3390/en18123002

Jui, T. D., & Rivas, P. (2024). Fairness issues, current approaches, and challenges in machine learning models. *International Journal of Machine Learning and Cybernetics, 15*(8), 3095–3125. DOI: 10.1007/s13042-023-02083-2

Jumper, J. M., Evans, R. O., Pritzel, A., Green, T., Figurnov, M., Ronneberger, O., Tunyasuvunakool, K., Bates, R., Žídek, A., Potapenko, A., Bridgland, A., Meyer, C., Kohl, S. A. A., Ballard, A. J., Cowie, A., Romera-Paredes, B., Nikolov, S., Jain, R. D., Adler, J., & Hassabis, D. (2021). Highly accurate protein structure prediction with AlphaFold. *Nature, 596*(7873), 583–589. DOI: 10.1038/s41586-021-03819-2 PMID: 34265844

Jurion, B. (2004). *Tax expenditure distortions and effects on growth. External Research Report.* University of Liege.

K, S. A., Gururaj, V., Shankar, S. P., S., S., & Bharadwaj, A. (2022). Electrocardiogram based cardiovascular disease detection with Ensemble Learning Classifier. *2022 4th International Conference on Circuits, Control, Communication and Computing (I4C)*, 48-53. DOI: 10.1109/I4C57141.2022.10057719

Kaplan, R. S., & Norton, D. P. (1996). *The balanced scorecard: Translating strategy into action.* Harvard Business School Press.

Kassa, B. Y., & Worku, E. K. (2025). The impact of artificial intelligence on organizational performance: The mediating role of employee productivity. *Journal of Open Innovation, 11*(1), 100474. DOI: 10.1016/j.joitmc.2025.100474

Kaushik, A., Barcellona, C., Mandyam, N. K., Tan, S. Y., & Tromp, J. (2025). Challenges and opportunities for data sharing related to artificial intelligence tools in health care in low- and middle-income countries: Systematic review and case study from Thailand. *Journal of Medical Internet Research*, *27*, e58338. DOI: 10.2196/58338 PMID: 39903508

Kern, C., Gerdon, F., Bach, R. L., Keusch, F., & Kreuter, F. (2022). Humans versus machines: Who is perceived to decide fairer? Experimental evidence on attitudes toward automated decision-making. *Patterns (New York, N.Y.)*, *3*(10), 100591. DOI: 10.1016/j.patter.2022.100591 PMID: 36277823

Khalil, M., & Er, E. (2023, June). Will chatgpt g et you caught? rethinking of plagiarism detection. In *International conference on human-computer interaction* (pp. 475-487). Cham: Springer Nature Switzerland.

Khalil, L., Serhier, Z., & Bennani Othmani, M. (2024). Exploring patients' acceptance of mental health e-services in Morocco: A quantitative approach. *Cureus*, *16*(12), e76143. DOI: 10.7759/cureus.76143 PMID: 39835084

Khan, T., & Emon, M. M. H. (2025a). Preparing for the Next Disruption: Lessons Learned From Past Crises and Building a Playbook for Future Supply Chain Shocks. In Supply Chain Disruptions and Impact on Global Inflation (pp. 253–286). IGI Global Scientific Publishing. DOI: 10.4018/979-8-3373-1982-7.ch010

Khan, T., & Hasan Emon, M. M. (2024). Determinants of AI Image Generator Adoption Among Marketing Agencies: The Mediating Effects of Perceived Usefulness. 2024 IEEE 3rd International Conference on Robotics, Automation, Artificial-Intelligence and Internet-of-Things (RAAICON), 177–182. DOI: 10.1109/RAAICON64172.2024.10928548

Khan, T., Emon, M. M. H., & Nath, A. (2024). Quantifying the Effects of AI-Driven Inventory Management on Operational Efficiency in Online Retail. 2024 27th International Conference on Computer and Information Technology (ICCIT), 2092–2097. DOI: 10.1109/ICCIT64611.2024.11021996

Khan, T., Emon, M. M. H., & Rahman, S. (2024). Marketing Strategy Innovation via AI Adoption: A Study on Bangladeshi SMEs in the Context of Industry 5.0. 2024 6th International Conference on Sustainable Technologies for Industry 5.0 (STI), 1–6. DOI: 10.1109/STI64222.2024.10951050

Khan, T., & Emon, M. M. H. (2025b). The role of digital supply chain practices in enhancing firm performance: Insights from the manufacturing sector of Bangladesh. *Brazilian Journal of Operations & Production Management*, *22*(2), 2493. DOI: 10.14488/BJOPM.2493.2025

Khlaif, Z. N., Mousa, A., Hattab, M. K., Itmazi, J., Hassan, A. A., Sanmugam, M., & Ayyoub, A. (2023). The Potential and Concerns of Using AI in Scientific Research: ChatGPT Performance Evaluation. *JMIR Medical Education, 9*, e47049. Advance online publication. DOI: 10.2196/47049 PMID: 37707884

Kickbusch, I., Piselli, D., Agrawal, A., Balicer, R., Banner, O., Adelhardt, M., Capobianco, E., Fabian, C., Gill, A. S., Lupton, D., Medhora, R. P., Ndili, N., Ryś, A., Sambuli, N., Settle, D., Swaminathan, S., Vega Morales, J., Wolpert, M., Wyckoff, A. W., & Xue, L. (2021). The Lancet and Financial Times Commission on governing health futures 2030: Growing up in a digital world. *The Lancet, 398*(10312), 1727–1776. DOI: 10.1016/S0140-6736(21)01824-9

Kim, G. J., Namkoong, K., Kreps, G. L., Huh, J., Park, E., & Rimal, R. (2025). Developing the digital health communication maturity model. *Journal of Medical Internet Research, 27*, e68344. DOI: 10.2196/68344 PMID: 40228239

Kim, H., So, K. K. F., Shin, S., & Li, J. (2024). Artificial intelligence in hospitality and tourism: Insights from industry practices, research literature, and expert opinions. *Journal of Hospitality & Tourism Research (Washington, D.C.), 49*(2), 366–385. DOI: 10.1177/10963480241229235

Kirilenko, A., Kyle, A. S., Samadi, M., & Tuzun, T. (2017). The flash crash: High-frequency trading in an electronic market. *The Journal of Finance, 72*(3), 967–998. DOI: 10.1111/jofi.12498

Kleinberg, J., Lakkaraju, H., Leskovec, J., Ludwig, J., & Mullainathan, S. (2018). Human decisions and machine predictions. *The Quarterly Journal of Economics, 133*(1), 237–293. DOI: 10.1093/qje/qjx032 PMID: 29755141

Köchling, A., & Wehner, M. C. (2020). Discriminated by an algorithm: A systematic review of discrimination and fairness by algorithmic decision-making in the context of HR recruitment and HR development. *Business Research, 13*(3), 795–848. DOI: 10.1007/s40685-020-00134-w

Ko, J. S., El-Toukhy, S., Quintero, S. M., Wilkerson, M. J., Nápoles, A. M., Stewart, A. L., & Strassle, P. D. (2023). Disparities in telehealth access, not willingness to use services, likely explain rural telehealth disparities. *The Journal of Rural Health, 39*(3), 617–624. DOI: 10.1111/jrh.12759 PMID: 37042413

Kougkoulos, I., Cakir, M. S., Kunz, N., Boyd, D. S., Trautrims, A., Hatzinikolaou, K., & Gold, S. (2021). A Multi-Method Approach to Prioritize Locations of Labor Exploitation for Ground-Based Interventions. *Production and Operations Management, 30*(12), 4396–4411. DOI: 10.1111/poms.13496

Krauss, M. B., & Johnson, H. G. (1972). The Theory of Tax Incidence: A Diagrammatic Analysis. *Economica, 39*(156), 357–382. DOI: 10.2307/2552880

Kshetri, N. (2024). Economics of Artificial intelligence governance. *Computer, 57*(4), 113–118. DOI: 10.1109/MC.2024.3357951

Kutscher, S. (2025). The EU AI Act: Law of Unintended Consequences? *Technology and Regulation, 2025*, 316–334. DOI: 10.71265/krne7205

Kuznets, S. (1955). Economic Growth and Income Inequality. In *AMERICAN ECONOMIC REVIEW* (Vol. 45, pp. 1–28). DOI: 10.4324/9780429311208-4

L'MGHAIFRI, H., & ZAMMAR, R. (2025). Revue de littérature sur l'intégration de l'Intelligence Artificielle (IA) dans les pratiques du Contrôle de Gestion: Opportunités et Défis. Revue de l'Intelligence Artificielle et du développement territorial durable, 2(01).

Labadie, C., & Legner, C. (2023). Building data management capabilities to address data protection regulations: Learnings from EU-GDPR. *Journal of Information Technology, 38*(1), 16–44. DOI: 10.1177/02683962221141456

Lahlimi, Y. K., & Messaoudi, A. (2024). L'utilité du contrôle de gestion socio-économique dans le pilotage de la performance socio-économique des organisations publiques: cas des écoles primaires publiques en milieu rural (Région Marrakech-Safi). African Scientific Journal, 3(23), 135–176. https://hal.science/hal-04538078/document

Lamarre, E., Smaje, K., & Zemmel, R. (2023). *Rewired: the McKinsey Guide to Outcompeting in the Age of Digital and AI.* John Wiley & Sons.

Larsson, S. (2019). The socio-legal relevance of artificial intelligence. *Droit Social, 103*(3), 573–593.

Laukka, E., Pölkki, T., & Kanste, O. (2022). Leadership in the context of digital health services: A concept analysis. *Journal of Nursing Management, 30*(7), 2763–2780. DOI: 10.1111/jonm.13763 PMID: 35942802

Law, R., Ye, H., & Lei, S. S. I. (2025). Ethical artificial intelligence (AI): Principles and practices. *International Journal of Contemporary Hospitality Management, 37*(1), 279–295. DOI: 10.1108/IJCHM-04-2024-0482

Leal, P. H., Caetano, R. V., & Marques, A. C. (2021). Is the relocation of polluting industries prompted by FDI flow and stock, globalization, corruption and regulation? *Ecological Economics, 183*, 106–118.

Leavy, S., O'Sullivan, B., & Siapera, E. (2020). Data, power and bias in artificial intelligence. arXiv preprint arXiv:2008.07341.

Lee, K. F. (2016). Automation, Computerisation and Future Employment In Singapore. *Social Science Research Network*. https://doi.org/DOI: 10.2139/SSRN.2995283

Léger-Rousseau, M. (2024). Les performances des grands modèles de la famille des transformateurs génératifs préentraînés en soutien aux services de référence virtuelle automatisés en…. *Documentation et Bibliothèques*.

Leiner, B. M., Cerf, V. G., Clark, D. D., Kahn, R. E., Kleinrock, L., Lynch, D. C., Postel, J., Roberts, L. G., & Wolff, S. (2009). *A brief history of the internet. 39*(5), 22–31. DOI: 10.1145/1629607.1629613

LEONARD, T. (2024). La compréhension et la prédiction des préférences des clients dans le commerce en détail grâce à l'ia: Une revue de littérature. Int. J. Econ. Stud. Manag.

Lepri, B., Oliver, N., & Pentland, A. (2021). Ethical machines: The human-centric use of artificial intelligence. *iScience*, *24*(3), 102249. DOI: 10.1016/j.isci.2021.102249 PMID: 33763636

Leventhal, G. S. (1980). What should be done with equity theory? In Gergen, K. J., Greenberg, M. S., & Willis, R. H. (Eds.), *Social Exchange* (pp. 27–55). Springer US. DOI: 10.1007/978-1-4613-3087-5_2

Li, D., & Shen, S. (2022). Social environment and healthy investment behavior: Joint influence of culture and institution on China. *China Economic Journal, 15*(2), 203–220.

Liehr Storm, K. I., Reiss, L. K., Guenther, E. A., Clar-Novak, M., & Muhr, S. L. (2023). Unconscious bias in the HRM literature: Towards a critical-reflexive approach. *Human Resource Management Review*, *33*(3), 100969. DOI: 10.1016/j.hrmr.2023.100969

Lipton, Z. C. (2018). The mythos of model interpretability: In machine learning, the concept of interpretability is both important and slippery. *Communications of the ACM*, *61*(10), 36–43. DOI: 10.1145/3233231

Lobacheva, A., & Kashtanova, E. (2022). Social discrimination in the epoch of artificial intelligence. *Wisdom*, *2*(1, SI), 97+. DOI: 10.24234/wisdom.v2i1.767

LOEFFEN, A. (2021). L'impact de l'intelligence artificielle sur la gestion des ressources humaines: Cas des entreprises marocaines. *The Interdisciplinary Journal of Human and Social Studies-The Interdisciplinary Journal of Human and Social Studies, 1*(1), 85–99.

Long, D., & Magerko, B. (2020, April). What is AI literacy? Competencies and design considerations. In *Proceedings of the 2020 CHI conference on human factors in computing systems* (pp. 1-16). DOI: 10.1145/3313831.3376727

Lum, K., & Isaac, W. (2016). To predict and serve? *Significance, 13*(5), 14–19. DOI: 10.1111/j.1740-9713.2016.00960.x

Lundberg, S. M., & Lee, S. I. (2017). A unified approach to interpreting model predictions. *Advances in Neural Information Processing Systems, 30*, 4765–4774.

Maailif, S., & Lkhoyaali, B. (2023). Le contrôle de gestion dans le secteur public: quelles spécificités? Revue du contrôle, de la comptabilité et de l'audit, 7(4), 129–150. https://www.revuecca.com/index.php/home/article/download/981/865

MacFarlane, J. M. (2020). The Techno-Centred Imagination. In *Transhumanism as a New Social Movement* (pp. 205–233). Springer International Publishing., DOI: 10.1007/978-3-030-40090-3_8

Mackey, J. W. (2024). *The Industrial Revolution. The Modernization of the Western World*. Routledge., DOI: 10.4324/9781003467328-12

Mahdaoui, A., Seffar, M., El Kettani, A., Belayachi, J., & Abouqal, R. (2023). Morocco's healthcare system: Achievements, challenges, and perspectives. *Cureus, 15*(7), e42706. DOI: 10.7759/cureus.41143 PMID: 37519608

Mandl, K. D., Gottlieb, D., & Mandel, J. C. (2024). Integration of AI in healthcare requires an interoperable digital data ecosystem. *Nature Medicine, 30*(3), 631–634. DOI: 10.1038/s41591-023-02783-w PMID: 38291298

Mangla, S. K., Kazançoğlu, Y., Yıldızbaşı, A., Öztürk, C., & Çalık, A. (2022). A conceptual framework for blockchain-based sustainable supply chain and evaluating implementation barriers: A case of the tea supply chain. *Business Strategy and the Environment, 31*(8), 3693–3716. DOI: 10.1002/bse.3027

Mankiw, N. G. (2007). *Principles of microeconomics* (p. 0). Mason, OH: Thomson South-Western.

Mantelero, A. (2018). AI and Big Data: A Blueprint for a Human Rights, Social and Ethical Impact Assessment. *Computer Law & Security Review, 34*(4), 754–772. DOI: 10.1016/j.clsr.2018.05.017

Manyika, J., Chui, M., Brown, B., Bughin, J., Dobbs, R., Roxburgh, C., & Hung Byers, A. (2011). Big data: The next frontier for innovation, competition, and productivity.

Marelli, L., Stevens, M., Sharon, T., Van Hoyweghen, I., Boeckhout, M., Colussi, I., & Shabani, M. (2023). The European Health Data Space: Too big to succeed? *Health Policy (Amsterdam)*, *135*, 104861. DOI: 10.1016/j.healthpol.2023.104861 PMID: 37399677

Mariani, M., & Wamba, S. F. (2020). Exploring how consumer information systems produce information and effect consumer behavior: The role of artificial intelligence and big data. *Information & Management*, *57*(8), 103370.

Marini, J. L., Patissier, C., & Toublanc, J. (2024). Booster sa PME grâce à l'intelligence artificielle: Choisir et intégrer des solutions IA à vraie valeur ajoutée.

Marozva, G., & Makoni, L. (2018). Foreign direct investment, infrastructure development and economic growth in African economies. *African Journal of Economic Development, 10*(1), 33–52.

Marshall, D. T., & Naff, D. B. (2024). The ethics of using artificial intelligence in qualitative research. *Journal of Empirical Research on Human Research Ethics; JERHRE*, *19*(3), 92–102. DOI: 10.1177/15562646241262659 PMID: 38881315

Martinez, A. L. (2025). Artificial Intelligence in Tax Administration: Enhancing Compliance, Transparency, and Ethical Governance. *SSRN*. 1-18. DOI: 10.2139/ssrn.5285760

Martin, K. (2019). Ethical implications and accountability of algorithms. *Journal of Business Ethics*, *160*(4), 835–850. DOI: 10.1007/s10551-018-3921-3

Martin, K., Shilton, K., & Smith, J. (2019). Business and the ethical implications of technology: Introduction to the symposium. *Journal of Business Ethics*, *160*(2), 307–317. DOI: 10.1007/s10551-019-04213-9

Ma, S. (2024). Enhancing tourists' satisfaction: Leveraging artificial intelligence in the tourism sector. *Pacific International Journal*, *7*(3), 89–98. DOI: 10.55014/pij.v7i3.624

Masiya, M., Hall, S., Murray, S., Etter-Phoya, R., Hannah, E., & O'Hare, B. (2024). Tax expenditures and progress to the Sustainable Development Goals. *Sustainable Development (Bradford)*, *32*(6), 6144–6162. DOI: 10.1002/sd.3016

Maskus, K. E. (1998). The role of intellectual property rights in encouraging foreign direct investment and technology transfer. *Duke J. Comp. & Int'l L.*, *9*, 109.

McCarthy, J. J., Minsky, M., Rochester, N., & Shannon, C. E. (2006). A Proposal for the Dartmouth Summer Research Project on Artificial Intelligence. *AI Magazine*, *27*(4), 12. DOI: 10.1609/AIMAG.V27I4.1904

McCarthy, J., Minsky, M. L., Rochester, N., & Shannon, C. E. (2006). A proposal for the dartmouth summer research project on artificial intelligence, august 31, 1955. *AI Magazine*, *27*(4), 12–12.

McGrath, P., McCarthy, L., Marshall, D., & Rehme, J. (2021). Tools and Technologies of Transparency in Sustainable Global Supply Chains. *California Management Review*, *64*(1), 67–89. DOI: 10.1177/00081256211045993

Mehrabi, N., Morstatter, F., Saxena, N., Lerman, K., & Galstyan, A. (2021). A survey on bias and fairness in machine learning. [REMOVED HYPERLINK FIELD]. *ACM Computing Surveys*, *54*(6), 1–35. DOI: 10.1145/3457607

Meijer, A., & Bolívar, M. P. R. (2016). Governing the smart city: A review of the literature on smart urban governance. *International Review of Administrative Sciences*, *82*(2), 392–408. DOI: 10.1177/0020852314564308

Meng, Z., Ao, B., Wang, S., Niu, T., Chen, Y., Ma, X., & Huang, Y. (2024). Relationships between screen time and childhood attention deficit hyperactivity disorder: A Mendelian randomization study. *Frontiers in Psychiatry*, *15*, 1441191. Advance online publication. DOI: 10.3389/fpsyt.2024.1441191 PMID: 39376970

Menissier, T. (2023). The four ethics of artificial intelligence. *Revue D'anthropologie Des Connaissances*, *17*(2).

Mention, A. L. (2019). The future of fintech. *Research Technology Management*, *62*(4), 59–63. DOI: 10.1080/08956308.2019.1613123

Mevorach, I. (2015). Beyond the search for certainty: Addressing the cross-border resolution gap. *Brook. J. Corp. Fin. & Com. L.*, *10*, 183.

Meyer, J. W., & Rowan, B. (1977). Institutionalized organizations: Formal structure as myth and ceremony. *American Journal of Sociology*, *83*(2), 340–363. DOI: 10.1086/226550

Mihai Gabriel, C. R. I. S. T. I. A. N., & Cosmin, T. I. L. E. A. G. Ă. (2024). Challenges and perspectives of AI in sustainable tourism. *Management of Sustainable Development, 16*(2).

Mihan, A., Pandey, A., & Van Spall, H. G. C. (2024). Mitigating the risk of artificial intelligence bias in cardiovascular care. *The Lancet. Digital Health*, *6*(10), e749–e754. DOI: 10.1016/S2589-7500(24)00155-9 PMID: 39214762

Mikalef, P., Krogstie, J., Pappas, I. O., & Pavlou, P. (2020). Exploring the Relationship between Big Data Analytics Capability and Competitive Performance: The Mediating Roles of Dynamic and Operational Capabilities. *Information & Management, 57*(2), 103169. Advance online publication. DOI: 10.1016/j.im.2019.05.004

Miller, F. A., Katz, J. H., & Gans, R. (2018). The OD is imperative to add inclusion to artificial intelligence algorithms. *OD Practitioner, 50*(1), 8.

Miller, G. J. (2022). Stakeholder roles in artificial intelligence projects. *Project Leadership and Society, 3*, 100068. DOI: 10.1016/j.plas.2022.100068

Milot, J. P. (2022). La comptabilité nationale: l'information économique au service du pilotage des finances publiques. Revue française d'administration publique, 183(3), 723–734. https://doi.org/DOI: 10.3917/rfap.183.0082

Ministry of Health and Social Protection of the Kingdom of Morocco (MSPS). (2020). Wiqaytna mobile application for the fight against. *COVID, 19*, •••. https://www.sante.gov.ma

Ministry of Health and Social Protection of the Kingdom of Morocco (MSPS). (2023). Santé en chiffres. https://www.sante.gov.ma

Minsky, M., & Papert, S. (1969). Perceptron: an introduction to computational geometry. *The MIT Press, Cambridge, expanded edition, 19*(88), 2. DOI: 10.7551/mitpress/11301.001.0001

Mirrlees, J. A. (1971). An Exploration in the Theory of Optimum Income Taxation. *The Review of Economic Studies, 38*(2), 175–208. DOI: 10.2307/2296779

Mitchell, M., Wu, S., Zaldivar, A., Barnes, P., Vasserman, L., Hutchinson, B., & Gebru, T. (2019). Model cards for model reporting. *Proceedings of the Conference on Fairness, Accountability, and Transparency*, 220–229. DOI: 10.1145/3287560.3287596

Mittelstadt, B. D. (2019). Principles alone cannot guarantee ethical AI. *Nature Machine Intelligence, 1*(11), 501–507. DOI: 10.1038/s42256-019-0114-4

Mkik, M., & Mkik, S. (2023). Acceptability aspects of artificial intelligence in Morocco: Managerial and theoretical contributions. In Motahhir, S., & Bossoufi, B. (Eds.), *Digital Technologies And Applications, ICDTA 2023, VOL 2* (Vol. 669, pp. 65–74)., DOI: 10.1007/978-3-031-29860-8_7

Mohan, M. P. (2021). Intellectual Property licenses in cross-border insolvency: Lessons from In Re Qimonda. *Hastings Bus. LJ, 18*, 181.

Mökander, J. (2023). Auditing of AI: Legal, ethical and technical approaches. *Digital Society : Ethics, Socio-Legal and Governance of Digital Technology*, *2*(49), 49. Advance online publication. DOI: 10.1007/s44206-023-00074-y

Mökander, J., Morley, J., Taddeo, M., & Floridi, L. (2021). Ethics-based auditing of automated decision-making systems: Nature, scope, and limitations. *Science and Engineering Ethics*, *27*(4), 44. DOI: 10.1007/s11948-021-00319-4 PMID: 34231029

Morandín-Ahuerma, F. (2023). Ten UNESCO recommendations on the ethics of artificial intelligence.

Morley, J., Floridi, L., Kinsey, L., & Elhalal, A. (2021). From what to how: An initial review of publicly available AI ethics tools, methods and research to translate principles into practices. *Science and Engineering Ethics*, *27*(1), 1–31. PMID: 33492516

Mosallanezhad, B., Hajiaghaei-Keshteli, M., Smith Cornejo, N. R., & Rodríguez Calvo, E. Z. (2023). An IoMT platform for an integrated sustainable energy-efficient disaster relief supply chain to prevent severity-driven disruptions during pandemics. *Journal of Industrial Information Integration*, *35*, 100502. DOI: 10.1016/j.jii.2023.100502

Moynihan, D. P., & Kroll, A. (2016). Performance management routines that work? An early assessment of the GPRA Modernization Act. *Public Administration Review*, *76*(2), 314–323. DOI: 10.1111/puar.12434

Mügge, D. (2024). EU AI sovereignty: For whom, to what end, and with what effect? *Journal of European Public Policy*, *31*(8), 2200–2225. Advance online publication. DOI: 10.1080/13501763.2024.2318475

Murikah, W., Nthenge, J. K., & Musyoka, F. M. (2024). Bias and ethics of AI systems applied in auditing - A systematic review. *Scientific African*, *25*, e02281. DOI: 10.1016/j.sciaf.2024.e02281

Murphy, J., Hofacker, C. F., & Gretzel, U. (2017). Dawning of the age of robots in hospitality and tourism: Challenges for teaching and research. *European Journal of Tourism Research*, *15*, 104–111. DOI: 10.54055/ejtr.v15i.265

Murugesan, U., Subramanian, P., Srivastava, S., & Dwivedi, A. (2023). A study of artificial intelligence impacts on human resource digitalization in Industry 4.0. *Decision Analytics Journal*, *7*, 100249. DOI: 10.1016/j.dajour.2023.100249

Musgrave, R. A., & Musgrave, P. B. (1989). PUBLIC FINANCE IN THEORY AND PRACTICE. *McGraw-HILL INTERNATIONAL EDITIONS. ISBN 0-07-1 00227-8*

Musib, M., Wang, F., Tarselli, M. A., Yoho, R., Yu, K.-H., Andrés, R. M., Greenwald, N. F., Pan, X., Lee, C.-H., Zhang, J., Dutton-Regester, K., Johnston, J. W., & Sharafeldin, I. M.Musib Feng. (2017). Artificial intelligence in research. *Science*, *357*(6346), 28–30. DOI: 10.1126/science.357.6346.28 PMID: 28684488

Mvile, B. N., & Bishoge, O. K. (2024). Mining industry's potential for community development, challenges, and way forward in the East Africa community. A review. *Local Development & Society*, •••, 1–24. DOI: 10.1080/26883597.2024.2423948

Nadeau, P. & Jobin, K. (2024). Intelligence artificielle: Génération Générative: chatgpt, Midjourney... S'approprier les nouvelles IA qui révolutionnent le monde professionnel.

Nannelli, M., Capone, F., & Lazzeretti, L. (2023). Artificial intelligence in hospitality and tourism. State of the art and future research avenues. *European Planning Studies*, *31*(7), 1325–1344. DOI: 10.1080/09654313.2023.2180321

Nasser, T. (2013). The effects of the economy, perception of business law, external anchors, and ethical behavior on corporate social responsibility in developing countries. *Journal of Business and Society, 19*(2), 44–62.

Nawaz, N., Arunachalam, H., Pathi, B. K., & Gajenderan, V. (2024). The adoption of artificial intelligence in human resources management practices. *International Journal of Information Management Data Insights*, *4*(1), 100208. DOI: 10.1016/j.jjimei.2023.100208

Nebie, E. & Tbitbi, E. (2022). La résistance au changement organisationnel dans la gouvernance des organisations publiques: meilleures pratiques pour réussir les transformations. Revue de Management et Cultures. 397–412.

Negri, S. M. C. A. (2021). *Robot As Legal Person: Electronic Personhood in Robotics and Artificial Intelligence*. DOI: 10.3389/frobt.2021.789327

Neuwirth, J. (2023). Prohibited artificial intelligence practices in the proposed eu artificial intelligence act (aia). *Computer Law & Security Review*, *48*, 105798. DOI: 10.1016/j.clsr.2023.105798

Newaz, A. I., Sikder, A. K., Rahman, M. A., & Uluagac, A. S. (2020). A survey on security and privacy issues in modern healthcare systems: Attacks and defenses. arXiv preprint arXiv:2005.07359. https://arxiv.org/abs/2005.07359

Ngai, E. W., Hu, Y., Wong, Y. H., Chen, Y., & Sun, X. (2011). The application of data mining techniques in financial fraud detection: A classification framework and an academic review of literature. *Decision Support Systems*, *50*(3), 559–569. DOI: 10.1016/j.dss.2010.08.006

Nguyen, L. T., & Zuidwijk, R. (2025). Sustainable supply chain governance: A literature review. *Business Ethics, the Environment & Responsibility*, *34*(2), 541–564. DOI: 10.1111/beer.12668

Niang, M., Fofana, F., Faye, C., & Tall, L. (2024). Lignes directrices pour l'élaboration d'un document national sur l'éthique de l'ia dans les pays d'Afrique de l'Ouest.

Nikolinakos, N. T. (2023). *EU Policy and Legal Framework for Artificial Intelligence*. Robotics and Related Technologies - The AI Act., DOI: 10.1007/978-3-031-27953-9

Nishant, R., Schneckenberg, D., & Ravishankar, M. (2024). The formal rationality of artificial intelligence-based algorithms and the problem of bias. *Journal of Information Technology*, *39*(1), 19–40. DOI: 10.1177/02683962231176842

Niu, Y., Wang, S., & Liu, Z. (2022). Managing brand competition with consumer fairness concerns via manufacturer incentives. *European Journal of Operational Research*, *300*(2), 661–675. DOI: 10.1016/j.ejor.2021.08.026

Nurik, C. L., Gauthier, M., Sawchuk, K., DeJong, S., Gómez, S., Canzonetta, J., & Cox, C. M. (2021). *Intersectional automations: robotics, AI, algorithms, and Equity*. Rowman & Littlefield.

O'Rourke, K. H., Rahman, A. S., & Taylor, A. M. (2013). Luddites, the industrial revolution, and the demographic transition. *Journal of Economic Growth*, *18*(4), 373–409. DOI: 10.1007/s10887-013-9096-y

Obal, W., & Merriman, K. K. (2025). Extending the geographic scope of corporate social responsibility: Remote work and housing affordability. *Journal of Business Ethics*. Advance online publication. DOI: 10.1007/s10551-025-05956-4

Obermeyer, Z., Powers, B., Vogeli, C., & Mullainathan, S. (2019). Dissecting racial bias in an algorithm used to manage the health of populations. *Science*, *366*(6464), 447–453. DOI: 10.1126/science.aax2342 PMID: 31649194

OECD. (2010). *Tax Expenditures in OECD Countries*. OECD Publishing., DOI: 10.1787/9789264076907-

OECD. (2016). *Technologies for Better Tax Administration: A Practical Guide for Revenue Bodies*. OECD Publishing.

OECD. (2019). *F. T. C. A. N.* Principles on Artificial Intelligence.

OECD. (2021). *OECD AI Principles*.

OECD. (2022). *Tax Administration 2022: Comparative Information on OECD and other Advanced and Emerging Economies*. OECD Publishing., DOI: 10.1787/1e797131-

OECD. (2023). *Tax Policy Reforms 2023 OECD and Selected Partner Economies*. OECD Publishing., DOI: 10.1787/d8bc45d9-

Olutimehin, D. O., Ofodile, O. C., Ejibe, I., Odunaiya, O. G., & Soyombo, O. T. (2024). Implementing AI in business models: Strategies for efficiency and innovation. *International Journal of Management & Entrepreneurship Research, 6*(3), 863–877. DOI: 10.51594/ijmer.v6i3.940

Omrana, H., Nassiri, S., Belouadha, F.-Z., & Roudiés, O. (2012). Design for distributed Moroccan hospital pharmacy information environment with service-oriented architecture. arXiv preprint arXiv:1207.3277. https://arxiv.org/abs/1207.3277

Organisation for Economic Co-operation and Development (OECD). (2019). OECD principles on artificial intelligence. OECD Publishing. https://www.oecd.org/en/topics/ai-principles.html

Organisation for Economic Co-operation and Development. (2018). *Digital government review of Morocco*. https://www.oecd.org/content/dam/oecd/fr/publications/reports/2018/07/digital-government-review-of-morocco_g1g8d20f/9789264299917-fr.pdf

ORMACHEA. A. (n.d.). L'intégration de l'ia dans la gestion de la chaîne d'approvisionnement et ses effets sur la relation client. Publication-theses.unistra.fr.

Osborne, D., & Gaebler, T. (Eds.). (1992). *Reinventing government. how the entrepreneurial spirit is transforming the public sector*. Penguin Books USA Inc.

Otley, D. T. (1980). The contingency theory of management accounting: Achievement and prognosis. *Accounting, Organizations and Society, 5*(4), 413–428. DOI: 10.1016/0361-3682(80)90040-9

Oubibi, M., Zhou, Y., Oubibi, A., Fute, A., & Saleem, A. (2022). The challenges and opportunities for developing the use of data and artificial intelligence (AI) in North Africa: case of Morocco. *International Conference on Digital Technologies and Applications*, 80–90. DOI: 10.1007/978-3-031-02447-4_9

Paci, B. (2024). The hidden abode of artificial intelligence production: Stretching the limits of artificial intelligence ethics and critique. *Thesis ELEVEN, 184–185*(1, SI), 152–166. DOI: 10.1177/07255136241292366

Padmapriya, S. S., & Jayamala, C. (2025). *Strategic Policy Recommendations and Implications for Effective Management in Organizational and Regulatory Frameworks*., DOI: 10.1007/978-3-031-84909-1_12

Pae, J. (2003). The EU Regulation on Insolvency Proceedings: The Need for a Modified Universal Approach. *Hastings Int'l & Comp. L. Rev., 27*, 555.

Pagé, J. (2025). *Validation du questionnaire proactif: un outil d'auto-repérage pour permettre aux personnes aînées d'évaluer elles-mêmes leurs risques de déclin fonctionnel. Manuscrit inédit présenté au congrès annuel du North American Primary Care Research Group*. NAPCRG.

Palacios-González, C. (2024). *Robotic Persons and Asimov's Three Laws of Robotics*. 185–194. DOI: 10.1093/oso/9780198876434.003.0018

Palm, G. (2012). *Neural assemblies: An alternative approach to artificial intelligence*. Springer Science & Business Media.

Papernot, N., McDaniel, P., Goodfellow, I., Jha, S., Celik, Z. B., & Swami, A. (2017). *Practical black-box attacks against deep learning systems using adversarial examples*. In *Proceedings of the 2017 ACM on Asia Conference on Computer and Communications Security* (pp. 506–519).

Parasuraman, R., & Manzey, D. H. (2010). Complacency and bias in human use of automation: An attentional integration. *Human Factors, 52*(3), 381–410. DOI: 10.1177/0018720810376055 PMID: 21077562

Parks, R., Wigand, R. T., Bennani Othmani, M., Serhier, Z., & Bouhaddou, O. (2019). Electronic health records implementation in Morocco: Challenges of silo efforts and recommendations for improvements. *International Journal of Medical Informatics, 129*, 430–437. DOI: 10.1016/j.ijmedinf.2019.05.026 PMID: 31445288

Parliament, E. U. (2017). *Civil law rules on robotics*.

Patel, D., Sahu, C. K., & Rai, R. (2024). Security in modern manufacturing systems: Integrating blockchain in artificial intelligence-assisted manufacturing. *International Journal of Production Research, 62*(3), 1041–1071. DOI: 10.1080/00207543.2023.2262050

Patil, D. (2024). Explainable Artificial Intelligence (XAI) for industry applications: Enhancing transparency, trust, and informed decision-making in business operation. *Trust, And Informed Decision-Making In Business Operation (December 03, 2024)*.

Pauleen, D. J., Rooney, D., & Intezari, A. (2017). Big data, little wisdom: Trouble brewing? Ethical implications for the information systems discipline. *Social Epistemology, 31*(4), 9–33. DOI: 10.1080/02691728.2016.1249436

Pawar, V., Chavan, P., Vhatkar, A., Khang, A., & Gawankar, S. (2025). Green Transportation and Moral Licensing: Navigating Ethical Challenges with Artificial Intelligence (AI) and Automation (pp. 527–562). DOI: 10.1007/978-3-031-72617-0_28

Pesaran, M. H., & Shin, Y. (1999). An autoregressive distributed lag modelling approach to cointegration analysis. In S. Strøm (Ed.), *Econometrics and economic theory in the 20th century: The Ragnar Frisch Centennial Symposium* (pp. 371–413). Cambridge University Press. DOI: 10.1017/CCOL521633230.011

Pesaran, M. H., Shin, Y., & Smith, R. J. (2001). Bounds testing approaches to the analysis of level relationships. *Journal of Applied Econometrics, 16*(3), 289–326. DOI: 10.1002/jae.616

Petticrew, M., & Roberts, H. (2006). *Systematic reviews in the social sciences: A practical guide*. Blackwell Publishing., DOI: 10.1002/9780470754887

Philippon, T. (2016). *The fintech opportunity* (No. w22476). National Bureau of Economic Research.

Phillips, P. (1987). Time series regression with a unit root. *Econometrica, 55*(2), 277–303. https://www.researchgate.net/publication/4814947_Time_Series_Regression_with_a_Unit_Root. DOI: 10.2307/1913237

Phillips, P., & Durlauf, S. N. (1986). Multiple time series regression with integrated processes. *The Review of Economic Studies, 53*(3), 473–495. DOI: 10.2307/2297602

Piteira, M., Aparicio, M., & Costa, C. J. (2019). Ethics of artificial intelligence: Challenges. *2019 14th Iberian Conference On Information Systems and Technologies (CISTI)*. DOI: 10.23919/CISTI.2019.8760826

Pollitt, C., & Bouckaert, G. (2011). *Public management reform: a comparative analysis—new public management, governance, and the neo-Weberian state* (3rd ed.). Oxford University Press.

Popescu, A., Polat-Erdeniz, S., Felfernig, A., Uta, M., Atas, M., Le, V.-M., Pilsl, K., Enzelsberger, M., & Tran, T. N. T. (2022). An overview of machine learning techniques in constraint solving. *Journal of Intelligent Information Systems, 58*(1), 91–118. DOI: 10.1007/s10844-021-00666-5

Porral, A. (2025). The protection of creditors in cross-border insolvency proceedings-International soft law, EU law and a comparative analysis of German and Australian law.

Prensky, M. (2001). Digital Natives, Digital Immigrants Part 1. *On the Horizon, 9*(5), 1–6. DOI: 10.1108/10748120110424816

Pria, S., Al Rubaie, I., & Prasad, V. (2024). *Enhancing Business Intelligence Through AI-Driven Integration of Sustainability Metrics via ESG Factors.*, DOI: 10.4018/979-8-3693-2185-0.ch004

Putera, N. S. F. M. S., Saripan, H., Bajury, M. S. M., & Yaâ, S. N. (2022). Artificial Intelligence in the Tourism Industry: A privacy impasse. *Environment-Behaviour Proceedings Journal, 7*(SI7), 433–440. DOI: 10.21834/ebpj.v7iSI7.3812

Pymetrics. (2021). Audit report on bias mitigation practices.

Raab, R., Küderle, A., Zakreuskaya, A., Stern, A. D., Klucken, J., Kaissis, G., Rueckert, D., Boll, S., Eils, R., Wagener, H., & Eskofier, B. M. (2023). Federated electronic health records for the European Health Data Space. *The Lancet. Digital Health, 5*(11), e840–e847. DOI: 10.1016/S2589-7500(23)00156-5 PMID: 37741765

Radanliev, P. (2025). AI Ethics: Integrating Transparency, Fairness, and Privacy in AI Development. *Applied Artificial Intelligence, 39*(1), 1–14. DOI: 10.1080/08839514.2025.2463722

Raji, I. D., & Buolamwini, J. (2022). Actionable auditing: Investigating the impact of publicly naming biased performance results of commercial AI products. *Proceedings of the 2022 ACM Conference on Fairness, Accountability, and Transparency*, 63–75.

Raji, I. D., Smart, A., White, R. N., Mitchell, M., Gebru, T., Hutchinson, B., & Barnes, P. (2020). Closing the AI accountability gap: Defining an end-to-end framework for internal algorithmic auditing. *Proceedings of the 2020 Conference on Fairness, Accountability, and Transparency*, 33–44. DOI: 10.1145/3351095.3372873

Rania, B. E. N. M. O. U. S. S. A., & Leila, K. I. S. S. I. (2025). Le marketing expérientiel à l'ère des technologies immersives: Une revue de la littérature sur le rôle des émotions dans l'engagement du consommateur. Journal of Economics [JEFM]. *Financial Management, 4*(3), 672–692.

Rashidov, A. (2024). Artificial intelligence in scientific research. *strategies for policy in science and education-strategy II NA Obrazovatelnata I Nauchnata Politika, 32*(5, S), 35–45. DOI: 10.53656/str2024-5s-3-ald

Rawas, S. (2024). AI: The future of humanity. *Discover Artificial Intelligence, 4*(25). Advance online publication. DOI: 10.1007/s44163-024-00118-3awe

Reisman, D., Schultz, J., Crawford, K., & Whittaker, M. (2018). *Algorithmic impact assessments: A practical framework for public agency.* AI Now Institute.

Resnik, D. B., & Hosseini, M. (2025). *Disclosing artificial intelligence use in scientific research and publication: When should disclosure be mandatory, optional, or unnecessary?* Accountability in Research-Ethics Integrity and Policy., DOI: 10.1080/08989621.2025.2481949

Reuel, A., & Undheim, T. A. (2024). Generative AI needs adaptive governance. *arXiv preprint arXiv:2406.04554.*

Rezaei, M., Pironti, M., & Quaglia, R. (2024). AI in knowledge sharing, which ethical challenges are raised in decision-making processes for organisations? *Management Decision.* Advance online publication. DOI: 10.1108/MD-10-2023-2023

Riad, M., Naimi, M., & Okar, C. (2024). Enhancing supply chain resilience through artificial intelligence: Developing a comprehensive conceptual framework for AI implementation and supply chain optimization. *Logistics (Basel)*, *8*(4), 111. DOI: 10.3390/logistics8040111

Riazul Islam, S. M., Kwak, D., Kabir, M. H., Hossain, M., & Kwak, K. S. (2015). The Internet of Things for health care: A comprehensive survey. *IEEE Access : Practical Innovations, Open Solutions*, *3*, 678–708. DOI: 10.1109/ACCESS.2015.2437951

Ribeiro Goraieb, M., Nascimento, M. R., & Verdu, F. C. (2019). Cultural influences on foreign direct investment. *Journal of International Management, 25*(4), 100–115.

Ribeiro, D., Rocha, T., Pinto, G., Cartaxo, B., Amaral, M., Davila, N., & Camargo, A. (2025). Toward Effective AI Governance: A Review of Principles. *arXiv preprint arXiv:2505.23417.*

Richey, R. G.Jr, Chowdhury, S., Davis-Sramek, B., Giannakis, M., & Dwivedi, Y. K. (2023). Artificial intelligence in logistics and supply chain management: A primer and roadmap for research. *Journal of Business Logistics*, *44*(4), 532–549. DOI: 10.1111/jbl.12364

Ridder, H.-G. (2014). *Book review: Qualitative data analysis. A methods sourcebook.* Sage publications Sage UK: London, England.

Riedmann-Streitz, C., Streitz, N., Antona, M., Marcus, A., Margetis, G., Ntoa, S., Rau, P.-L. P., & Rosenzweig, E. (2025). How to Create and Foster Sustainable Smart Cities? Insights on Ethics, Trust, Privacy, Transparency, Incentives, and Success. *International Journal of Human-Computer Interaction*, *41*(4), 2491–2522. DOI: 10.1080/10447318.2024.2325175

Rochette, C. (2022). L'évaluation, socle de l'organisation publique apprenante. Gestion et management public, 10(3), 6–9. https://shs.cairn.info/revue-gestion-et-management-public-2022-3-page-6?lang=fr

Rodriguez, A. L. T. (2022). Ethics for sustainable artificial Intelligence. *Arbor-Ciencia pensamiento Y cultura, 198*(806). DOI: 10.3989/arbor.2022.806013

Rodríguez-Mejías, S., Degli-Esposti, S., González-García, S., & Parra-Calderón, C. L. (2024). Toward the European Health Data Space: The IMPaCT-Data secure infrastructure for EHR-based precision medicine research. *Journal of Biomedical Informatics, 156*, 104670. DOI: 10.1016/j.jbi.2024.104670 PMID: 38880235

Roe, J., & Perkins, M. (2023). 'What they're not telling you about ChatGPT': Exploring the discourse of AI in UK news media headlines. *Humanities & Social Sciences Communications, 10*(1), 1–9.

Rouleau, G., Wu, K., Ramamoorthi, K., Boxall, C., Liu, R. H., Maloney, S., Zelmer, J., Scott, T., Larsen, D., Wijeysundera, H. C., Ziegler, D., Bhatia, S., Kishimoto, V., Steele Gray, C., & Desveaux, L. (2024). Mapping theories, models, and frameworks to evaluate digital health interventions: Scoping review. *Journal of Medical Internet Research, 26*, e51098. DOI: 10.2196/51098 PMID: 38315515

Roumate, F. (2021). Artificial intelligence, ethics and international human rights law. *International Journal of Information Ethics*, •••, 29.

ROUSSI, K. (2022). L'Intelligence Artificielle au service du métier de l'expert-comptable. Revue Internationale du Chercheur.

Royaume du Maroc. (2015). Dahir n° 1-15-62 du 14 chaabane 1436 (2 juin 2015) portant promulgation de la loi organique n° 130-13 relative à la loi de finances. Bulletin Officiel du Royaume du Maroc, n° 6370, 1 ramadan 1436 (18 juin 2015).

Royaume du Maroc. (2023). Décret n° 2-22-580 du 10 chaabane 1444 (3 mars 2023) relatif à la mise en place du dispositif de contrôle de gestion au sein des départements ministériels. Bulletin Officiel du Royaume du Maroc, n° 7200, 12 kaada 1444 (1er juin 2023).

Rudin, C. (2019). Stop explaining black box machine learning models for high stakes decisions and use interpretable models instead. *Nature Machine Intelligence, 1*(5), 206–215. DOI: 10.1038/s42256-019-0048-x PMID: 35603010

Rumbold, J. M. M., & Pierscionek, B. K. (2017). A critique of the regulation of data science in healthcare research in the European Union. *BMC Medical Ethics, 18*(1), 27. Advance online publication. DOI: 10.1186/s12910-017-0184-y PMID: 28388916

Ruotsalainen, P., & Blobel, B. (2025). A system model and requirements for transformation to human-centric digital health. *Journal of Medical Internet Research, 27*, e68661. DOI: 10.2196/68661 PMID: 40294402

Russell, S. (2022). Artificial intelligence and the problem of control. *Perspectives on Digital Humanism, 19*, 1–322. DOI: 10.1007/978-3-030-86144-5_3

Ryan, M. (2024). Will the real data sovereign please stand up? An EU policy analysis. *International Journal of Law and Information Technology*. Advance online publication. DOI: 10.1093/ijlit/eaae006

Ryll, L., Barton, M. E., Zhang, B. Z., McWaters, R. J., Schizas, E., Hao, R., ... & Yerolemou, N. (2020). Transforming paradigms: A global AI in financial services survey.

Saba, C. S., & Monkam, N. (2025). Artificial intelligence's (AI's) role in enhancing tax revenue, institutional quality, and economic growth in selected BRICS-plus countries. *Journal of Social and Economic Development*. Advance online publication. DOI: 10.1007/s40847-024-00401-0

Saberi, S., Kouhizadeh, M., Sarkis, J., & Shen, L. (2019). Blockchain technology and its relationships to sustainable supply chain management. *International Journal of Production Research, 57*(7), 2117–2135. DOI: 10.1080/00207543.2018.1533261

Saheb, T., Saheb, T., & Carpenter, D. O. (2021). Mapping research strands of ethics of artificial intelligence in healthcare: A bibliometric and content analysis. *Computers in Biology and Medicine, 135*, 104660. Advance online publication. DOI: 10.1016/j.compbiomed.2021.104660 PMID: 34346319

Salesforce. (2021). Ethical use of technology guidelines.

Salmandjee, Y. (2025). Utiliser l'ia en 90 cas concrets pour se faciliter la vie: Travail, santé, cuisine, sport, loisirs

Samara, D., Magnisalis, I., & Peristeras, V. (2020). Artificial intelligence and big data in tourism: A systematic literature review. *Journal of Hospitality and Tourism Technology, 11*(2), 343–367. DOI: 10.1108/JHTT-12-2018-0118

Samarawickrama, M. (2022). AI governance and ethics framework for sustainable AI and sustainability. *arXiv preprint arXiv:2210.08984.*

Sambala, L. G. (2016). The role of multinational business corporations in sustainable development: The case of Tanzania Tobacco Processing Ltd. *African Journal of Business and Sustainability, 12*(1), 77–94.

Sanday, A., & Nalweera, A. (2025). Leveraging Digital Public Infrastructure to Enhance Uganda's Tax Revenue Potential. IMF.

Sardi, A., Rizzi, A., Sorano, E., & Guerrieri, A. (2021). Cyber risk in health facilities: A systematic literature review. arXiv preprint arXiv:2102.04093. https://arxiv.org/abs/2102.04093

Sarker, I. H. (2021). Deep learning: A comprehensive overview of techniques, taxonomy, applications, and research directions. *SN Computer Science*, *2*(420), 420. DOI: 10.1007/s42979-021-00815-1 PMID: 34426802

Sauvant, K. P. (2011). The regulatory framework for investment: Where are we headed? *World Investment Review, 7*(1), 1–23.

Sauvant, K. P. (2021). Improving the distribution of FDI benefits: The need for policy-oriented research, advice, and advocacy. *Transnational Corporations Review, 13*(2), 89–105.

Sauvant, K. P., & Mann, H. K. (2017). Towards an indicative list of FDI sustainability characteristics. *Investment Policy Journal, 3*(1), 12–28.

Saveliev, A. M., Zhurenkov, D. A., Poikin, A. E., & Berkutova, T. A. (2021). Ethics of Artificial Intelligence and Post-non-classical Scientific Rationality. *IFAC-PapersOnLine*, *54*(13), 397–401. DOI: 10.1016/j.ifacol.2021.10.480

Savulescu, J., Giubilini, A., Vandersluis, R., & Mishra, A. (2024). Ethics of artificial intelligence in medicine. *Singapore Medical Journal*, *65*(3), 150–158. DOI: 10.4103/singaporemedj.SMJ-2023-279 PMID: 38527299

Scandura, T. A., & Williams, E. A. (2000). Research methodology in management: Current practices, trends, and implications for future research. *Academy of Management Journal*, *43*(6), 1248–1264. DOI: 10.2307/1556348

Schacherer, S. (2024). Building responsible and sustainable supply chain frameworks: Limits of international investment law and the CSR initiatives taken by the EU and China. *Asia Pacific Law Review*, *32*(2), 347–369. DOI: 10.1080/10192557.2024.2349825

Schär, F. (2021). Decentralized finance: On blockchain-and smart contract-based financial markets. *FRB of St. Louis Review*.

Scheibner, J., Raisaro, J. L., Troncoso-Pastoriza, J. R., Ienca, M., Fellay, J., Vayena, E., & Hubaux, J. P. (2020). Revolutionizing medical data sharing using advanced privacy enhancing technologies: Technical, legal and ethical synthesis. arXiv preprint arXiv:2010.14445. https://arxiv.org/abs/2010.14445 DOI: 10.2196/preprints.25120

Schiff, D., Biddle, J., Borenstein, J., & Laas, K. (2020, February). What's next for AI ethics, policy, and governance? A global overview. In *Proceedings of the AAAI/ACM Conference on AI, Ethics, and Society* (pp. 153–158). DOI: 10.1145/3375627.3375804

Schlinger, H. D.Jr. (2009). Theory of Mind: An Overview and Behavioral Perspective. *The Psychological Record*, *59*(3), 435–448. DOI: 10.1007/BF03395673

Schneider, J., Abraham, R., Meske, C., & Vom Brocke, J. (2023). Artificial intelligence governance for businesses. *Information Systems Management*, *40*(3), 229–249. DOI: 10.1080/10580530.2022.2085825

Schwartz, D. L., & Rogers, M. (2022). Inventorless Inventions? The Constitutional Conundrum of AI-Produced Inventions. *Social Science Research Network*. DOI: 10.2139/ssrn.4025434

Schwartz, R., Dodge, J., Smith, N. A., & Etzioni, O. (2020). Green AI. *Communications of the ACM*, *63*(12), 54–63. DOI: 10.1145/3381831

Seargeant, P., & Tagg, C. (2019). Social media and the future of open debate: A user-oriented approach to Facebook's filter bubble conundrum. *Discourse, Context & Media*, *27*, 41–48. DOI: 10.1016/j.dcm.2018.03.005

Selbst, A. D., Boyd, D., Friedler, S. A., Venkatasubramanian, S., & Vertesi, J. (2019). Fairness and abstraction in sociotechnical systems. FAT 2019 - Proceedings of the 2019 Conference on Fairness, Accountability, and Transparency*, 1(1), 59–68. DOI: 10.1145/3287560.3287598

Seppälä, P., & Małecka, M. (2024). AI and discriminative decisions in recruitment: Challenging the core assumptions. *Big Data & Society*, *11*(1), 20539517241235872. Advance online publication. DOI: 10.1177/20539517241235872

Shams, R. A., Zowghi, D., & Bano, M. (2023). AI and the quest for diversity and inclusion: A systematic literature review. *AI and Ethics*. Advance online publication. DOI: 10.1007/s43681-023-00362-w

Shamsuddoha, M., Kashem, M. A., Nasir, T., Hossain, A. I., & Ahmed, M. F. (2025). Quantum Computing Applications in Supply Chain Information and Optimization: Future Scenarios and Opportunities. *Information (Basel)*, *16*(8), 693. DOI: 10.3390/info16080693

Sharadhi, A. K., (2025). Scouring through the crowd simulation dynamics in urban environments. *International Journal of Innovative Science and Research Technology*, 1439–1459. https://doi.org/DOI: 10.38124/ijisrt/25mar1176

Sharadhi, A. K., Gururaj, V., Shankar, S. P., Supriya, M. S., & Chogule, N. S. (2022). Face mask recogniser using image processing and computer vision approach. *Global Transitions Proceedings*, *3*(1), 67–73. DOI: 10.1016/j.gltp.2022.04.016

Sharma, S. (2024). Benefits or concerns of AI: A multistakeholder responsibility. *Futures, 157*, 103328. DOI: 10.1016/j.futures.2024.103328

Sheets, L. R., Wallach, E., Khairat, S., Mutrux, R., Edison, K., & Becevic, M. (2020). Similarities and differences between rural and urban telemedicine utilization. *Perspectives in Health Information Management, 18*(Winter), 1e. https://www.ncbi.nlm.nih.gov/pmc/articles/PMC7883358/ PMID: 33633515

Sheikh, H., Prins, C., & Schrijvers, E. (2023). Artificial intelligence: Definition and background. In *Mission AI. Research for Policy* (pp. 13-30). Springer.

Shen, Y., & Zhang, X. (2024). The impact of artificial intelligence on employment: The role of virtual agglomeration. *Humanities & Social Sciences Communications, 11*(122), 122. DOI: 10.1057/s41599-024-02647-9

Shortliffe, E. H. (1974). *A rule-based computer program for advising physicians regarding antimicrobial therapy selection.* 739. DOI: 10.1145/1408800.1408906

Shoup, C. S. (1975). Surrey's Pathways to Tax Reform—A Review Article. *The Journal of Finance, 30*(5), 1329–1341. DOI: 10.2307/2326658

Shrivastava, U., Song, J., Han, B. T., & Dietzman, D. (2021). Do data security measures, privacy regulations, and communication standards impact the interoperability of patient health information? *International Journal of Medical Informatics, 148*, 104401. DOI: 10.1016/j.ijmedinf.2021.104401 PMID: 33571743

Siau, K., & Wang, W. (2020). Artificial intelligence (AI) ethics: Ethics of AI and ethical AI. *Journal of Database Management, 31*(2), 74–87. DOI: 10.4018/JDM.2020040105

Siddique, S., Haque, M. A., George, R., Gupta, K. D., Gupta, D., & Faruk, M. J. H. (2024). Survey on machine learning biases and mitigation techniques. *Digital, 4*(1), 1–68. DOI: 10.3390/digital4010001

Siham, N. (2021). *Du New Public Management à l'émergence du dogme de la performance et du contrôle de gestion dans les Établissements et Entreprises Publics Marocains.* Alternatives Managériales Economiques.

Sikstrom, L., Maslej, M. M., Hui, K., Findlay, Z., Buchman, D. Z., & Hill, S. L. (2022). Conceptualising fairness: Three pillars for medical algorithms and health equity. *BMJ Health & Care Informatics, 29*(1), e100459. DOI: 10.1136/bmjhci-2021-100459 PMID: 35012941

Simen Nana, S. F., Flindjoa, D., Tankpe, T. A., & Kaka, Z. Y. (2024). Diriger l'entreprise africaine à l'ère de la transformation numérique et de l'intelligence artificielle.

Simons, R. (1995). *Levers of control: How managers use innovative control systems to drive strategic renewal.* Harvard Business School Press.

Singh, R. K., Agarwal, A., Dwivedi, D. K., & Yadav, S. (2024). Ethical Claims and Enforcement in Legal Metrology in India. In Handbook of Quality System, Accreditation and Conformity Assessment (pp. 1–23). Springer Nature Singapore. DOI: 10.1007/978-981-99-4637-2_51-1

Sirignano, J., & Giesecke, K. (2019). Risk analysis for large pools of loans. *Management Science*, *65*(1), 107–121. DOI: 10.1287/mnsc.2017.2947

Sironi, P. (2016). *FinTech innovation: from robo-advisors to goal based investing and gamification.* John Wiley & Sons. DOI: 10.1002/9781119227205

Smith, H. (2021). Clinical AI: Opacity, accountability, responsibility and liability. *AI & Society*, *36*(2), 535–545. DOI: 10.1007/s00146-020-01019-6

Soled, J. A., & Thomas, K. D. (2019). Automation and the Income Tax. *Columbia Journal of Tax Law, 10*(1), 1–48. DOI: 10.7916/CJTL.V10I1.2864

Soori, M., Arezoo, B., & Dastres, R. (2023). Artificial intelligence, machine learning and deep learning in advanced robotics: A review. *Cognitive Robotics*, *3*, 54–70. DOI: 10.1016/j.cogr.2023.04.001

Sreenivasan, A., & Suresh, M. (2022). Modeling the enablers of sourcing risks faced by startups in COVID-19 era. *Journal of Global Operations and Strategic Sourcing*, *15*(2), 151–171. DOI: 10.1108/JGOSS-12-2020-0070

Srikanth, G. (2025). Transforming governance: Exploring the intersection of e-governance and artificial intelligence in India. *Electronic Government-. International Journal (Toronto, Ont.)*, *21*(3), 300–312. Advance online publication. DOI: 10.1504/EG.2025.145913

Srivastava, S., & Sinha, K. (2023). From bias to fairness: A review of ethical considerations and mitigation strategies in artificial intelligence. *International Journal for Research in Applied Science and Engineering Technology*, *11*(3), 2247–2251. DOI: 10.22214/ijraset.2023.49990

Stefanelli, V., Manta, F., & Toma, P. (2022). *Digitalization and competition in the EU banking sector: Regulatory strategies and market evolution. arXiv.*

Stroumpoulis, A., Kopanaki, E., & Varelas, S. (2022). Role of artificial intelligence and big data analytics in smart tourism: A resource-based view approach. *WIT Transactions on Ecology and the Environment, 256*(2022), 99-108. method. *Mathematics*, *10*(13), 2190.

Strubell, E., Ganesh, A., & McCallum, A. (2020, April). Energy and policy considerations for modern deep learning research. *Proceedings of the AAAI Conference on Artificial Intelligence*, *34*(09), 13693–13696. DOI: 10.1609/aaai.v34i09.7123

Subrahmanyam, S. (2024). Role of data analytics and big data in operational decision-making and performance improvement. In Martínez-Falcó, J., Sánchez-García, E., Marco-Lajara, B., & Visser, G. (Eds.), *Convergence of technology and operations management in modern businesses* (pp. 105–140). IGI Global Scientific Publishing. DOI: 10.4018/979-8-3693-6200-6.ch003

Subrahmanyam, S. (2024). The evolution of human resources in the age of technology. In Kaur, J. (Ed.), *Technological enhancements for improving employee performance, safety, and well-being* (pp. 229–256). IGI Global Scientific Publishing. DOI: 10.4018/979-8-3693-9631-5.ch011

Subrahmanyam, S. (2025). Role of Artificial Intelligence in Re-Inventing Human Resource Management (HRM) 2.0. In Kapoor, T., Singh, A., Sahay, K., Kaur, H., & Verma, R. (Eds.), *Sustainable Management Practices for Employee Retention and Recruitment* (pp. 243–264). IGI Global Scientific Publishing. DOI: 10.4018/979-8-3693-8679-8.ch012

Sugimoto, S. (2018). Ethics in responsible investment: How to incorporate ethics into investment analysis. *Journal of Business Ethics, 152*(3), 709–726.

Suler, J. R. (2004). The Online Disinhibition Effect. *Cyberpsychology, Behavior, and Social Networking*, *7*(3), 321–326. DOI: 10.1089/1094931041291295 PMID: 15257832

Sun, M., & Ji, N. (2024). Erosion and Remodeling: New Alienation of User Subjects Led by Intelligent Algorithms. *Communications in Humanities Research*, *36*(1), 50–61. https://doi.org/10.54254/2753-7064/36/2024BJ0035DOI: 10.54254/2753-7064/50/20242548

Sun, P. (2020). Security and privacy protection in cloud computing: Discussions and challenges. *Journal of Network and Computer Applications*, *160*, 102642. DOI: 10.1016/j.jnca.2020.102642

Sun, T. Q., & Medaglia, R. (2019). Mapping the challenges of artificial intelligence in the public sector: Evidence from public healthcare. *Government Information Quarterly*, *36*(2), 368–383. DOI: 10.1016/j.giq.2018.09.008

Suresh, H., & Guttag, J. V. (2021). A framework for understanding unintended consequences of machine learning. *Communications of the ACM*, *64*(7), 62–71.

Syed, R., Suriadi, S., Adams, M., Bandara, W., Leemans, S. J., Ouyang, C., ter Hofstede, A. H. M., van de Weerd, I., Wynn, M. T., & Reijers, H. A. (2020). Robotic process automation: Contemporary themes and challenges. *Computers in Industry*, *115*, 103162. DOI: 10.1016/j.compind.2019.103162

Szafran, D., & Bach, R. (2024). "The human must remain the central focus": Subjective fairness perceptions in automated decision-making. *Minds and Machines*, *34*(1), 24. DOI: 10.1007/s11023-024-09684-y

Taeihagh, A. (2021). Governance of artificial intelligence. *Policy and Society*, *40*(2), 137–157. DOI: 10.1080/14494035.2021.1928377

Tafenzi, H. A., Essaadi, I., & Belbaraka, R. (2025). Digital oncology in Morocco: Embracing artificial intelligence in a new era. *JCO Global Oncology*, *11*(11), e2400583. DOI: 10.1200/GO-24-00583 PMID: 39787448

Taj, S., Imran, A. S., Kastrati, Z., Daudpota, S. M., Memon, R. A., & Ahmed, J. (2023). IoT-based supply chain management: A systematic literature review. *Internet of Things : Engineering Cyber Physical Human Systems*, *24*, 100982. DOI: 10.1016/j.iot.2023.100982

Tamanine, R., Fahim, I., Houssas, M. B., Elkandoussi, F., & Baazizi, Y. (2024). Assessing the Determinants of AI Integration in Tourism SMEs. International Journal of Business and Management Research Tuo, Y., Wu, J., Zhao, J., & Si, X. (2025). Artificial intelligence in tourism: Insights and future research agenda. *Tourism Review*, *80*(4), 793–812.

Tao, R., Wu, J., & Zhao, H. (2023). Do Corporate Customers Prefer Socially Responsible Suppliers? An Instrumental Stakeholder Theory Perspective. *Journal of Business Ethics*, *185*(3), 689–712. DOI: 10.1007/s10551-022-05171-5

Tapscott, D., & Tapscott, A. (2016). *Blockchain revolution: how the technology behind bitcoin is changing money, business, and the world*. Penguin.

Teixeira, A. R., Ferreira, J. V., & Ramos, A. L. (2025). Intelligent supply chain management: A systematic literature review on artificial intelligence contributions. *Information (Basel)*, *16*(5), 399. DOI: 10.3390/info16050399

Tergan, S.-O. (1997). Conceptual and Methodological Shortcomings in Hypertext/Hypermedia Design and Research. *Journal of Educational Computing Research*, *16*(3), 209–235. DOI: 10.2190/2R2J-NXB9-0MD3-T603

Thakor, A. V. (2020). Fintech and banking: What do we know? *Journal of Financial Intermediation*, *41*, 100833. DOI: 10.1016/j.jfi.2019.100833

Thomas, J., Patil, R. S., Patil, M., & John, J. (2023). Addressing the Sustainability Conundrums and Challenges within the Polymer Value Chain. *Sustainability (Basel), 15*(22), 15758. DOI: 10.3390/su152215758

Thompson, A. (2018). Robotic Process Automation (RPA) in Banking Operations. *International Journal of Artificial Intelligence and Machine Learning, 1*(2).

Thuraisingham, B., & Thomas, T. (2024). Social media governance and fake news detection integrated with artificial intelligence governance. *2024 IEEE International Conference on Information Reuse and Integration for Data Science, IRI 2024*, 190–197. DOI: 10.1109/IRI62200.2024.00048

Timmons, A. C., Duong, J. B., Simo Fiallo, N., Lee, T., Vo, H. P. Q., Ahle, M. W., Comer, J. S., Brewer, L. C., Frazier, S. L., & Chaspari, T. (2023). A call to action on assessing and mitigating bias in artificial intelligence applications for mental health. *Perspectives on Psychological Science, 18*(5), 1062–1096. DOI: 10.1177/17456916221134490 PMID: 36490369

Tirkolaee, E. B., Sadeghi, S., Mooseloo, F. M., Vandchali, H. R., & Aeini, S. (2021). Application of machine learning in supply chain management: A comprehensive overview of the main areas. *Mathematical Problems in Engineering, 2021*(1), 1476043. DOI: 10.1155/2021/1476043

Tolan, S. (2019). Fair and unbiased algorithmic decision making: Current state and future challenges. arXiv preprint arXiv:1901.04730.

Toma, C. L. (2021). Analyse comparative des différentes méthodes de gestion intégrée des rejets miniers permettant de minimiser l'impact des opérations minières. https://savoirs.usherbrooke.ca/handle/11143/18420

Touicher, O., & Loulid, M. (2023). Gestion Budgétaire: un levier clé pour la prise de décisions stratégiques. revue de l'entrepreneuriat et de l'innovation, 5(20).

Truera. (2022). Explainable and responsible AI for enterprises.

Tula, S. T., Kess-Momoh, A. J., Omotoye, G. B., Bello, B. G., & Daraojimba, A. I. (2024). AI-enabled customer experience enhancement in business. Computer Science & IT Research Journal, 5(2), 365-389.

Turing, A. M. (1950). Computing Machinery and Intelligence. *Mind, LIX*(236), 433–460. DOI: 10.1093/mind/LIX.236.433

Turkle, S. (2011). The Tethered Self: Technology Reinvents Intimacy and Solitude. *Continuing Higher Education Review, 75*, 28–31. https://files.eric.ed.gov/fulltext/EJ967807.pdf

Tussyadiah, I. P. (2020). A review of research into automation in tourism: Launching the Annals of Tourism Research Curated Collection on Artificial Intelligence and Robotics in Tourism. *Annals of Tourism Research, 81*, 102883. DOI: 10.1016/j.annals.2020.102883

Tussyadiah, I. P., & Griffin, T. (2018). The implications of AI for tourism: A discussion. In *Information and communication technologies in tourism 2018* (pp. 637–641). Springer.

UMAB. (2019). Applying the ARDL methodology for economic measurement: An EViews V.10 case study. *Université de Mostaganem*. Algeria. https://coilink.org/20.500.12592/2n7bzjc

UNESCO. (2021). *Recommendation on the ethics of artificial intelligence.*

UNESCO. (2021). Recommendation on the ethics of artificial intelligence. https://www.unesco.org/en/artificial-intelligence/recommendation-ethics?hub=32618

Union, E. (2021). Proposal for a regulation of the european parliament and of the council laying down harmonised rules on artificial intelligence (artificial intelligence act) and amending certain union legislative acts. *COM/2021/206final.*

UNWTO. (2022). Tourism and COVID-19: Unprecedented economic impacts. World Tourism Organization. https://www.unwto.org/covid-19-and-tourism-2022

Uvet, H., Park, A., Dickens, J., Oh, J., & Hazen, B. (2025). How blockchain technology utilization influences corporate social responsibility through supply chain transparency and the role of supplier risk. *International Journal of Logistics Management, 36*(4), 1195–1221. Advance online publication. DOI: 10.1108/IJLM-04-2024-0245

Vaccari, E. (2018). The Ammanati Affair: Seven Centuries Old, and Not Feeling the Age. Chi.-. *Kent L. Rev., 93*, 831.

Vakkuri, V., & Kemell, K.-K. (2019). Implementing artificial intelligence ethics: A tutorial. In S. Hyrynsalmi, M. Suoranta, A. NguyenDuc, P. Tyrvainen, & P. Abrahamsson (Eds.), *Software Business (ICSOB 2019)* (Vol. 370, pp. 439–442). DOI: 10.1007/978-3-030-33742-1_38

van Kolfschooten, H., & Shachar, C. (2023). The Council of Europe's AI Convention (2023-2024): Promises and pitfalls for health protection. *Health Policy (Amsterdam), 138*, 104935. DOI: 10.1016/j.healthpol.2023.104935 PMID: 37925880

Varela, J. Felix,R.,Mas, L.,Olivier, C., Ivan, k., Yksic, C., Omar, V., & Paola,A.,R., al. (2022). Digital Transformation of Tax and Customs Administrations. Equitable Growth, *Finance & Institutions Insight. The World Bank.* http://hdl.handle.net/10986/37629

Vatinel, T. (2024). *Lutte contre la fraude documentaire: état des lieux et avancées technologiques grâce à l'ia.* Archimag.

Veale, M., & Binns, R. (2017). Fairer machine learning in the real world: Mitigating discrimination without collecting sensitive data. *Big Data & Society*, *4*(2). DOI: 10.1177/2053951717743530

Veale, M., & Borgesius, F. Z. (2021). Demystifying the draft EU Artificial Intelligence Act. *Computer Law Review International*, *22*(4), 97–112. DOI: 10.9785/cri-2021-220402

Vicente, L., & Matute, H. (2023). Humans inherit artificial intelligence biases. *Scientific Reports*, *13*(1), 15737. DOI: 10.1038/s41598-023-42384-8 PMID: 37789032

Vijaykumar, V., Mercy, P., Lucia Agnes Beena, T., Leena, H. M., & Savarimuthu, C. (2024). Convergence of IoT, Artificial Intelligence and Blockchain Approaches for Supply Chain Management. In Blockchain, IoT, and AI Technologies for Supply Chain Management (pp. 45–89). Apress. DOI: 10.1007/979-8-8688-0315-4_2

Vinuesa, R., Azizpour, H., Leite, I., Balaam, M., Dignum, V., Domisch, S., Felländer, A., Langhans, S. D., Tegmark, M., & Fuso Nerini, F. (2020). The role of artificial intelligence in achieving the sustainable development Goals. *Nature Communications*, *11*(1), 233. DOI: 10.1038/s41467-019-14108-y PMID: 31932590

Voigt, P., & Von dem Bussche, A. (2017). *The EU General Data Protection Regulation (GDPR): A Practical Guide*. Springer. DOI: 10.1007/978-3-319-57959-7

von Eschenbach, W. J. (2021). Transparency and the Black Box Problem: Why We Do Not Trust AI. *Philosophy & Technology*, *34*(4), 1607–1622. DOI: 10.1007/s13347-021-00477-0

Votto, A. M., Valecha, R., Najafirad, P., & Rao, H. R. (2021). Artificial intelligence in tactical human resource management: A systematic literature review. *International Journal of Information Management Data Insights*, *1*(2), 100047. DOI: 10.1016/j.jjimei.2021.100047

Walker-Munro, B., & Assaad, Z. (2023). The Guilty (Silicon) Mind: Blameworthiness and Liability in Human-Machine Teaming. *Cambridge L. Rev.*, *8*, 1.

Waltl, B. (2021). Explainability and transparency of machine learning in ADM systems. In *Research handbook on big data law* (pp. 341–356). Edward Elgar Publishing. DOI: 10.4337/9781788972826.00023

Wamba, S. F., Akter, S., Edwards, A., Chopin, G., & Gnanzou, D. (2015). How 'big data'can make big impact: Findings from a systematic review and a longitudinal case study. *International Journal of Production Economics, 165*, 234–246. DOI: 10.1016/j.ijpe.2014.12.031

Wanjohi Ngeng, A. (2018). Unlocking the quandary: Towards embedding sustainable development into the Sino outbound FDIs. *Journal of International Business and Economics, 6*(2), 45–57.

Warner, R., & Sloan, R. H. (2021). Making artificial intelligence transparent: Fairness and the problem of proxy variables. *Criminal Justice Ethics, 40*(1), 23–39. DOI: 10.1080/0731129X.2021.1893932

Wasi, A. T., Anik, M. A., Rahman, A., Hoque, M. I., Islam, M. D., & Ahsan, M. M. (2025). A Theoretical Framework for Graph-based Digital Twins for Supply Chain Management and Optimization. *arXiv preprint arXiv:2504.03692*. DOI: 10.2139/ssrn.5293091

Weber, M. (1922). *Economy and society: An outline of interpretive sociology*. University of California Press.

Webster, P. (2019). Patient data in the cloud. *The Lancet. Digital Health, 1*(9), e385. DOI: 10.1016/S2589-7500(19)30202-X

Weizenbaum, J. (1966). ELIZA — A computer program for the study of natural language communication between man and machine. *Communications of the ACM, 9*(1), 36–45. DOI: 10.1145/365153.365168

Wellbrock, W., Malinovska, M., & Ludin, D. (2025). Ethical implications and potential opportunities and risks of artificial intelligence in supply chain management. *Discover Sustainability, 6*(1), 886. DOI: 10.1007/s43621-025-01808-3

Weller, A. (2019). Transparency: Motivations and challenges. *Proceedings of the 2019 ICML Workshop on Human Interpretability in Machine Learning.*

Weller, F. (2024). Blockchain Technology for Secure and Transparent Supply Chain Management. *International Journal of Computing and Engineering, 6*(3), 15–28. DOI: 10.47941/ijce.2138

Westerstrand, S. (2024). Reconstructing AI Ethics Principles: Rawlsian Ethics of Artificial Intelligence. *Science and Engineering Ethics, 30*(5), 46. Advance online publication. DOI: 10.1007/s11948-024-00507-y PMID: 39384600

West, S. M., Whittaker, M., & Crawford, K. (2019). *Discriminating systems: Gender, race and power in AI.* AI Now Institute.

Whittaker, M., Alston, J., & Crawford, K. (2021). *The steep cost of capture: Examining the impact of corporate influence on AI ethics research.* AI Now Institute. DOI: 10.1145/3488666

Willcocks, L., Lacity, M., & Craig, A. (2017). Robotic process automation: Strategic transformation lever for global business services. *Journal of Information Technology Teaching Cases*, *7*(1), 17–28. DOI: 10.1057/s41266-016-0016-9

Winfield, A. F., Michael, K., Pitt, J., & Evers, V. (2019). Machine ethics: The design and governance of ethical AI and autonomous systems [Scanning the issue]. *Proceedings of the IEEE, 107*(3), 509–517. DOI: 10.1109/JPROC.2019.2900622

Winter, J. S., & Davidson, E. (2019). Governance of artificial intelligence and personal health information. *Digital Policy Regulation and Governance, 21*(3, SI), 280–290. DOI: 10.1108/DPRG-08-2018-0048

Wirtz, B. W., Weyerer, J. C., & Geyer, C. (2019). Artificial intelligence and the public sector—Applications and challenges. *International Journal of Public Administration, 42*(7), 596–615. DOI: 10.1080/01900692.2018.1498103

Wirtz, J., Patterson, P. G., Kunz, W. H., Gruber, T., Lu, V. N., Paluch, S., & Martins, A. (2018). Brave new world: Service robots in the frontline. *Journal of Service Management, 29*(5), 907–931. DOI: 10.1108/JOSM-04-2018-0119

World Bank. (2024). Tax Expenditure Manual. *Washington, DC 20433.* https://documents1.worldbank.org/curated/en/099062724151636908/pdf/P174543148ba880bb188fd1ce06f588a6aa.pdf

World Economic Forum. (2023). *Global Cybersecurity Outlook 2023.* Retrieved from https://www.weforum.org/reports/global-cybersecurity-outlook-2023/

World Health Organization. (2023). 2023 emerging technologies and scientific innovations: A global public health perspective. https://www.who.int/publications/i/item/9789240073876

WTTC. (2024). *AI Set to Shape the Future of Travel & Tourism, Says WTTC.* Retrieved from https://wttc.org/news/ai-set-to-shape-the-future-of-travel-and-tourism-says-wttc

Xu, J., Guo, S., Xie, D., & Yan, Y. (2020). Blockchain: A new safeguard for agri-foods. *Artificial Intelligence in Agriculture, 4*, 153–161. DOI: 10.1016/j.aiia.2020.08.002

Yamani, F. Z., & El Merouani, M. (2020). Review the enterprise resource planning in Moroccan healthcare organizations. arXiv preprint arXiv:2010.06989. https://arxiv.org/abs/2010.06989

Yang, Y., Lin, M., Zhao, H., Peng, Y., Huang, F., & Lu, Z. (2024). A survey of recent methods for addressing AI fairness and bias in biomedicine. *Journal of Biomedical Informatics, 154*, 104646. DOI: 10.1016/j.jbi.2024.104646 PMID: 38677633

Zeng, Z., Chen, P. J., Lew, A. A., & Lin, H. (2020). From high-touch to high-tech: COVID-19 drives robotics adoption. *Tourism Geographies, 22*(3), 724–734. DOI: 10.1080/14616688.2020.1762118

Zest AI. (2021). Responsible credit underwriting with AI.

Zhao, L., Zhang, L., Wu, Z., Chen, Y., Dai, H., Yu, X.-X., Liu, Z., Zhang, T., Hu, X., Jiang, X., Li, X., Zhu, D., Shen, D., & Liu, T. (2023). When Brain-inspired AI Meets AGI. *Meta-Radiology, 1*(1), 100005. DOI: 10.1016/j.metrad.2023.100005

Zheng, H., & Peng, C. (2022). The impact of public health emergency governance based on artificial intelligence. *Journal of Intelligent Systems, 31*(1), 891–901. DOI: 10.1515/jisys-2022-0065

Zhu, T., & Liu, X. (2023). The future is now? Consumers' paradoxical expectations of human-like service robots. *Technological Forecasting and Social Change, 196*, 122830. Advance online publication. DOI: 10.1016/j.techfore.2023.122830

Ziegler, T., Shneor, R., Wenzlaff, K., Wang, B., Kim, J., Paes, F. F. D. C., & Adams, N. (2021). The global alternative finance market benchmarking report. *Available at SSRN* 3771509.

Zirar, A., Ali, S. I., & Islam, N. (2023). Worker and workplace artificial intelligence (AI) coexistence: Emerging themes and research agenda. *Technovation, 124*, 102747. DOI: 10.1016/j.technovation.2023.102747

Zuiderwijk, A., Chen, Y. C., & Salem, F. (2021). Implications of the use of artificial intelligence in public governance: A systematic literature review and a research agenda. *Government Information Quarterly, 38*(3), 101577. DOI: 10.1016/j.giq.2021.101577

About the Contributors

Ouail El Imrani is a Research Professor with specialization in Economics, Intelligent systems in Management, Optimization, port logistics and international trade ; he provides courses at the Faculty of Economic,Legal and Social Sciences of Tetouan, the Faculty of sciences, The Polydisciplinary Faculty of Larache, the institute for professional training in logistics and other institutions… His work applies in an evolutionary perspective in the field of economics and management. He is member of The European Organization of Scientific Research since its creation (Eursed) . Pr. Ouail EL IMRANI obtained his doctoral degree PhD in Economics, Management and Sustainable Development at the Abdelmalek Essaadi University of Tetouan and his master degree in project management of platforms at the National School of Commerce and Management in Tangier. His areas of interest include: Economics, Management, International trade, Cost optimization, Performance, Logistics, Project management, Social sciences, international economic development... Thus, he is present at various local, regional, national and international meetings. He has more than 10 years of experience in higher education. He supervised more than 100 student researchers and other researchers for specialized training. He has published lot of research articles in national and international journals. He has also presented more than 30 research papers at international symposia, conferences and scientific seminars at the national and international levels. He has participated in many training programs, include summer courses from lot of Universities and countries as well as participated in the organization of several scientific events nationally and internationally. He is a founding member and organizer of various International Symposiums and congress. He acted as resource person, moderator and examiner for many international events. He is the recipient of the "versatility exceptional award" at the Polydisciplinary Faculty of Tetouan.

***

Sharadhi A. K. is a graduate student researcher currently pursuing her Master of Science in Information Science (MSIS) from the College of Information Science, University of Arizona, Tucson, USA. She completed her B.Tech in Computer Science and Engineering from MS Ramaiah University of Applied Sciences in Bangalore, Karnataka, India. She has published several research papers in prestigious IEEE and Elsevier journals, as well as at international conferences, covering topics such as crowd simulation, cardiovascular disease detection, software defect prediction, and computer vision applications. Her research interests include human-centered computing, Human-Computer Interaction (HCI), simulation, cyber psychology, healthcare informatics, and digital health interventions, particularly at the intersection of human well-being and technology. he is currently exploring AI ethics and responsible AI design, examining how ethical considerations can be embedded in the development and deployment of intelligent systems. With interdisciplinary expertise spanning computer science, information systems, and healthcare applications, she actively contributes to academic literature as a reviewer for Elsevier journals, including Intelligent Systems with Applications and the Journal of Information and Intelligence.

Ahmed Aftiss is a professor and researcher at Sidi Mohamed Ben Abdellah University, Faculty of Legal, Economic and Social Sciences of Fez, Morocco.

Hamid Ait Lemqeddem is a Research Professor at the National School of Commerce and Management (ENCG) of Ibn Tofail University in Kenitra, Morocco. He holds a PhD in Economic Sciences from the Faculty of Economic Sciences, Rabat-Agdal. His research interests include environmental governance, sustainable development, logistics, digitalization of public procurement, territorial economic intelligence, big data, customer relationship management, SME performance, social entrepreneurship, crowdfunding, and the application of artificial intelligence in management. He has published numerous articles in these fields and participated in several international conferences.

Yassin Allammari, a Doctor in Management and Entrepreneurship, is affiliated with the Management, Finance, Digitization, and Applied Statistics Research Laboratory at the Faculty of Economic and Social Sciences, Abdelmalek Essaadi University. He is a researcher specializing in the fields of entrepreneurship, strategy, market dynamics, and SME performance. His work focuses on studying entrepreneurial and strategic dynamics that influence the competitiveness and sustainability of small and medium-sized enterprises.

Badih Baz, MS in Computer Science, PhD in Management, is an Associate Professor in Business Computing at the USEK Business School. He joined

USEK in 1994 as an Assistant Professor. He currently serves as the Head of the Business Computing program and is a member of the USEK University Senate. He has held several academic and administrative leadership positions throughout his career in higher education, including Chairperson of the Business Computing Department, Director of the Registrar's Office, and Dean of the Faculty of Sciences at USEK. He currently teaches undergraduate and graduate courses related to ERP E-Business Systems, Decision Support Models, Operations Research, and Database Programming. His research interests span both computer networking and optimization. Baz has published several papers in peer-reviewed indexed journals on networking and computer systems and holds a U.S. patent in the field of location tracking systems. Beyond academia, he has actively engaged with industry, collaborating on projects that bridge the gap between research and real-world application. His industry partnerships and consultations have enriched his understanding of practical challenges and opportunities in mobile app programming. He has worked on several progressive web apps and mobile Android/iOS apps currently published and active on multiple international platforms. With a robust presence at over thirty national and international events, Baz has been an active participant and contributor to numerous conferences and symposiums, showcasing expertise in IT and management through roles including speaker, VIP guest, session chair, and committee member. In addition, he has served as an IT consultant and executive board member for several national and international companies and institutions, including Al Houkair (Saudi Arabia), Retail Group (Lebanon), and Bull - Zenith Data Systems (France).

Nouhaila Bentatou is a doctoral researcher at Ibn Tofail University, Kenitra, Morocco. She holds a Master's degree in International Trade and Logistics and is currently pursuing a PhD in Management Sciences, focusing on the impact of digital transformation on the optimization of logistics costs in the Moroccan automotive sector. Her research interests include digital transformation, Industry 4.0, sustainable logistics, and ethical artificial intelligence applied to supply chain management. She has contributed to several academic publications and international conference presentations, reflecting her commitment to research on sustainability and responsibility in contemporary logistics systems.

Nadir El Morabit is an EFL teacher and Ph.D. candidate at Abdelmalek Essaâdi University, Morocco. His research focuses on postcolonial discourse, cultural studies, educational technology, and AI. He has published in Scopus and Web of Science-indexed journals and presented at international conferences. El Morabit also teaches at Imam El Ghazali High School and has lectured in AI-related programs at National School of Applied Science of Tetouan.

Md Mehedi Hasan Emon is a research practitioner in Business and Management, with specialized in Supply Chain Management, Digital Transformation, Marketing, Information Systems, and Educational Technology (Edutech). His work bridges a diverse range of domains including Artificial Intelligence in Business, Human-Computer Interaction, Cognitive Behaviour, E-commerce, and Smart Logistics. His research extensively explores the integration of emerging technologies in business ecosystems focusing on AI-driven innovation, behavioral responses to technology adoption, and digital marketing strategies. He has also contributed significantly to sustainability-focused fields such as GreenTech, Green Finance, Renewable Energy, and Blockchain-enabled financial innovation. Additionally, his work in Smart Cities and IoT adoption emphasizes the importance of user trust, security, and operational efficiency. Emon has authored numerous peer-reviewed publications on topics including AI adoption intention, supply chain visibility, digital learning acceptance, and crisis-resilient supply chains. Many of his studies adopt behavioral and cognitive perspectives to understand how professionals and consumers interact with technology highlighting the psychological and user-centric aspects of digital transformation. He has participated in several research initiatives funded by the University Grants Commission of Bangladesh and the European Union. His work aligns closely with the United Nations Sustainable Development Goals (SDGs), particularly in areas such as poverty reduction, access to quality education, clean and affordable energy, inclusive economic growth, sustainable industrialization, responsible consumption, climate resilience, and ocean sustainability. His work also reflects a commitment to reducing inequalities and fostering international partnerships for sustainable development. In addition to his research contributions, he serves as a peer reviewer for several high-impact academic journals and is widely recognized for promoting academic rigor and ethical scholarship. As a prolific author, he continues to shape the discourse on sustainability, innovation, and digital advancement through his research, writing, and knowledge dissemination.

Aruditya Jasrotia is working as an Assistant Professor in Amity Institute of Travel and Tourism, Amity University Uttar Pradesh. He completed his Ph.D. and M. Phil. from Central University of Jammu. He has published 23 Scopus Publications, 2 Patent and more than 30 Publications, including research papers and book chapters. His broad research areas of interest are: Smart Tourism, Tourism Marketing, Sustainable Tourism, Tourism Policy and Planning, Tourism Products and Services. He is also associated as a reviewer for various Scopus-indexed and ABDC-listed international journals like International Journal of Tourism Cities, Journal of Hospitality, Leisure, Sport & Tourism Education, International Journal of Tourism Policy and Kybernets.

Bouchra Merrahi holds a Diploma of Advanced Studies (DESA) in Money, Finance, and Banking from Mohammed V University in Rabat. She has built a solid career bridging the academic and banking sectors. Currently a PhD candidate at the ENCG, Ibn Tofail University. Her doctoral thesis focuses on the transmission channels of monetary policy and their effectiveness during periods of crisis, with a particular emphasis on the case of Bank Al-Maghrib, Morocco's central bank. Bouchra Merrahi participates in national and international academic conferences, where she presents papers related to monetary policy, financial stability, and banking governance.

El Hassania Rahou holds a PhD in Economics and Management from Abdelmalek Essaâdi University. She has extensive experience in the field of teaching and education. Currently, she holds the position of Assistant Professor of Higher Education at the Faculty of Economics and Management at Hassan I University. Ms. Rahou's research interests cover a wide range of topics, such as entrepreneurship, marketing and management, human resources and accounting.

Pramendra Singh is an Assistant Professor at Faculty of Management Studies, Marwadi University, India. He has more than eight years of teaching and research experience. He has done three post graduate degrees, two in Tourism & Travel Management and one in History. He did his doctorate (PhD) from Jiwaji University, India in 2018. He has in his credit more than thirty research papers and book chapters published by internationally reputed publishers. He has been a part of various academic activities as resource person and speaker. He also published five books on tourism interlinking it with sustainability and technology. His research interest areas are travel & tourism management, marketing, entrepreneurship, higher education and sustainability.

Satya Subrahmanyam is a distinguished academic leader, researcher, and institutional strategist with over 27 years of experience in higher education, leadership, and research management. He holds two PhDs and a Postdoctoral qualification from the Polytechnic Institute of Leiria, Portugal. Currently serving as Professor and Dean (Research) at Holy Spirit University of Kaslik (USEK), Lebanon, he spearheads research initiatives that enhance academic excellence and institutional reputation. He played a pivotal role in the development of the Catholic University in Erbil, Kurdistan, where he established new departments, led research programs, and collaborated on international educational projects. As a founding member and Managing Partner of Vignan Institute of Technology and Management, India, he contributed to its transformation into a premier institution. With expertise in AI-driven business strategies, HR analytics, and leadership in technology adoption, he has published extensively, guided PhD research, and led training programs in corporate and academic settings.

Samuel Tshikuta is a dedicated health science professional with a global academic and research footprint. He earned his undergraduate degree from Rajiv Gandhi University of Health Sciences in India, where he also presented groundbreaking research at a public health conference in Bengaluru on the prevention of cholera and Ebola in the DR Congo. While back home in Kolwezi, he shared insights on the psychological stress experienced by international students, reflecting his deep concern for student well-being. Driven by a passion for community health, he collaborated with local NGOs to support HIV/AIDS prevention initiatives and sanitation efforts across Congo. His academic journey continued in the United States at Coppin State University, where he pursued his master's degree. In 2024, he presented research on the challenges of the Internet of Things (IoT) in health data management at the MdHIMA Annual Conference, followed by a 2025 presentation on telemedicine in Central Africa, spotlighting the DR Congo. His core passions lie in research, medical science, and music disciplines through which he seeks to impact both local and global communities.

Index

A

B

C

D

E

F

G

H

I

L

www.ingramcontent.com/pod-product-compliance
Lightning Source LLC
LaVergne TN
LVHW062335300926
849161LV00007B/1699